HISTORIC PHOTOS OF
THE BROOKLYN BRIDGE

TEXT AND CAPTIONS BY JOHN B. MANBECK

TURNER
PUBLISHING COMPANY

Revised traffic patterns: Engineers build new traffic patterns as shown in this cross-section drawing of the bridge.

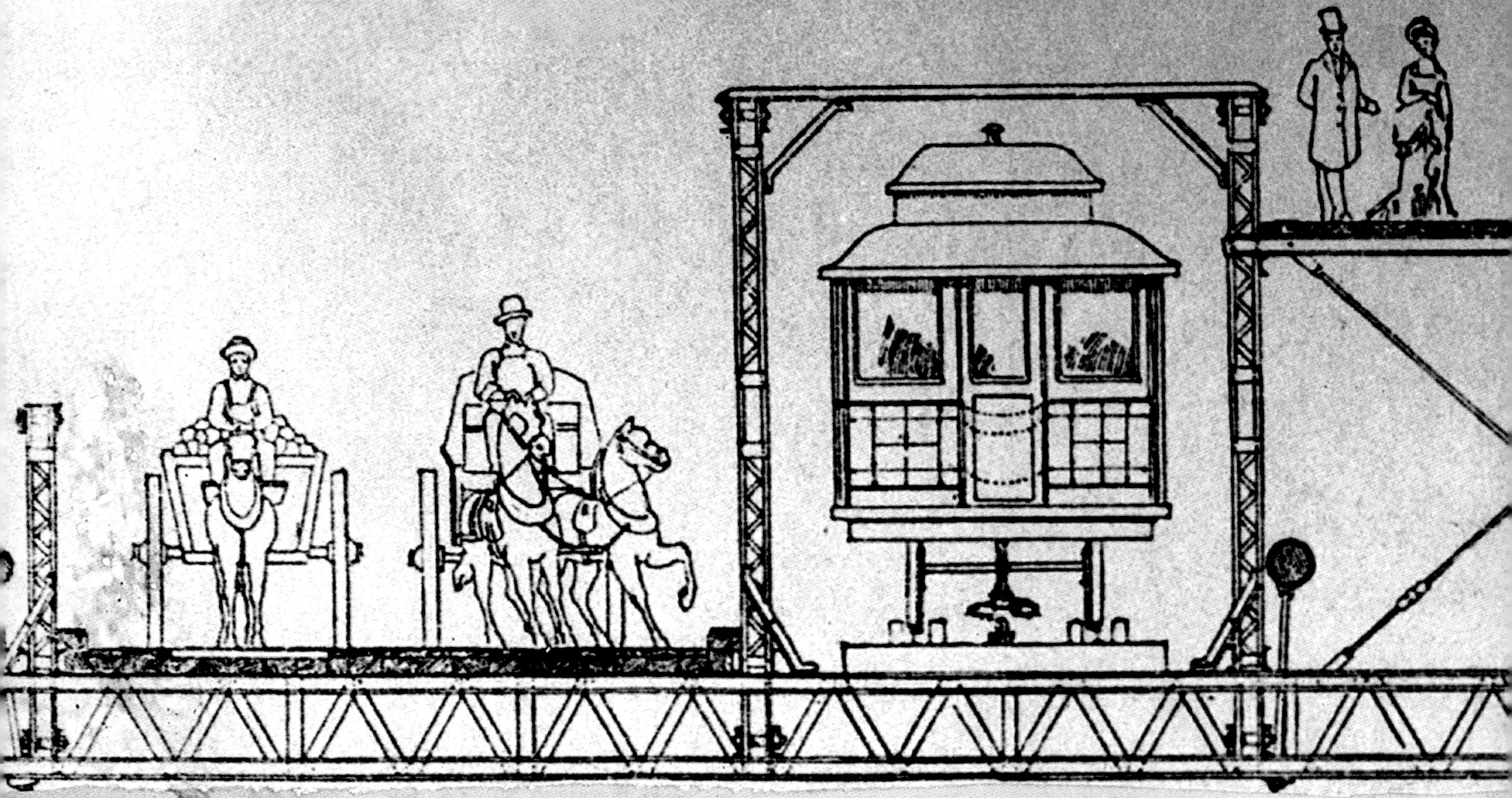

HISTORIC PHOTOS OF
THE BROOKLYN BRIDGE

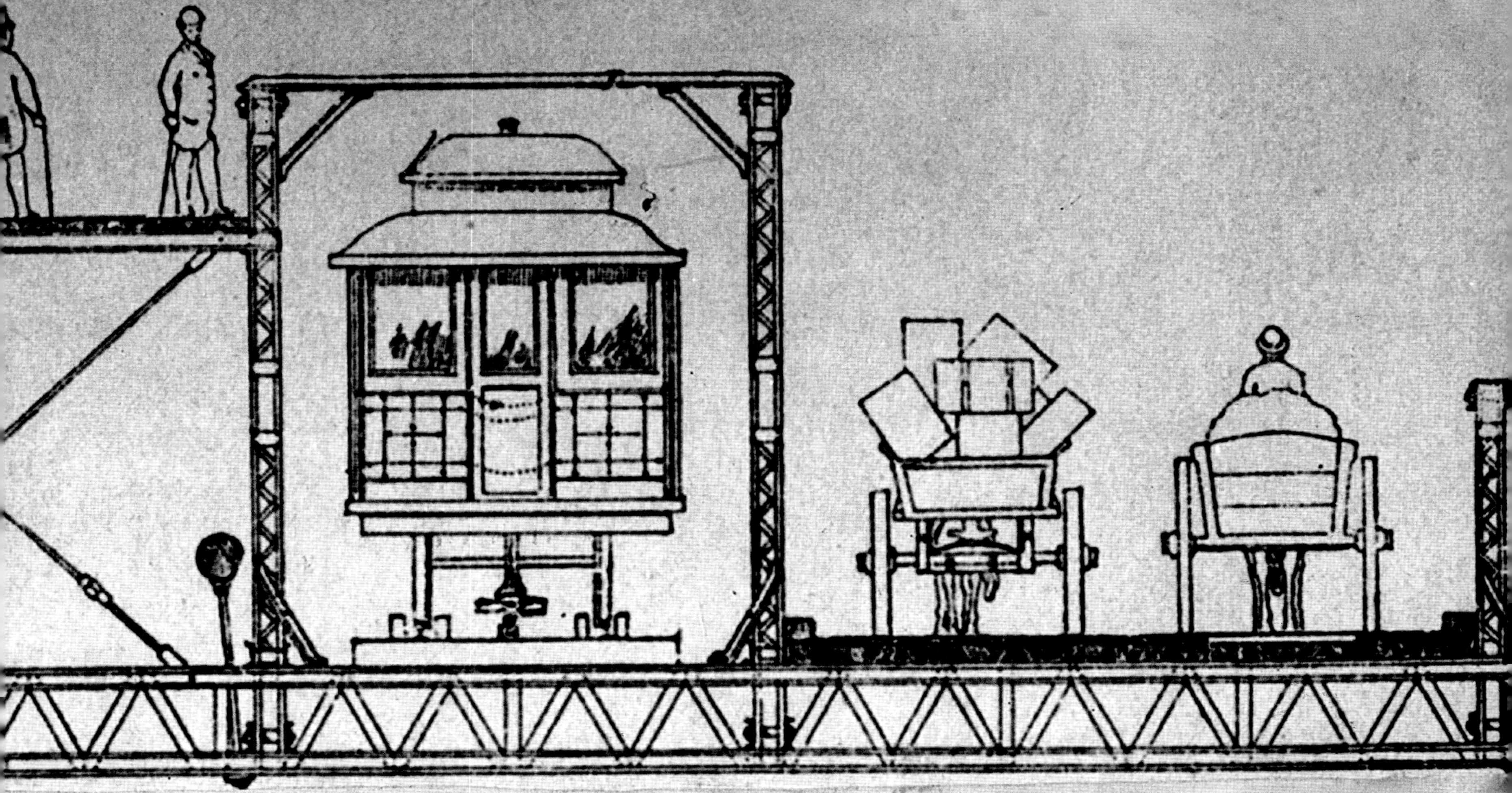

Turner Publishing Company
200 4th Avenue North • Suite 950
Nashville, Tennessee 37219
(615) 255-2665

www.turnerpublishing.com

Historic Photos of the Brooklyn Bridge

Library of Congress Control Number: 2008910983

ISBN-13: 978-1-59652-525-2

Printed in China

09 10 11 12 13 14 15 16—0 9 8 7 6 5 4 3 2 1

Contents

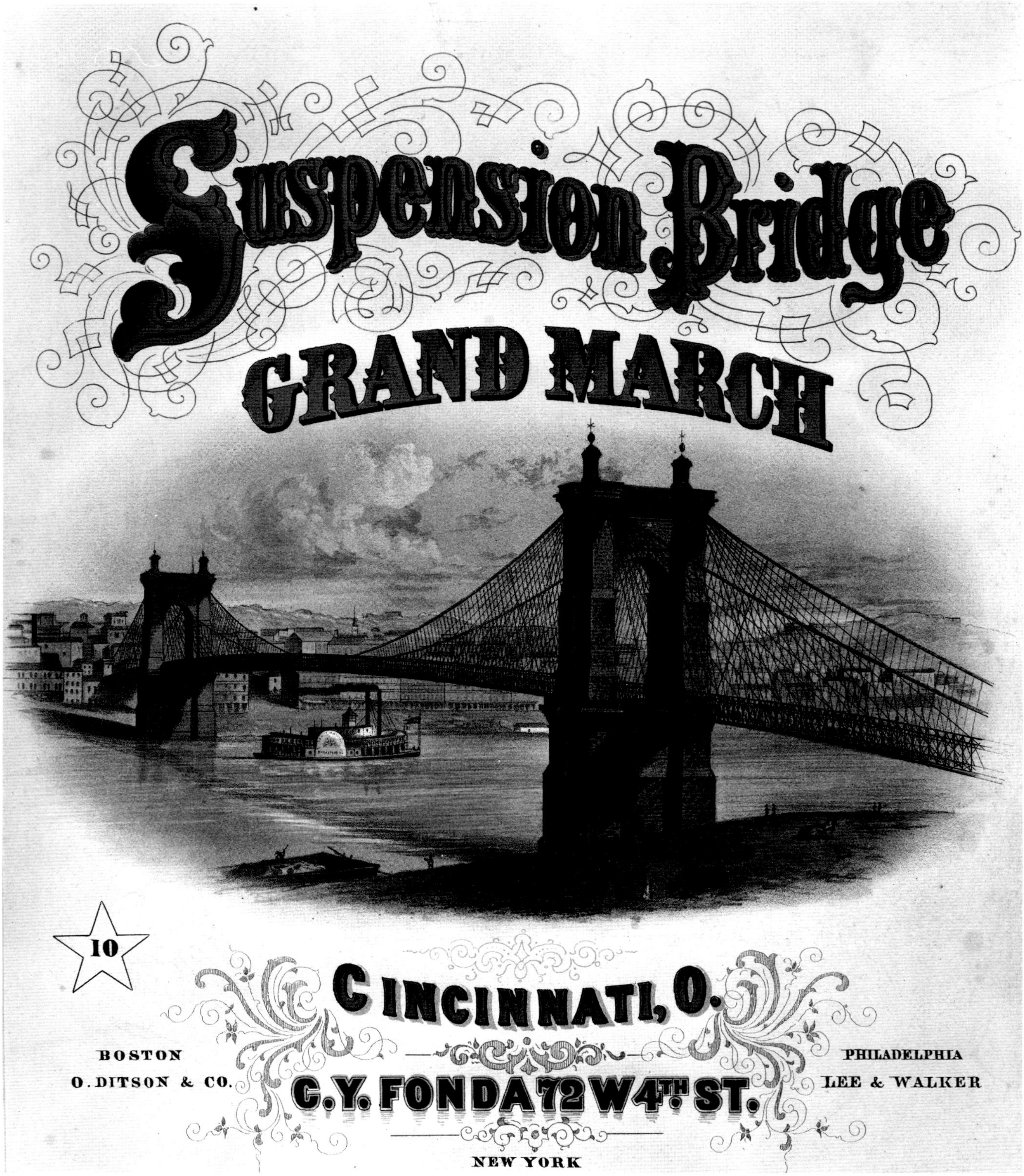

Suspension Bridge song sheet: This popular song appeared as a tribute to the Cincinnati-Covington Bridge, the closest model to Roebling's next assignment, the Brooklyn Bridge.

Acknowledgments

This volume, *Historic Photos of the Brooklyn Bridge,* is the result of the cooperation and efforts of many individuals, organizations, and corporations. It is with great thanks that we acknowledge the valuable contribution of the Brooklyn College Library Archives; the Brooklyn Historical Society; the Brooklyn Public Library, Brooklyn Collection; the Library of Congress; the New York State Archives; and the Queens Borough Public Library, Long Island Collection for their generous support.

We would also like to thank the following individuals for valuable contributions and assistance in making this work possible:

Joy Holland, Brooklyn Public Library
Marianne LaBatto and Michael Armstrong
Julie May, Brooklyn Historical Society, Brooklyn College Archives Library
My wife, Virginia, who helped shape my choice of words

I'm greatly indebted to David McCullough for the inspiration he provided in *The Great Bridge: The Epic Story of the Building of the Brooklyn Bridge* (New York: Simon & Schuster, 1972). Other useful references have been A&E's *Modern Marvels* series production about the Brooklyn Bridge; *Brooklyn Bridge: Fact and Symbol* by Alan Trachtenberg; and Ken Burns' first documentary, *The Brooklyn Bridge.*
Now that this book has re-created the pains and effort of building the daring enterprise called the Brooklyn Bridge, may it be a tribute to the many forces involved with its original creation.

With the exception of touching up imperfections that have accrued with the passage of time and cropping where necessary, no changes have been made to the images. The focus and clarity of many images are limited to the technology and the ability of the photographer at the time they were taken.

PREFACE

Within this city of towering steel skyscrapers, the most impressive structure was created from stone and steel over 100 years ago, stands only 274 feet tall and emerged from Brooklyn. In the city of New York with its constantly altering skyline, only the Brooklyn Bridge has remained constant for over a century.

The towers' Gothic arches evoke the majesty of another time, their peaks reminiscent of the crenellated turrets of a castle. Yet there the bridge stands, connecting the shores of Brooklyn and Manhattan, the first permanent link between two key segments of New York City. It started when Brooklyn and New York existed as individual cities. It commands attention still.

The bodies of water that swept around the islands of New York posed an obstacle to the mobility of the population, but the people resigned themselves to watercraft, from rowboats to steamers. As early as 1788, dreamers fantasized about bridges across the East River—a pontoon bridge by John Stevens in 1807, a wooden "Rainbow Bridge" by Thomas Pope in 1811, and a chain suspension bridge in 1829. But it took industrialization and Roebling engineering to succeed in suspending a steel bridge across the waters.

After experimenting with aqueducts in Pennsylvania and bridges at Niagara Falls and in Ohio, John Roebling, a manufacturer of steel wire and cable, drew up blueprints for the first realistically conceived East River bridge. With his son Washington, also an engineer, John studied construction possibilities and considered contingencies and alternatives. When John died in 1869, Washington took over; when he was injured, his wife Emily assisted. The bridge was a family affair.

After 14 years, multiple deaths, charges of corruption, and politically inspired delays, the bridge opened for traffic on May 24, 1883. That Brooklyn world appeared vastly different from the scene today. Life was simpler then. Consider that in 1869—the year that John Roebling started surveying for the Great East River Bridge—the steam engine was only 55 years old, suspension bridges were relatively unknown, and the tallest structures in urban communities were church steeples and masts of sailing ships.

In the city of Brooklyn, as elsewhere, the primary transportation was by horse, although the new bicycle had caught the public's fancy. Houses and streets relied on candles, gas, and kerosene for illumination. While suburban communities in Brooklyn expanded into independent villages, large chunks of agrarian property prevailed in southern Kings County.

By 1869, when Roebling started drawing up his plans, the Civil War had ended only four years earlier and Coney Island possessed one small hotel, several bathhouses, and much useless sand. But in 1883, when the bridge opened, Brooklyn had blossomed. As a visiting cabinet member observed, "I thought Brooklyn had but one hotel and a shipyard or two, but I see it's quite a town." The city had at least five major steam railroads running Brooklyn's length: at the railroads' terminals in Coney Island, about 25 hotels had been erected, and elevated trains crisscrossed northern Brooklyn. Brooklyn mayor Seth Low accepted national praise for the city's progress.

Today, with its image imbedded in our psyche, the "Eighth Wonder of the Modern World" has become as important to our American identity as Mickey Mouse. The Gothic arches, classic lines, and majestic design of the bridge have inspired artists, authors, photographers, playwrights and moviemakers, poets, songwriters, and commercial copywriters. In more than 22 films, the bridge has played a major role; over 15 popular songs have featured it. In 1964, the government declared it a National Historic Landmark. And the borough of Brooklyn esteems it as the symbol of its success. Outstanding among tributes to the span are *The Great Bridge* by David McCullough, *Brooklyn Bridge: Fact and Symbol* by Alan Trachtenberg, and the documentary film *Brooklyn Bridge* by Ken Burns.

During the fragile beginnings of the Brooklyn Bridge, the doubts of naysayers surrounded every step of its creation—yet the Roeblings never hesitated. The imagery and mystery surrounding this timeless creation remains a story of the ages. This book is a pictorial record of a struggle to build the great bridge—and of its success.

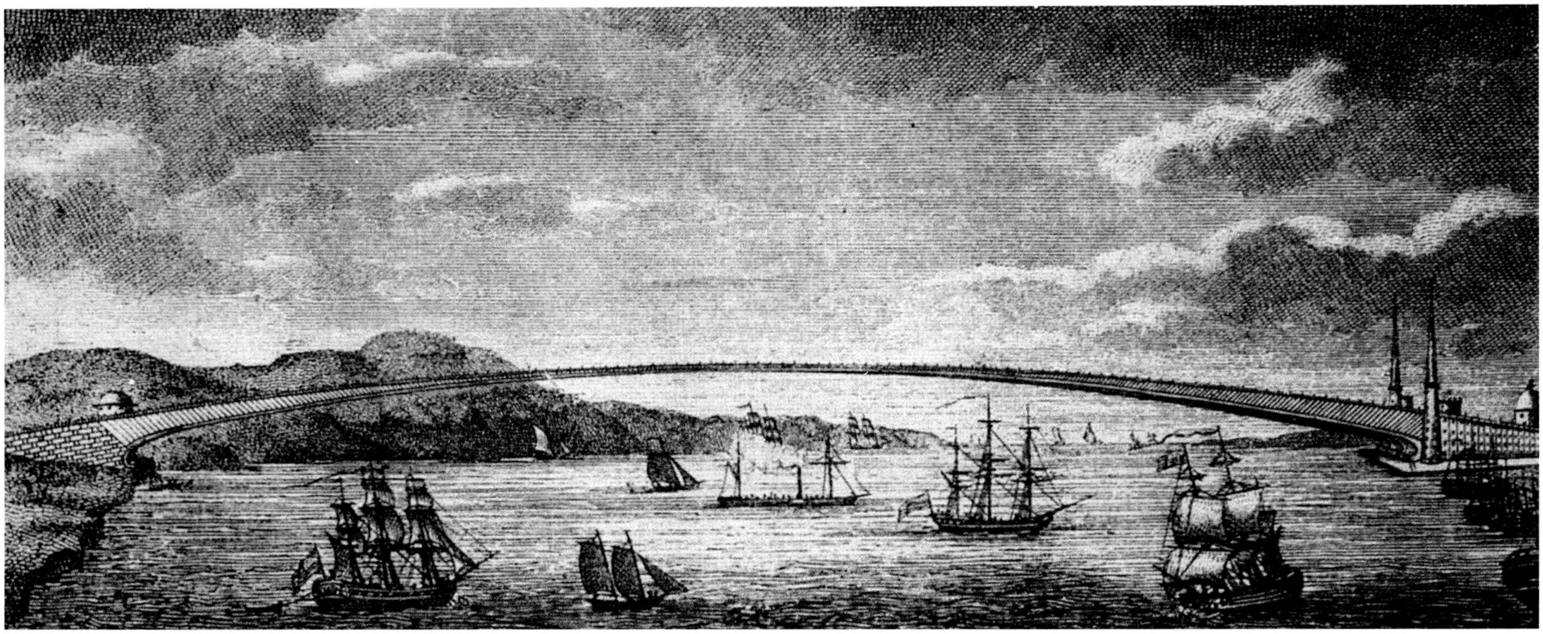

Rainbow Bridge: Among the earliest proposals for a bridge across the East River was this one by Thomas Pope. In 1811 he conceived of the Flying Pendant Lever Bridge made from wood: a 1,800-foot-long, 200-foot-high arch with no suspension and no supports. It looked like a rainbow. Pope created a model that was destroyed in a storm, just as the real bridge would have been.

Inspiration: Suspension

(1860s)

Four years before the 1806 birth of John Roebling in Mühlhausen, Thuringia, a section of Prussia destined to become the heart of Germany, the New York State legislature received a petition to construct a bridge across the East River. Without the know-how, though, the task was impossible.

Roebling, an architectural student who studied philosophy under Hegel, received a degree in civil engineering in 1826 from the Royal Polytechnic Institute of Berlin. While traveling through Bavaria, he brightened at the sight of his first suspension bridge, intrigued with the concept of suspension and what would make it foolproof. Fleeing Germany's turmoil, he immigrated to America in 1831 at the age of 25 and settled in Pennsylvania near Pittsburgh.

In 1837, while surveying railroad and canal routes, he designed steel wires to replace rope hawsers on canals. Out of these projects, Roebling received his first commission in 1844 to build a suspension aqueduct to carry the Pennsylvania Canal over the Allegheny River. By 1848, he had completed four more aqueducts and opened a profitable wire factory. He then moved to Trenton, New Jersey, and created Trenton Wire Rope plant, which manufactured bridge cables. Using his engineering knowledge, he designed suspension bridges in Cincinnati, St. Louis, Pittsburgh, and at Niagara Falls.

In 1865, John's son Washington, a Union colonel in the Civil War, was released from duty and joined his father in completing John's most ambitious project yet: the Covington-Cincinnati Suspension Bridge between Kentucky and Ohio. Opening in 1867, it was the longest suspended span in the world and most closely resembled the Brooklyn Bridge design.

Encouraged by a severe freeze-over of the East River during the winters of 1866 and 1867, William Kingsley, editor of the *Brooklyn Eagle,* persuaded state senator Henry Cruse Murphy to issue a charter authorizing a toll bridge between Brooklyn and New York City. Then the New York legislature passed a bridge bill. In April 1867, John Roebling—whose designs included three types of bridge towers—was named chief engineer and in September, the private New York Bridge Company held its first meeting. In 1869, Congress approved the construction and President Ulysses Grant signed the bill. Bridge construction began in 1870—but without Roebling, who had died in 1869 after sustaining an injury to his foot while surveying for the Brooklyn tower. His son Washington was appointed chief engineer.

John A. Roebling: The uncompromising father of the Brooklyn Bridge. A true "man of iron," his attention to detail guaranteed success of the Brooklyn Bridge project before a stone was turned. He died in a freak accident in 1869, a year before construction began. Nevertheless, his hand-built bridge immortalized Brooklyn.

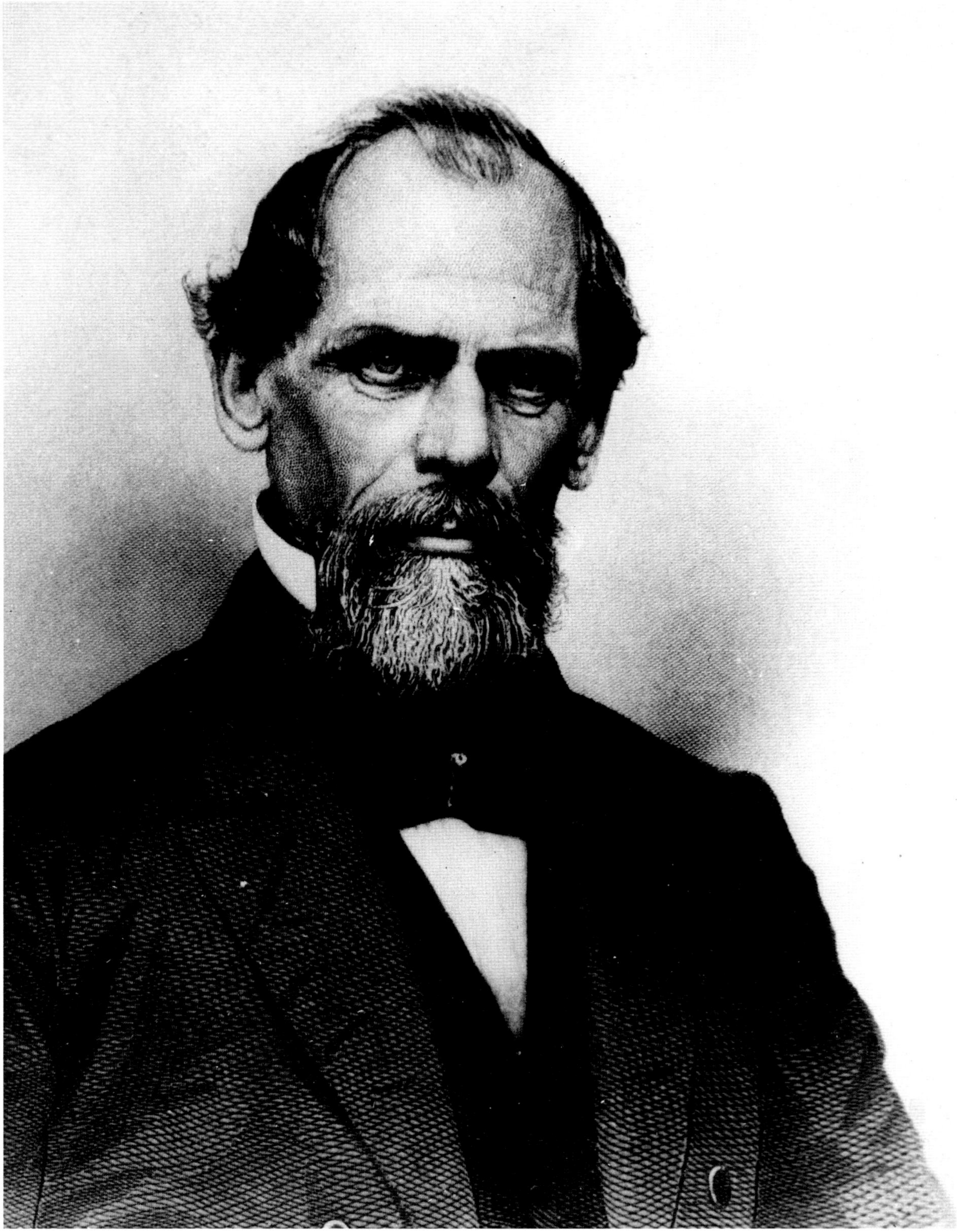

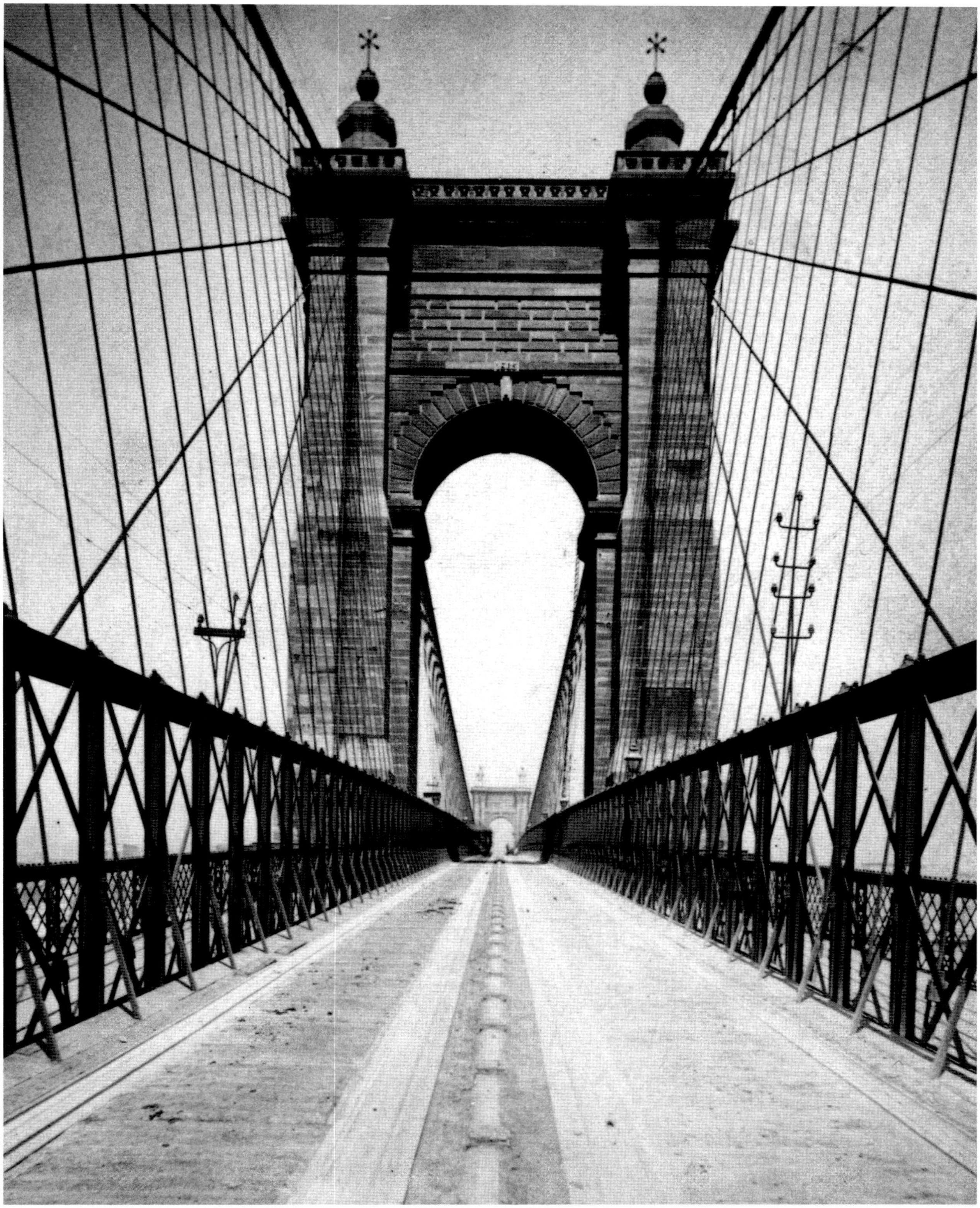

The Cincinnati-Covington Bridge: With success building aqueducts and a suspension bridge at Niagara Falls, John Roebling graduated to a full-sized bridge across the Ohio River, a suspension bridge crossing from Cincinnati, Ohio, to Covington, Kentucky. On his discharge from the army in 1865, John Roebling's son, Washington, joined him to complete the work in 1867. In 1975, the bridge was named a National Historic Landmark and renamed the John A. Roebling Bridge.

Following Spread: "Sectional View of Foundation": This sketch from an 1870 edition of *Harper's Weekly* shows a cross-section of the bridge when construction first began.

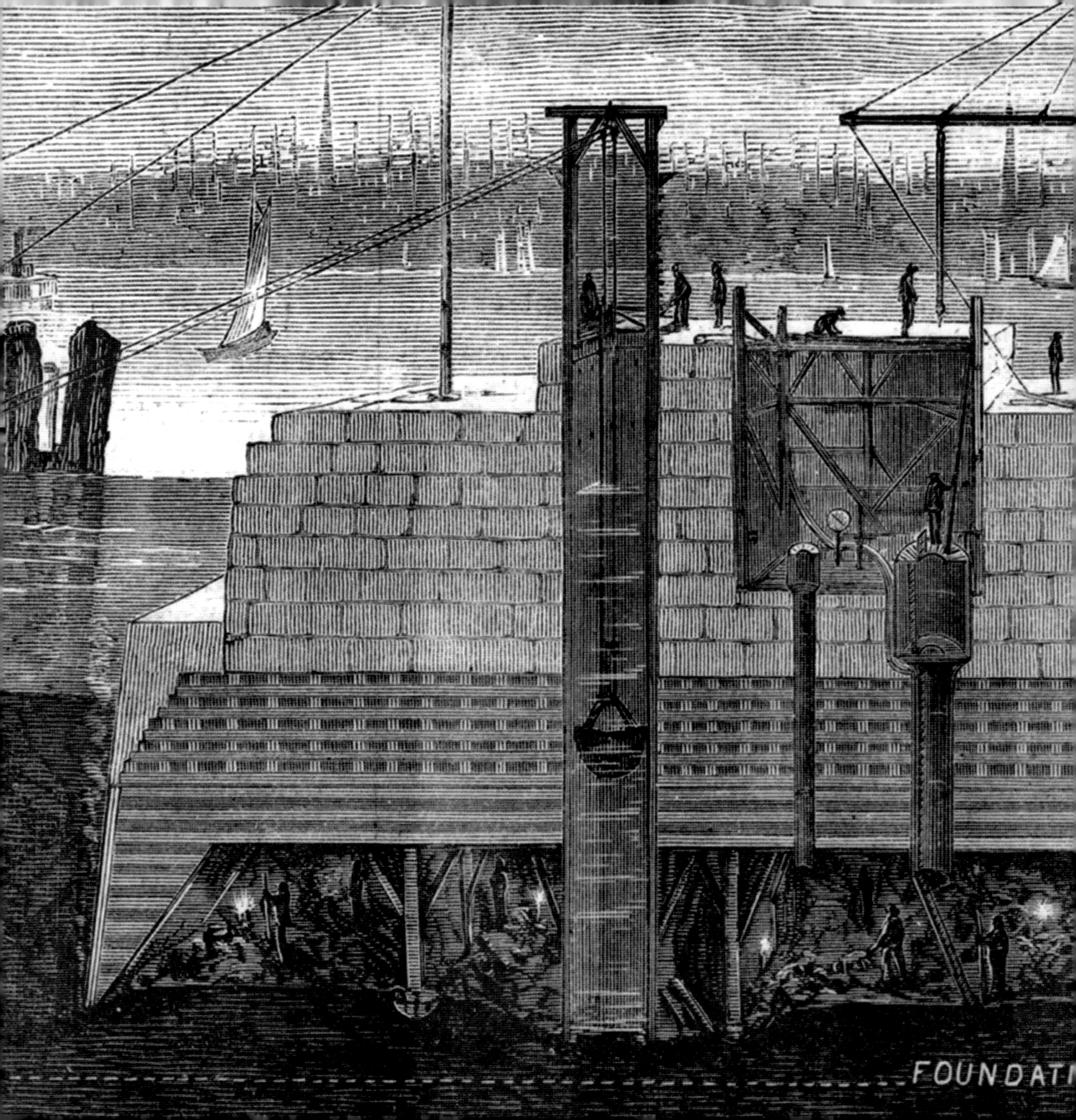
FOUNDATI

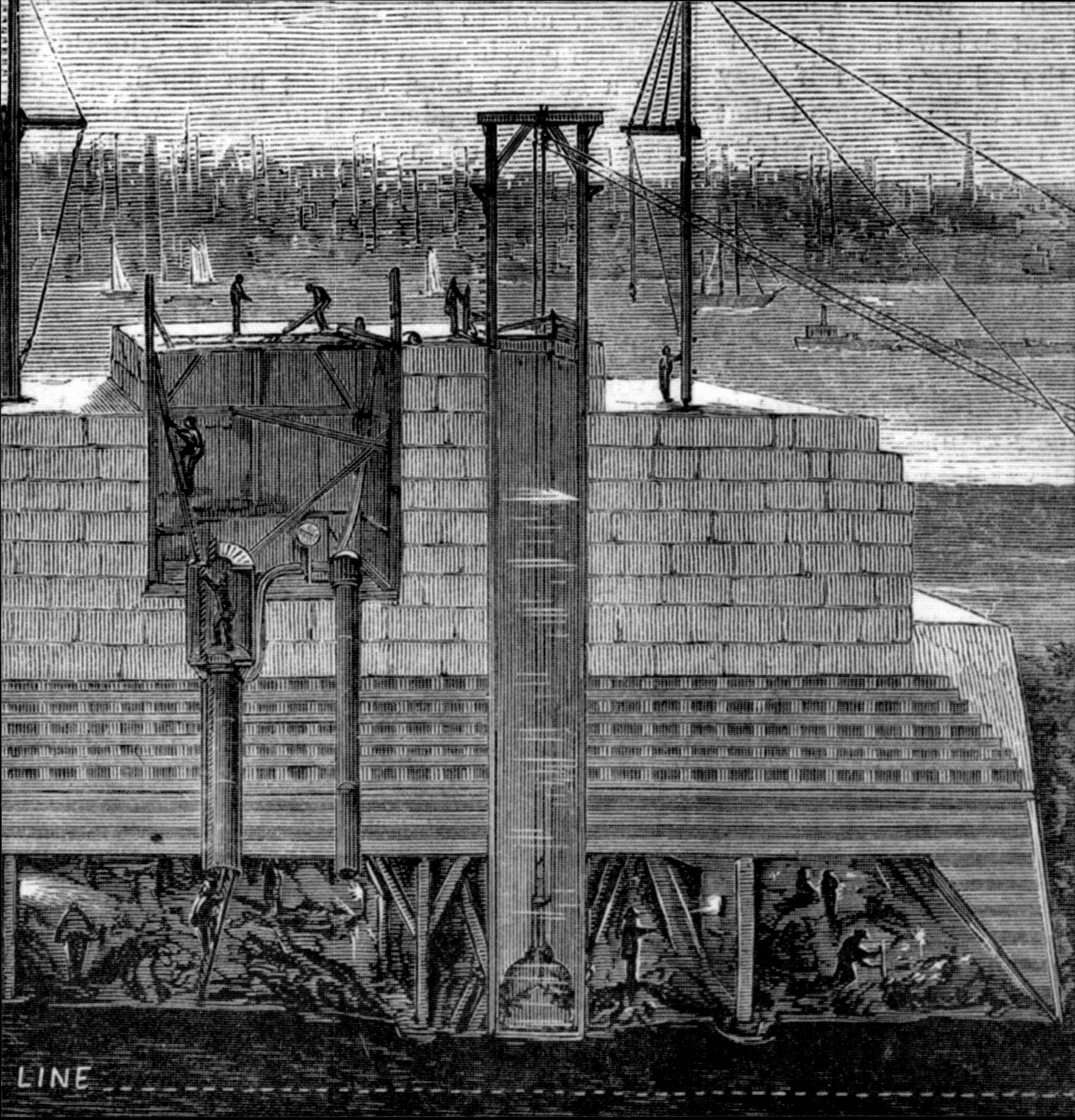
LINE

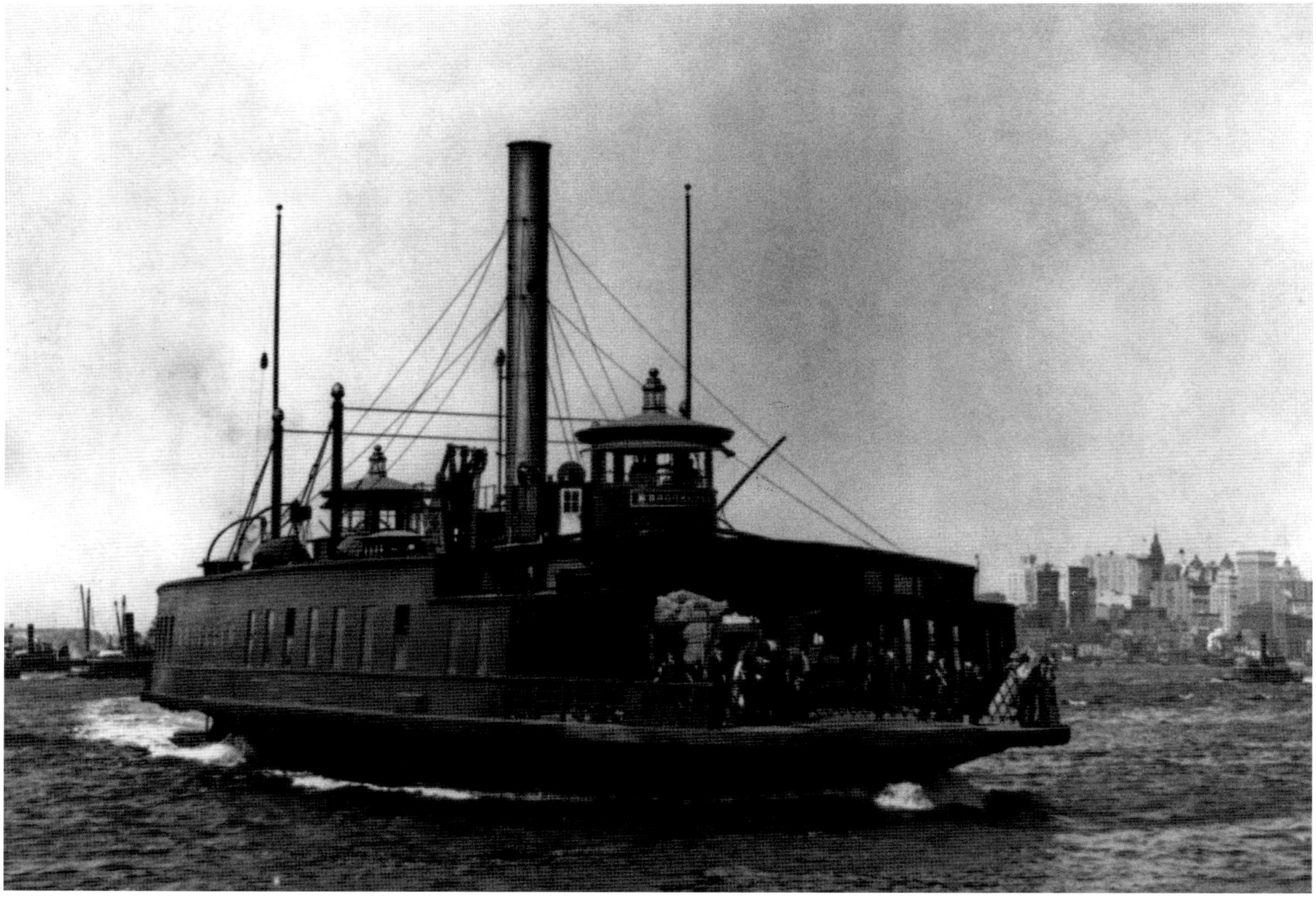

Ferryboat *Brooklyn*: Transportation across the East River for over 200 years was by boat, mostly ferryboats. In 1869, when work on the bridge started, Brooklyn—a city of 400,000—had over 5 ferry lines and 13 boats traversing the East River day and night. Ferries operated until 1924 even though three bridges crossed the East River by 1909.

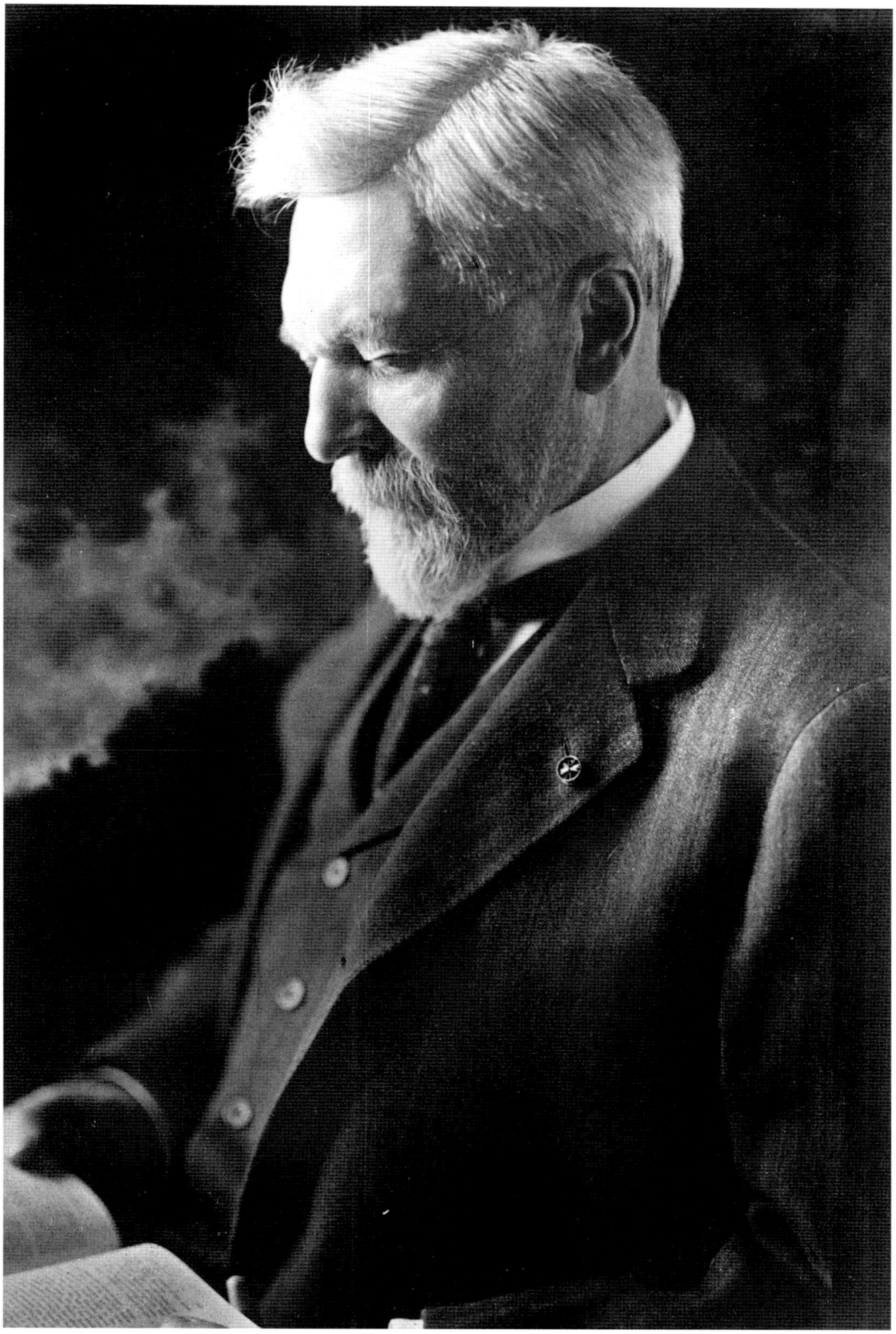

Washington A. Roebling: The first son of John and Johanna Roebling, also a civil engineer, inherited his father's sense of precision and thoroughness. He enlisted as a private in the Union Army and fought his way through eight major battles of the Civil War, including Gettysburg, retiring at the rank of colonel. It took the Brooklyn Bridge to wound him, but it never defeated him.

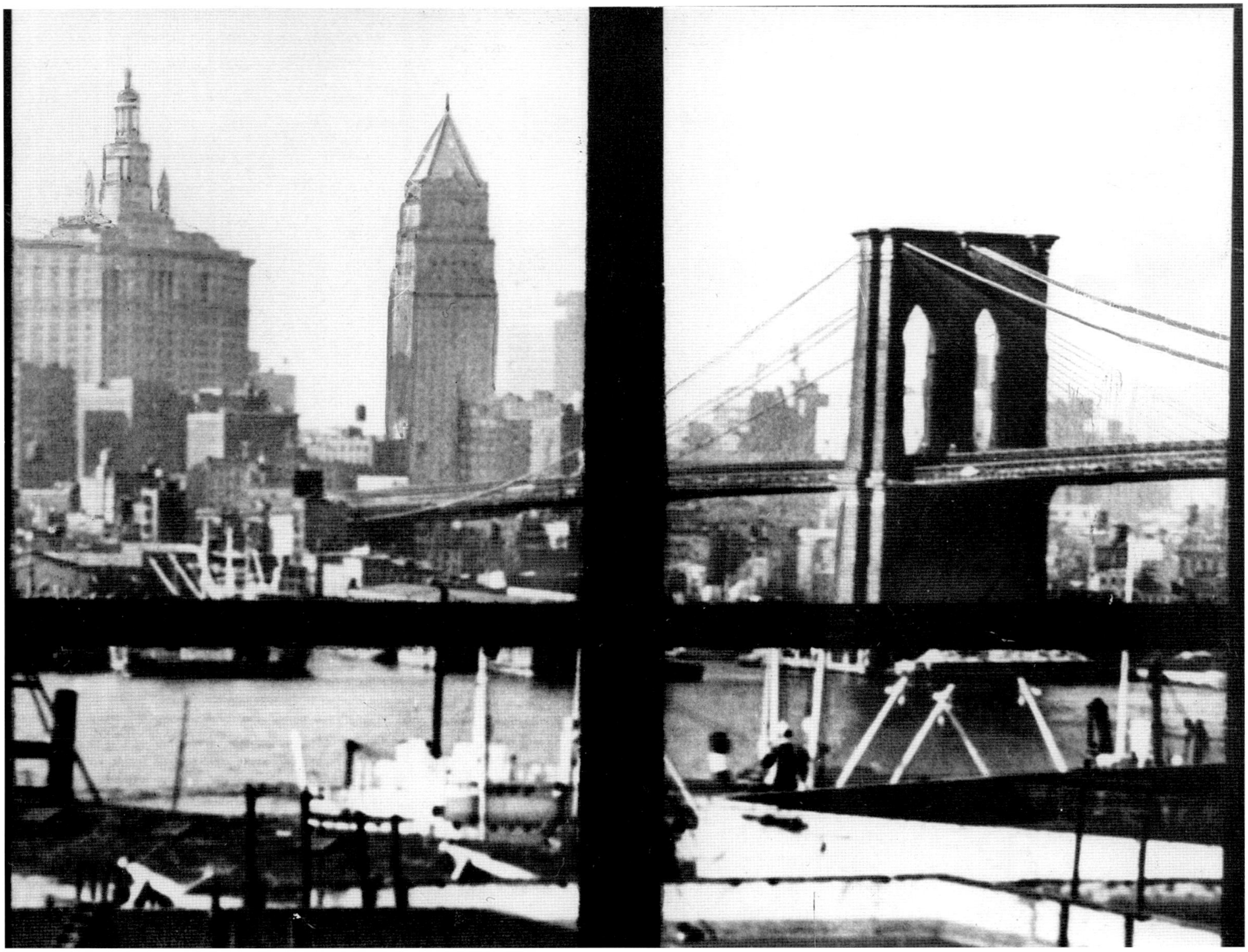

The Brooklyn Bridge from Columbia Heights: The view from the window of 110 Columbia Heights, now demolished, where Washington Roebling watched progress of the construction on the bridge after he was disabled by the "bends."

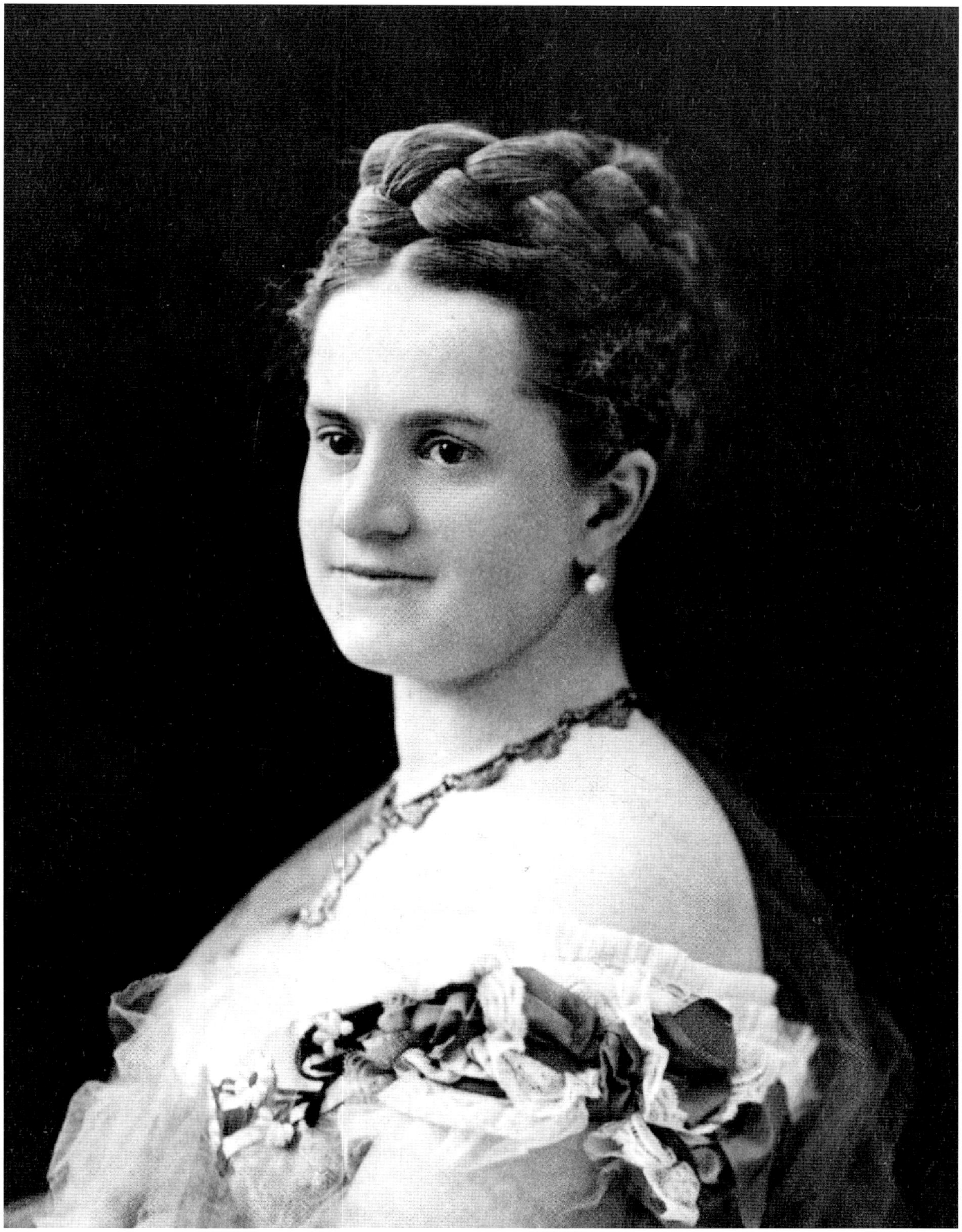

Emily Roebling: The sister of a Civil War general, Emily studied mathematics as a student, which enabled her to learn bridge construction and to assist and then to supervise construction of the Brooklyn Bridge when her husband, Washington, collapsed with the bends. Making daily rounds, she assumed the responsibility of completing the project and was first to cross the bridge and glory in its success.

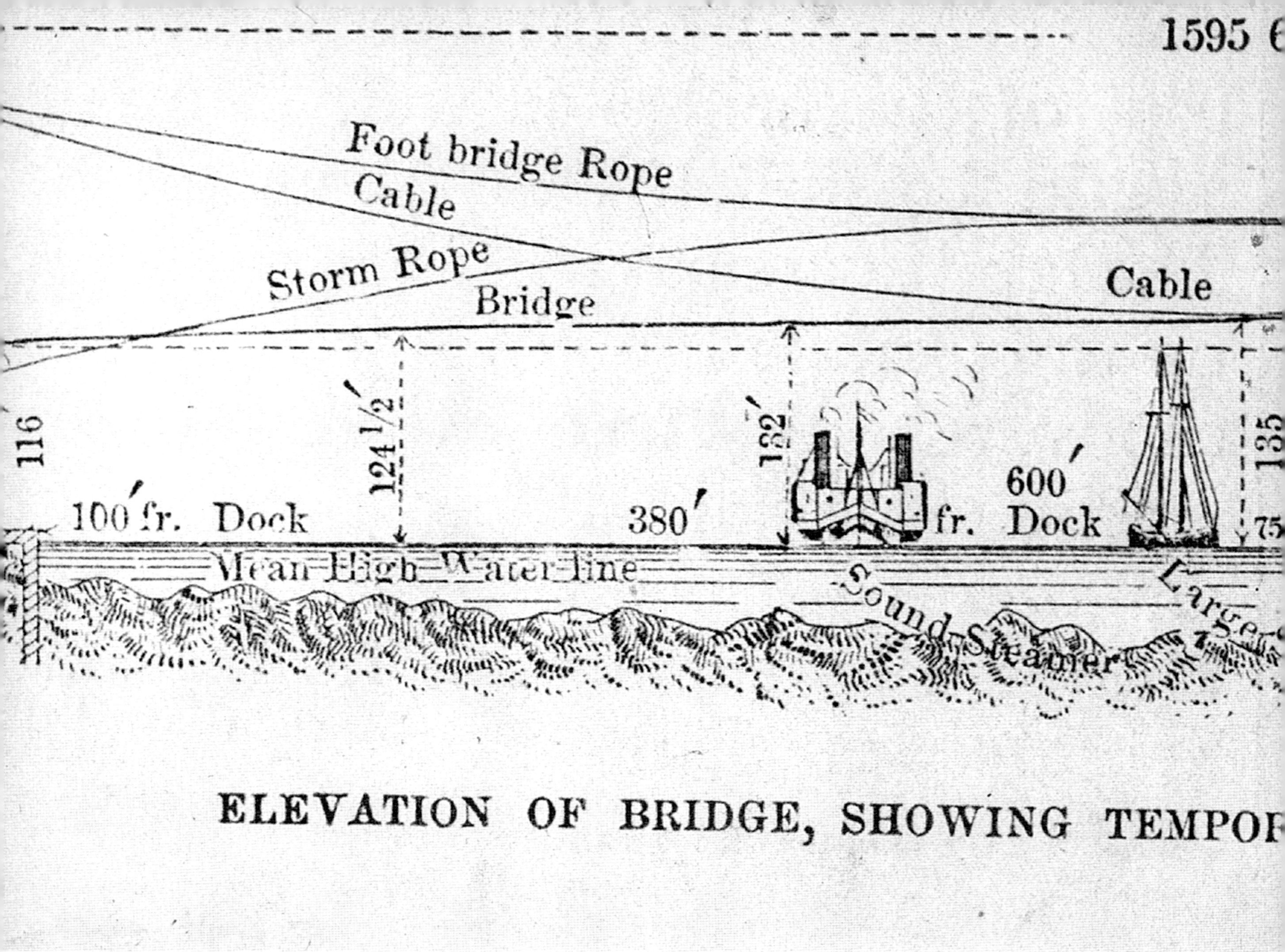

Elevation of the Brooklyn Bridge: An unidentified sketch shows measurements of the bridge with drawings of different-sized ships that might pass beneath it. Cables and ropes are indicated as well as the footbridge used before the roadbed became reality. This sketch from the 1870s may have been shown to visitors.

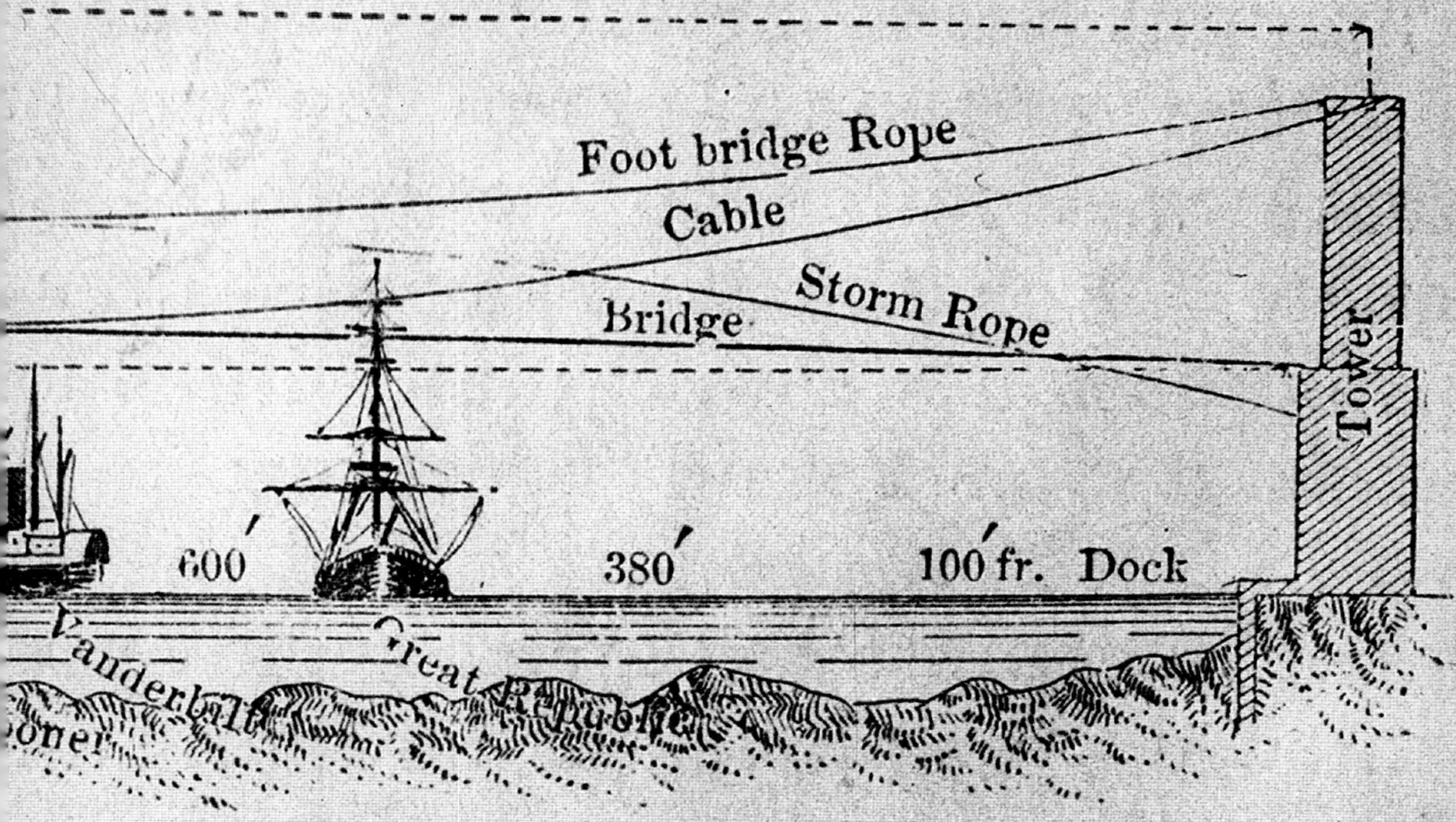

RY ROPES USED IN CABLE-MAKING.

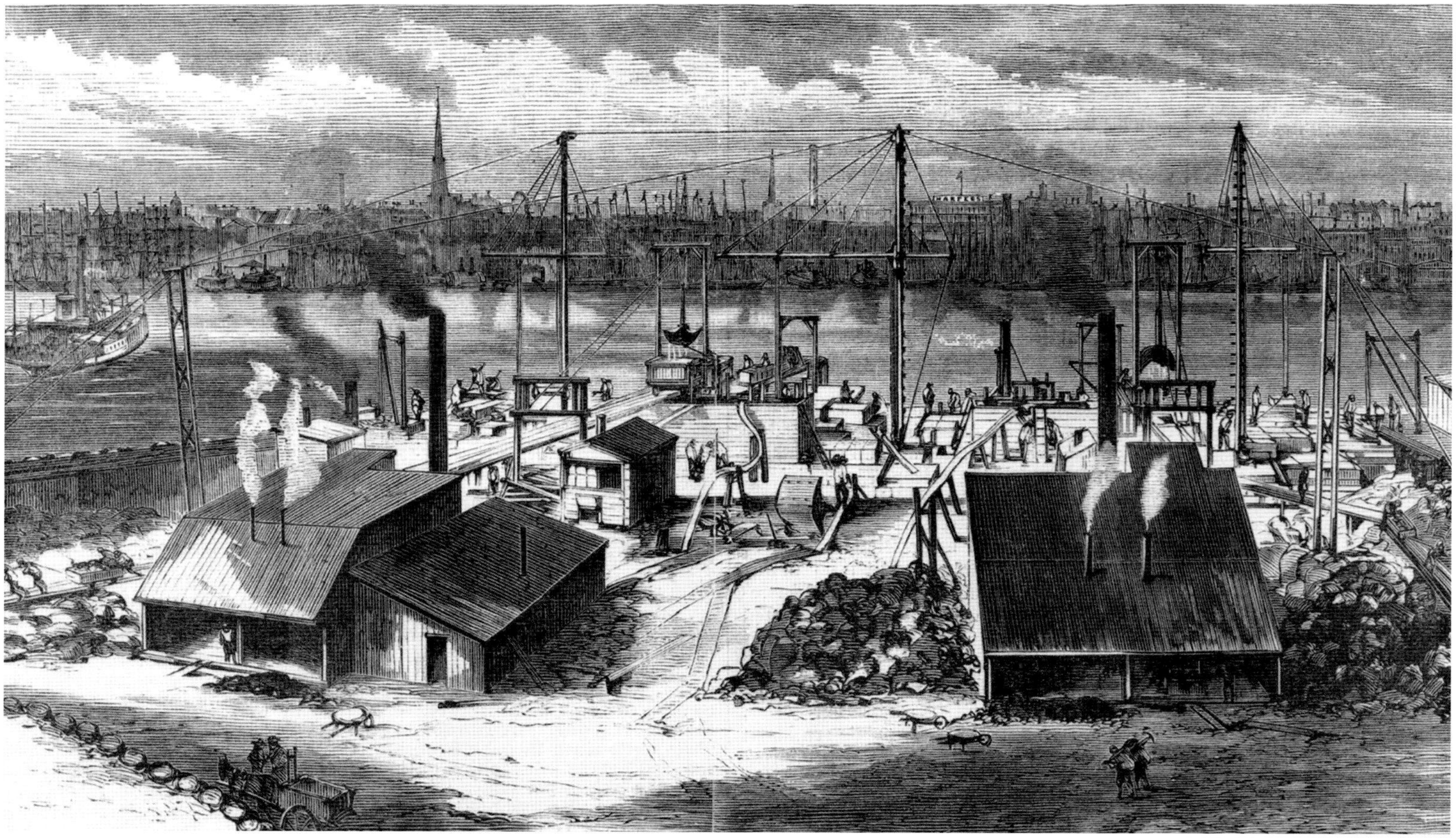

"The Brooklyn Pier—General View of Foundation": In this *Harper's Weekly* sketch, the East River Bridge appears during its foundation stage after Washington Roebling took over construction.

The Creative Years: Crossing the Divide (1869–1883)

After the New York Bridge Company elected Henry Cruse Murphy president, the building of the longest suspension bridge in the world began. When completed, the bridge would link the cities of Brooklyn and New York. In spite of the high temperatures, noise of explosions, blowouts, fires, flooding, and constant threat of caisson disease, workers continued to toil at daily wages of $2.25. The Brooklyn tower slowly but steadily rose.

One of the greatest calamities occurred in June 1872, when Washington Roebling descended into the Brooklyn caisson to inspect a fire and surfaced with the bends, which paralyzed him. That October, two cranes toppled from the Brooklyn tower. Investigations followed with accusations, counter-accusations, and demands to change the original charter, but the work progressed.

By 1875, the majestic Manhattan tower stood complete as the New York Bridge Company went public. Still paralyzed, Washington Roebling watched the bridge's progress from his window at 110 Columbia Heights as the New York tower reached completion and the first cable stretched from the Brooklyn tower to the bridge. His wife, Emily, had studied higher mathematics and then bridge engineering, enabling her to deliver and explain daily directives from the chief engineer to the supervisors.

In a show of daring, the master mechanic, E. Frank Farrington, rode a boatswain's chair between the towers in the first crossing of the East River in 1876. Midstream, he stood on his seat, doffed his cap, and blew kisses to the admiring crowd below. Soon after, a temporary footbridge from tower to tower opened to the public for those brave enough to cross.

Still, misfortunes continued to delay construction: faulty cables were supplied dishonestly by J. Lloyd Haigh, questions arose about bribes on a steel contract involving Henry Murphy, and—in 1878—a foreman was whipped from the New York tower by a snapped cable. The following year, a lawsuit sought to halt construction of the bridge for interfering with "navigable waters." It failed. In 1882, Mayor Seth Low of Brooklyn tried to force vacationing Roebling to resign for his "lack of leadership," to no avail. The bridge floor and cables eventually reached completion, and the train terminals took shape. Destiny beckoned.

Henry Cruse Murphy: One of several Brooklyn politicians who facilitated the construction of the bridge. Murphy had founded the *Brooklyn Daily Eagle* in 1841, invested in Coney Island with the Brighton Beach Railroad and Hotel, and served as mayor of Brooklyn, state senator, and a member of Congress. William Kingsley, editor of the *Eagle,* persuaded Murphy to create the New York Bridge Company, which oversaw construction of the Brooklyn Bridge. Murphy became its president.

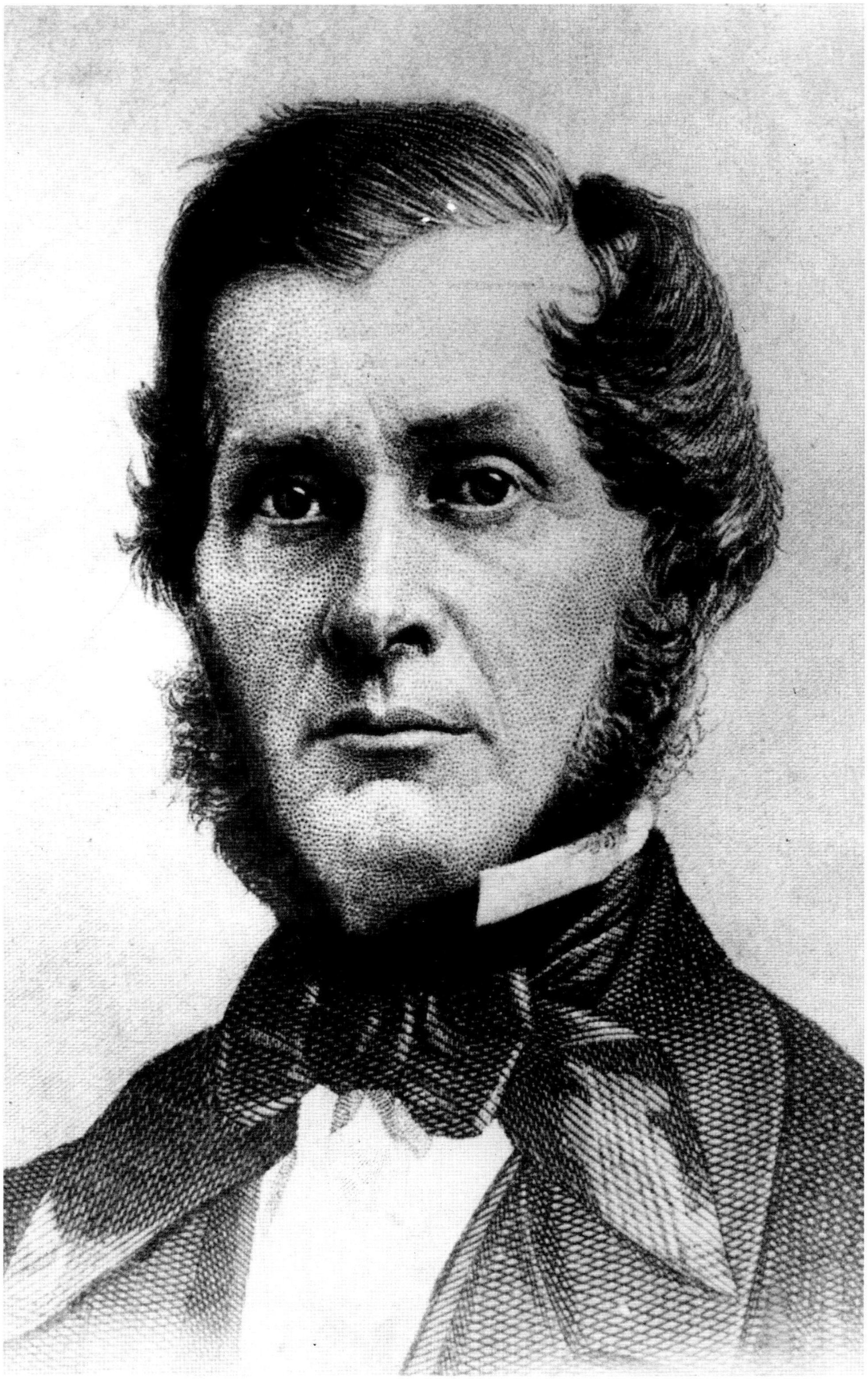

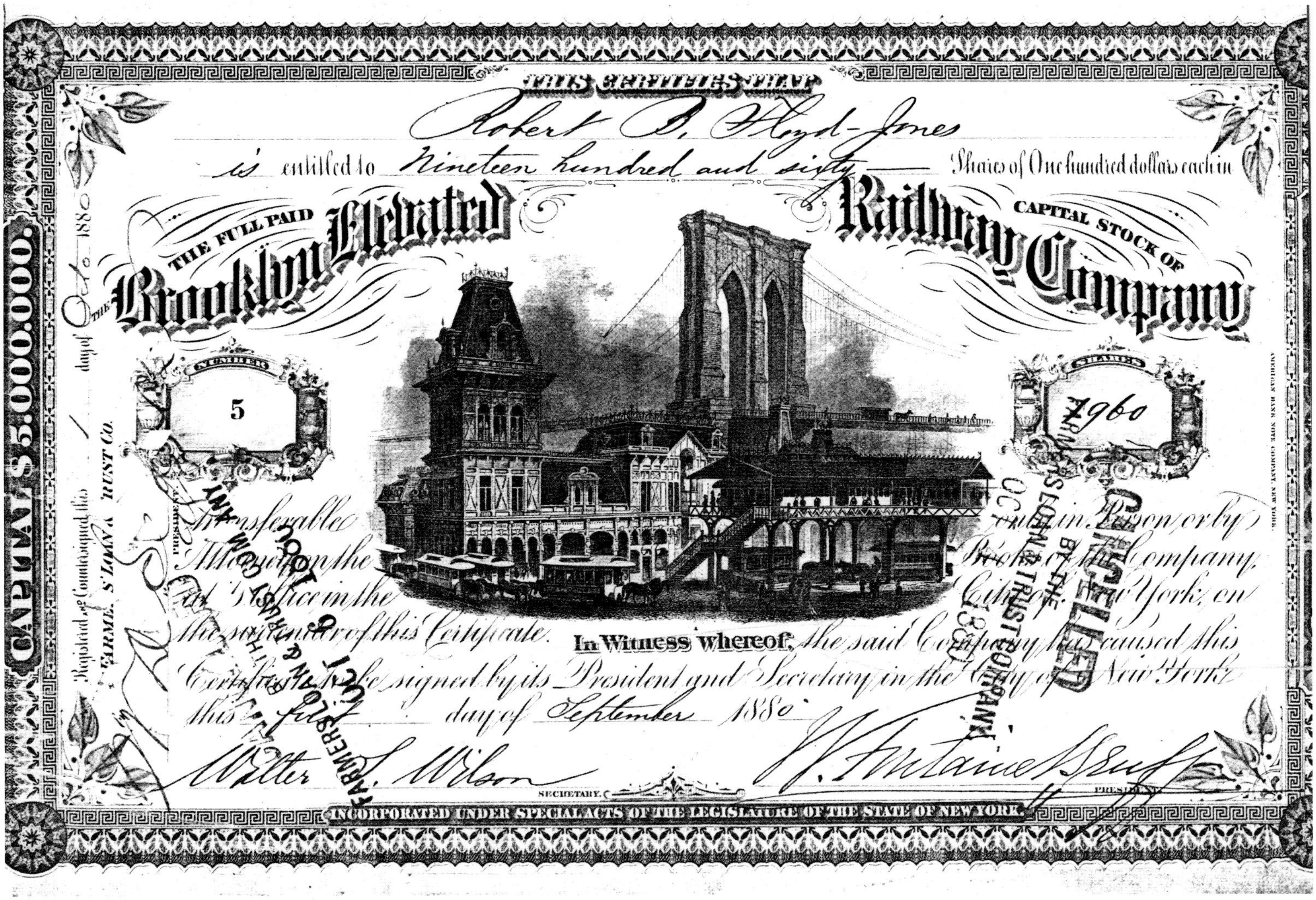

Stock certificate: A flamboyant certificate for stock was issued to finance construction of the railroad over the Brooklyn Bridge.

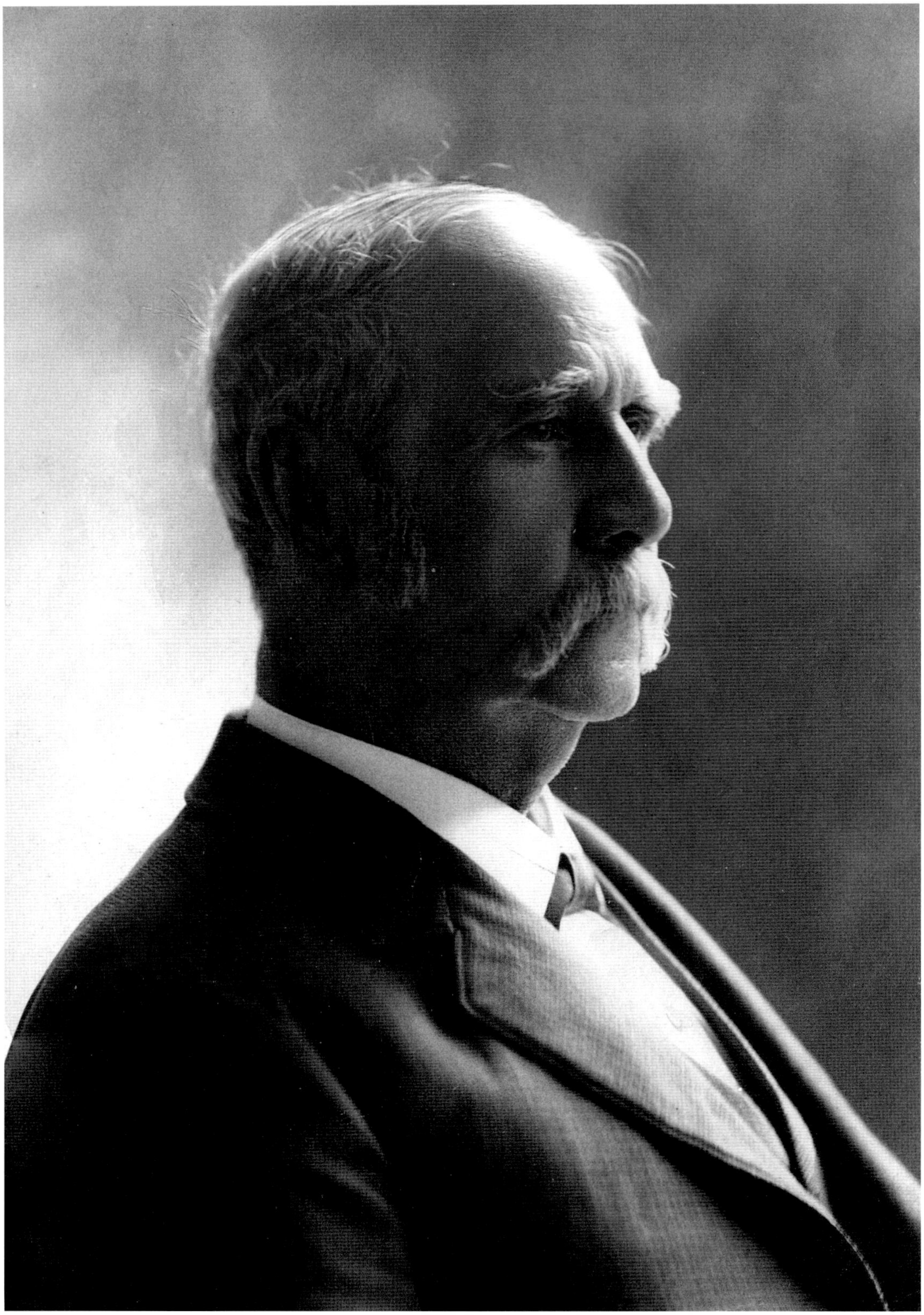

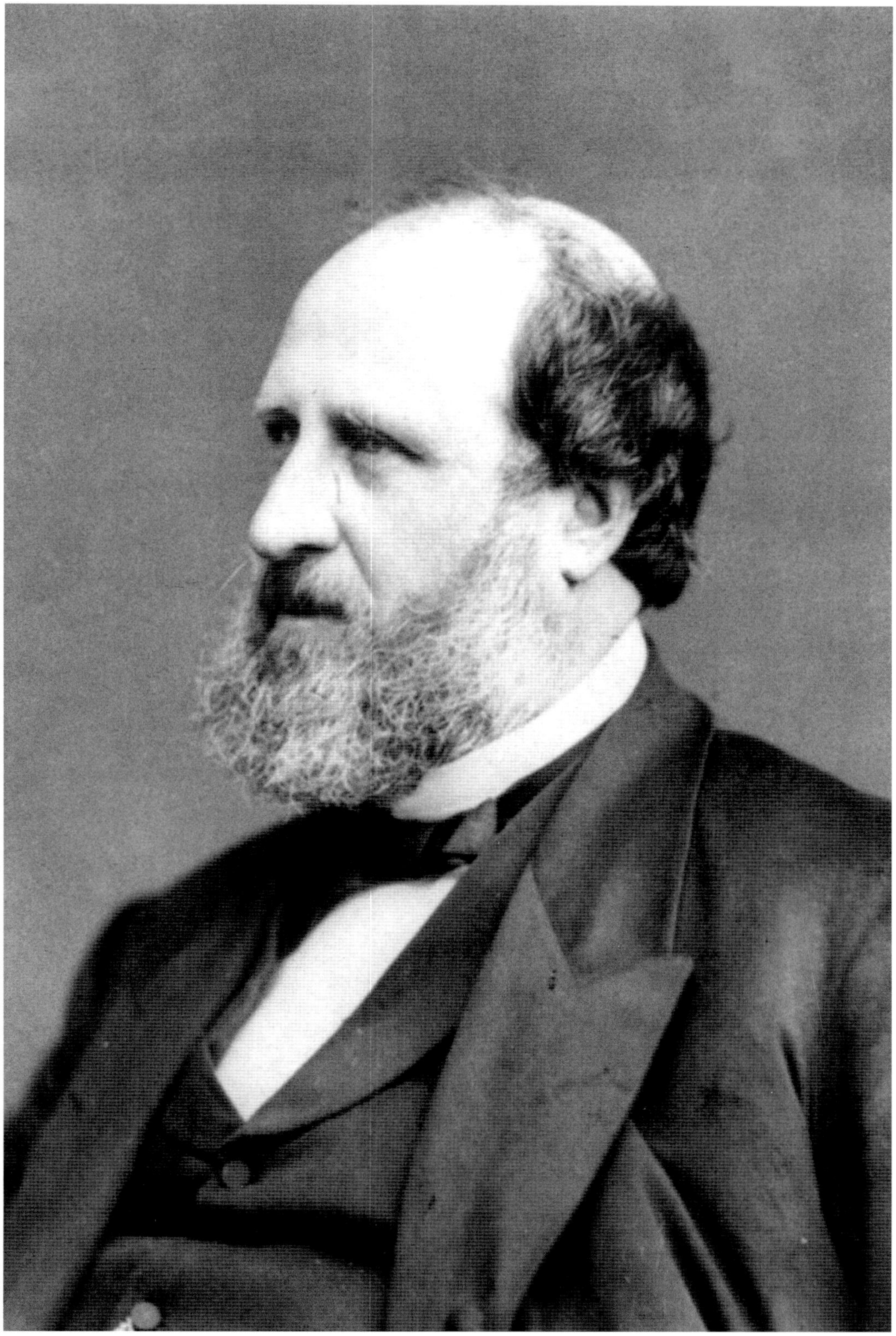

Hugh McLaughlin (left) and William Tweed: Two ward bosses (political supervisors who are not elected) saw the Brooklyn Bridge as a chance for personal profit. Boss McLaughlin of Brooklyn became a stockholder; Boss Tweed of New York sat on the board of directors and exerted political influence in Albany—until the Tweed Ring collapsed.

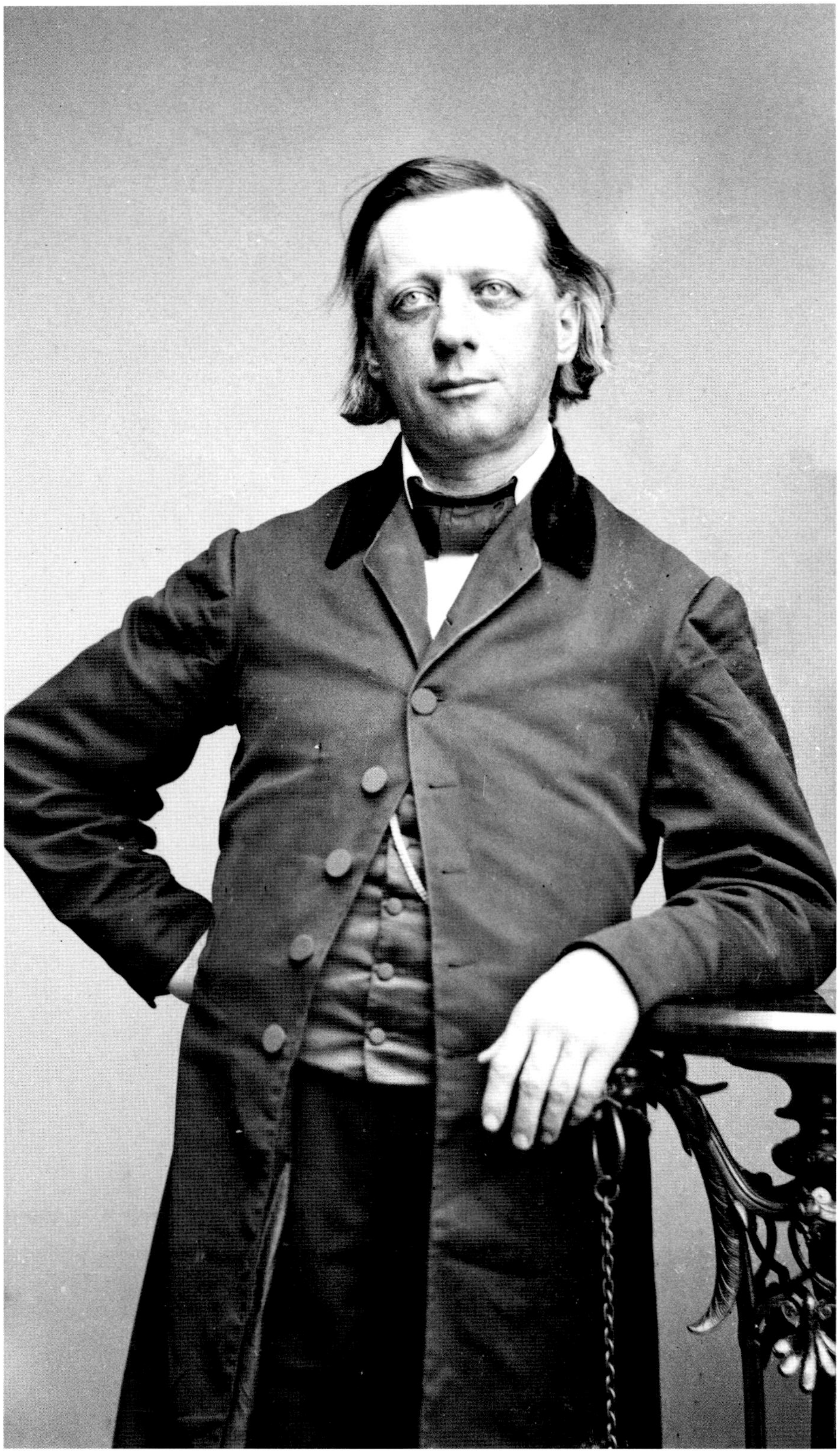

Henry Ward Beecher: The Reverend Beecher of Plymouth Church represented the moral Right of Brooklyn. His abolitionist stand before and during the Civil War earned him the respect of President Lincoln. "Beecher boats" of parishioners who crowded onto the ferries to hear his sermons meant that a bridge would facilitate their journeys. Beecher's support of Brooklyn translated into support of the bridge by the populace.

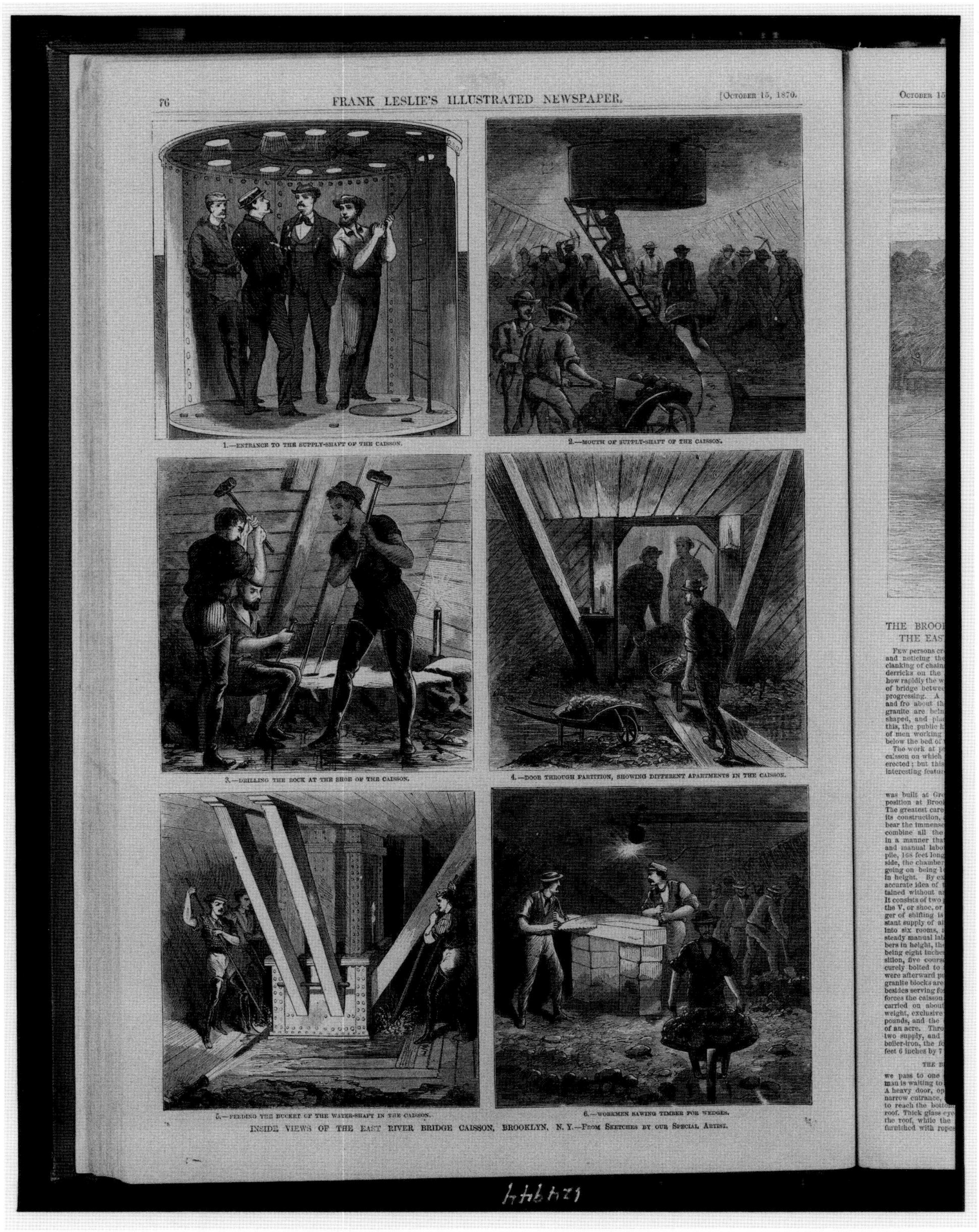

Inside the Brooklyn caisson: These six engravings of work inside a caisson appeared in an 1870 issue of *Frank Leslie's Illustrated Newspaper.* Included are pictures of three visitors entering through the air lock, the mouth of the shaft, men splitting boulders, a door showing different compartments, workmen stirring a pool beneath a water shaft, and men sawing timber for wedges. As workers removed debris, the caisson sank farther into the earth beneath the East River, creating a foundation and anchor for the tower.

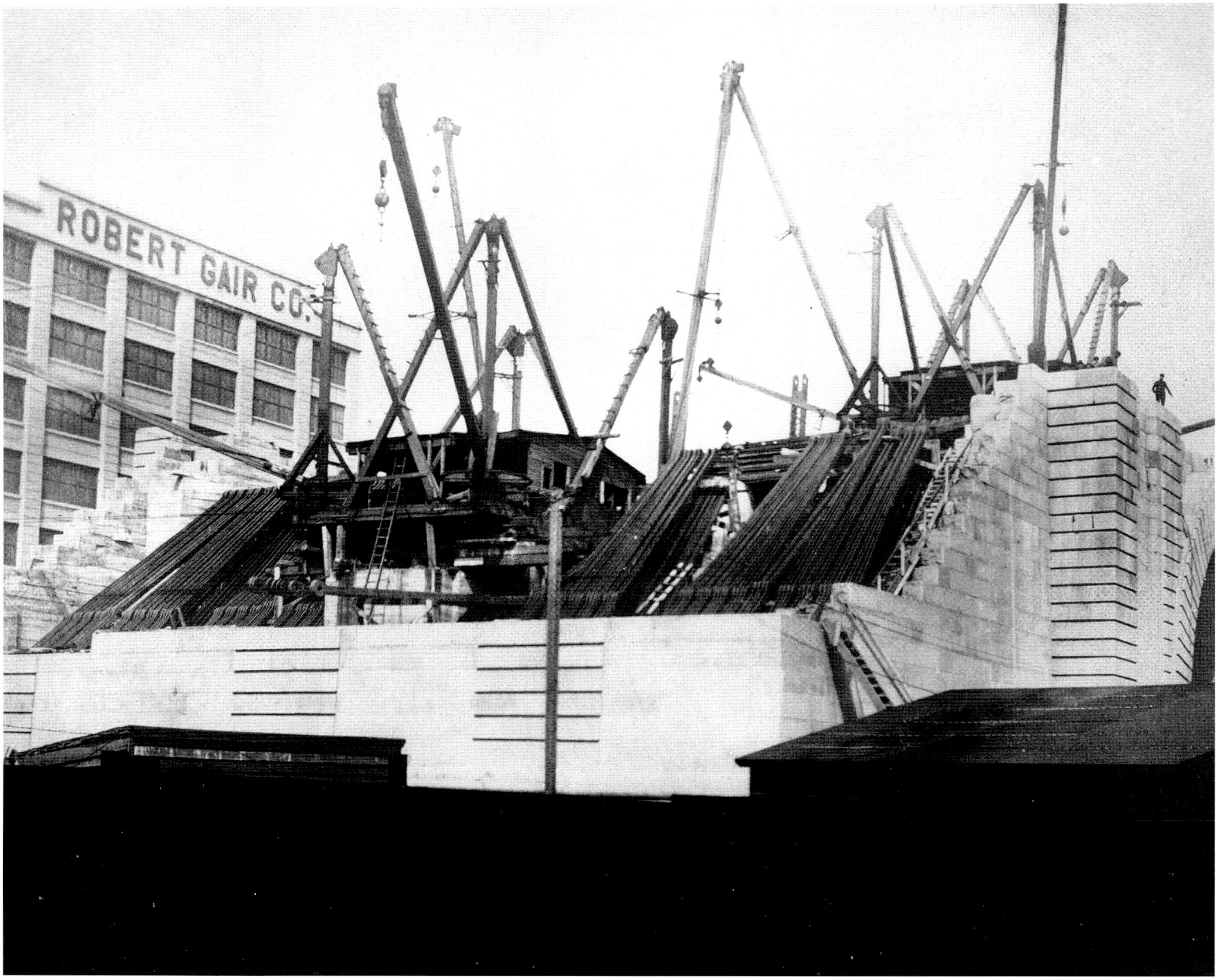

Foundation of Brooklyn tower: Steel derricks protrude from the anchorage base of the Brooklyn Bridge. On the top right, a man stands supervising. Behind the construction looms a Robert Gair warehouse, now in D.U.M.B.O. Gair, a printer who manufactured paper bags in the 1870s, went on to invent the corrugated box in 1890.

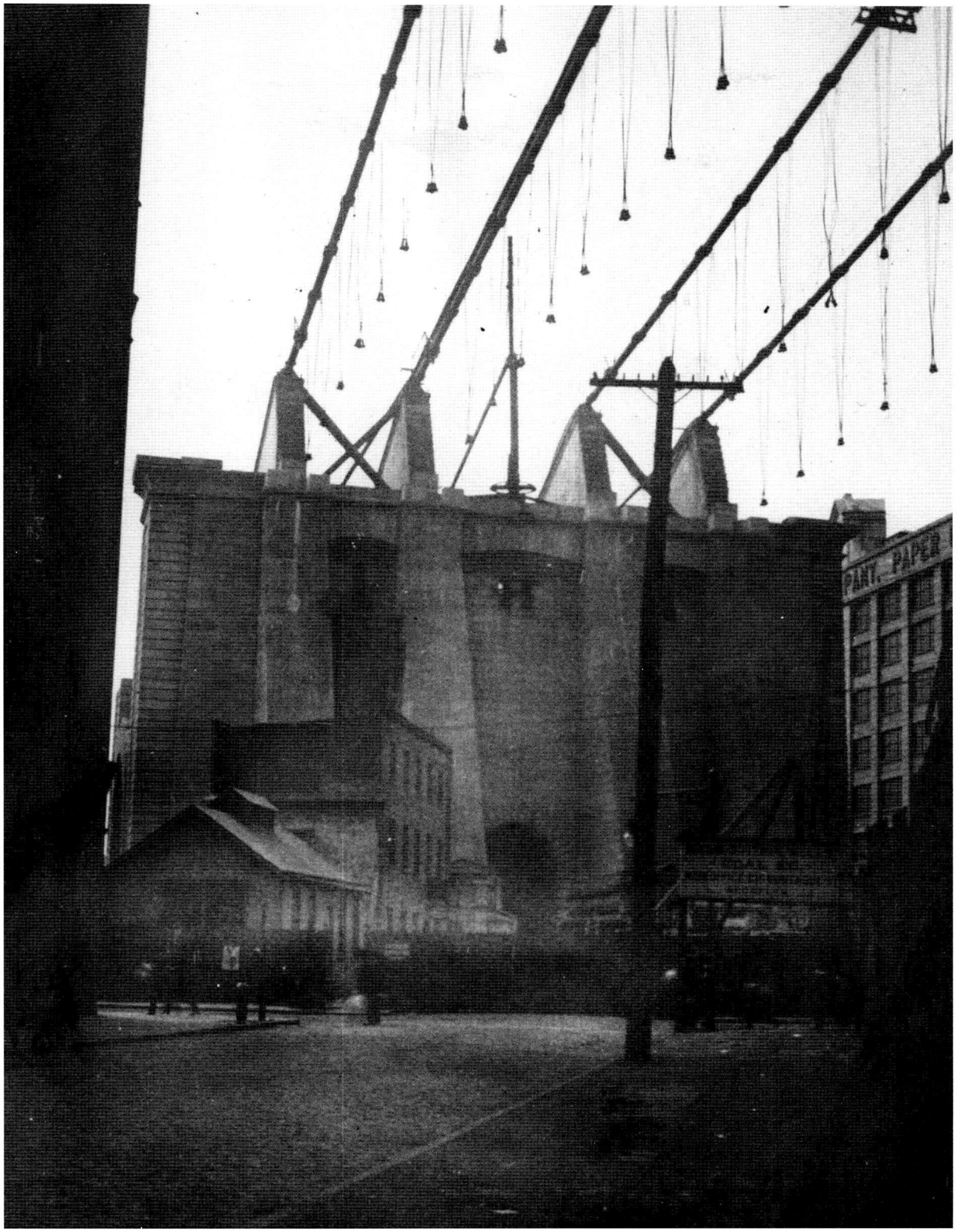

Rising elevation: Another view of the Brooklyn foundation with cables emerging from the tower's base. In the foreground are construction sheds and other buildings with the Gair building visible behind the bridge construction.

Following Spread: Bird's-eye view: Artist's concept of the Manhattan map in the 1870s as seen from Hoboken, New Jersey, originally published in *Illustrated London News,* 1876. Between Manhattan Island and Brooklyn can be seen the rising foundation of the Brooklyn Bridge.

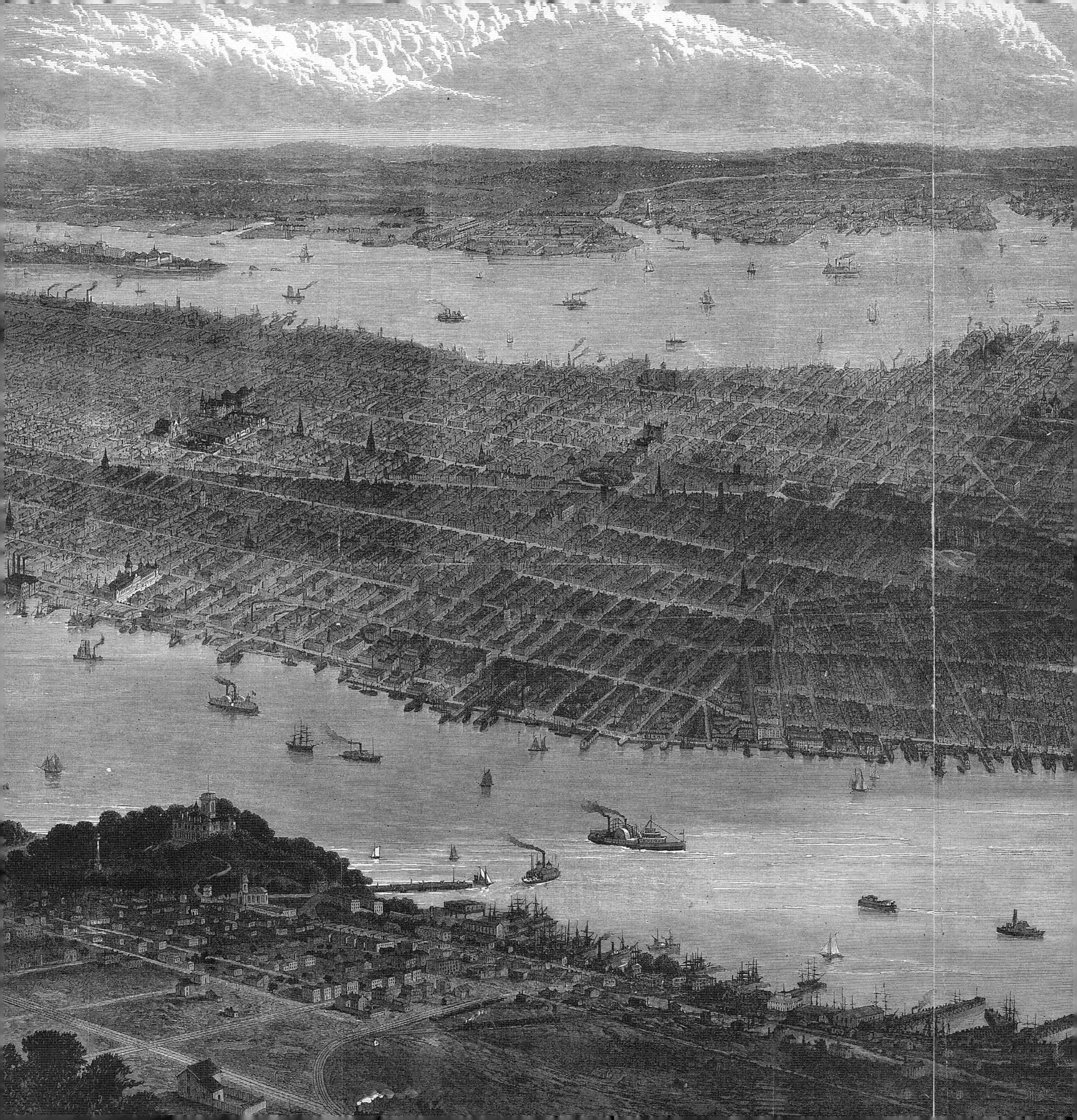

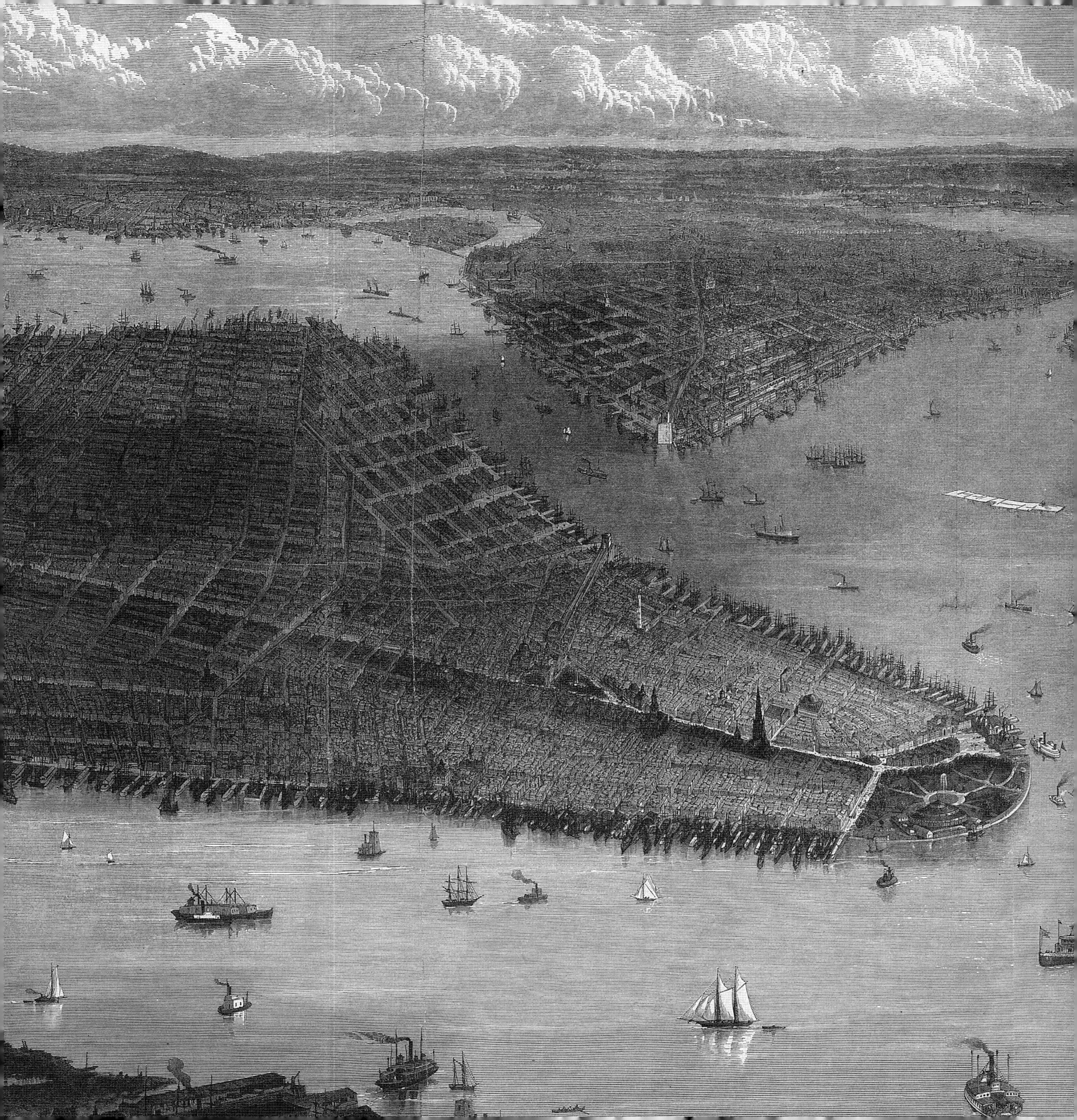

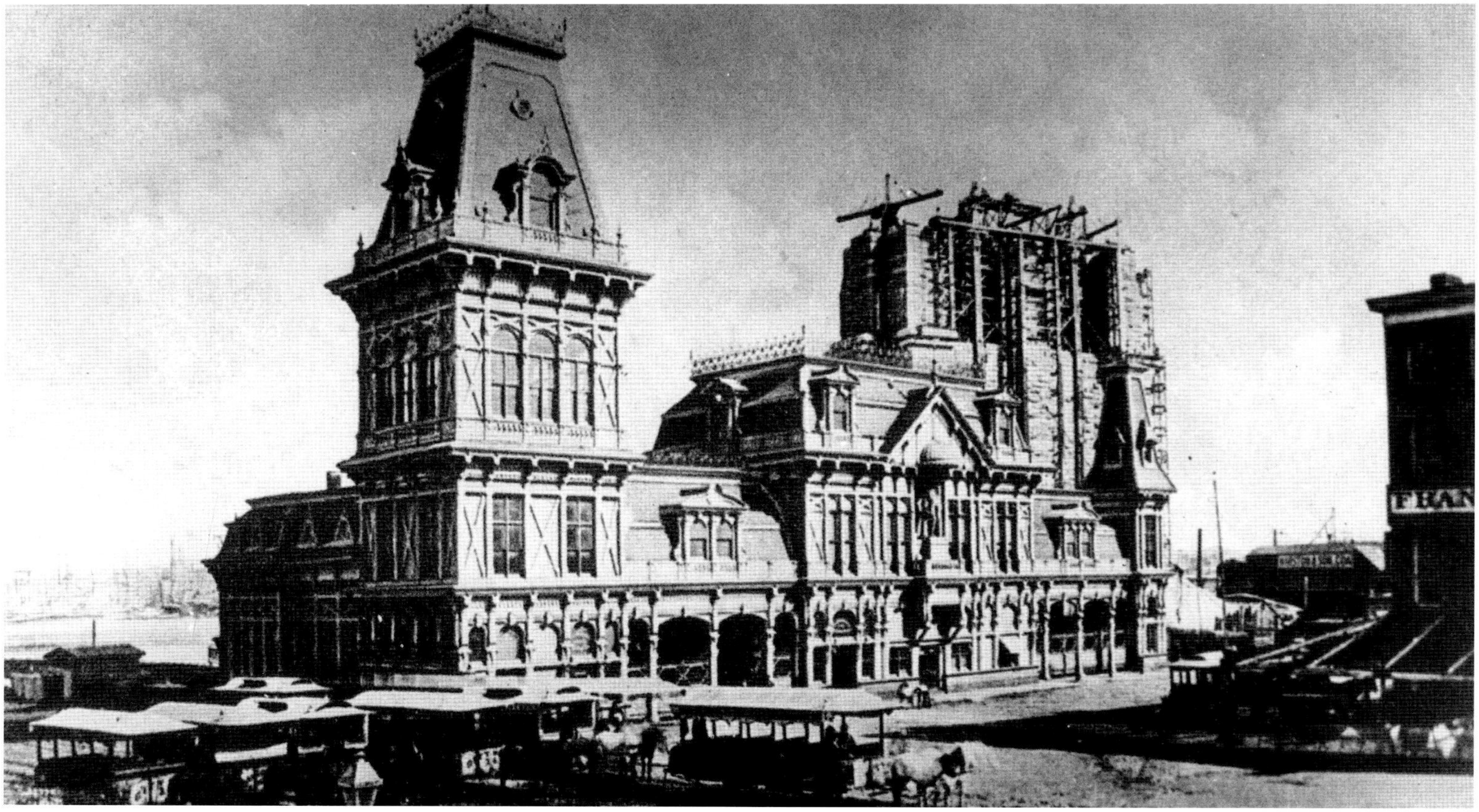

Brooklyn Ferry vs. Brooklyn Bridge: As the Brooklyn ferry busily transports passengers across the East River in 1874, the Brooklyn tower of the bridge rises behind the ferry building. Eventually one mode of transportation would replace the other.

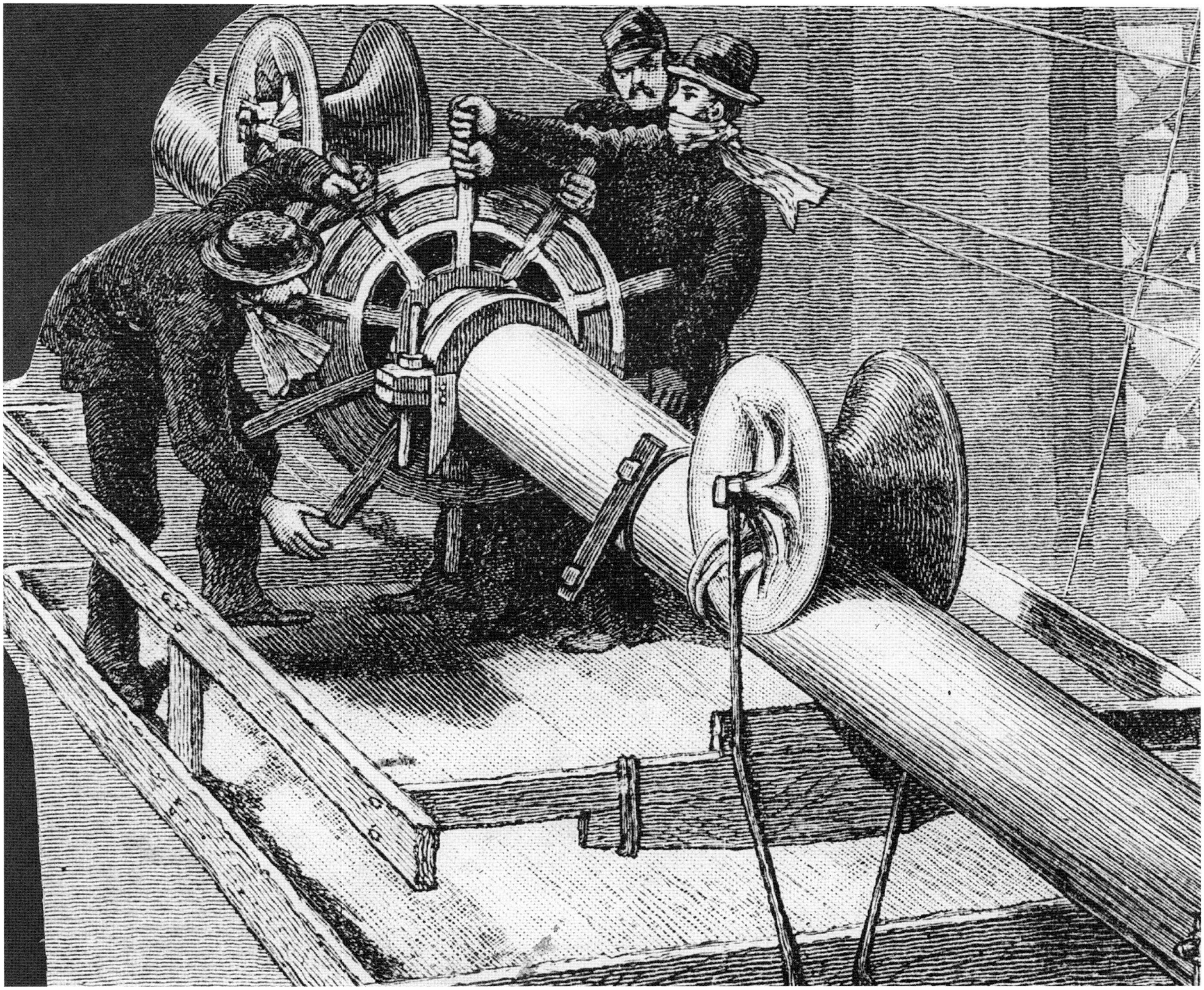

Engraving: Engraving of men working on the Brooklyn Bridge cabling.

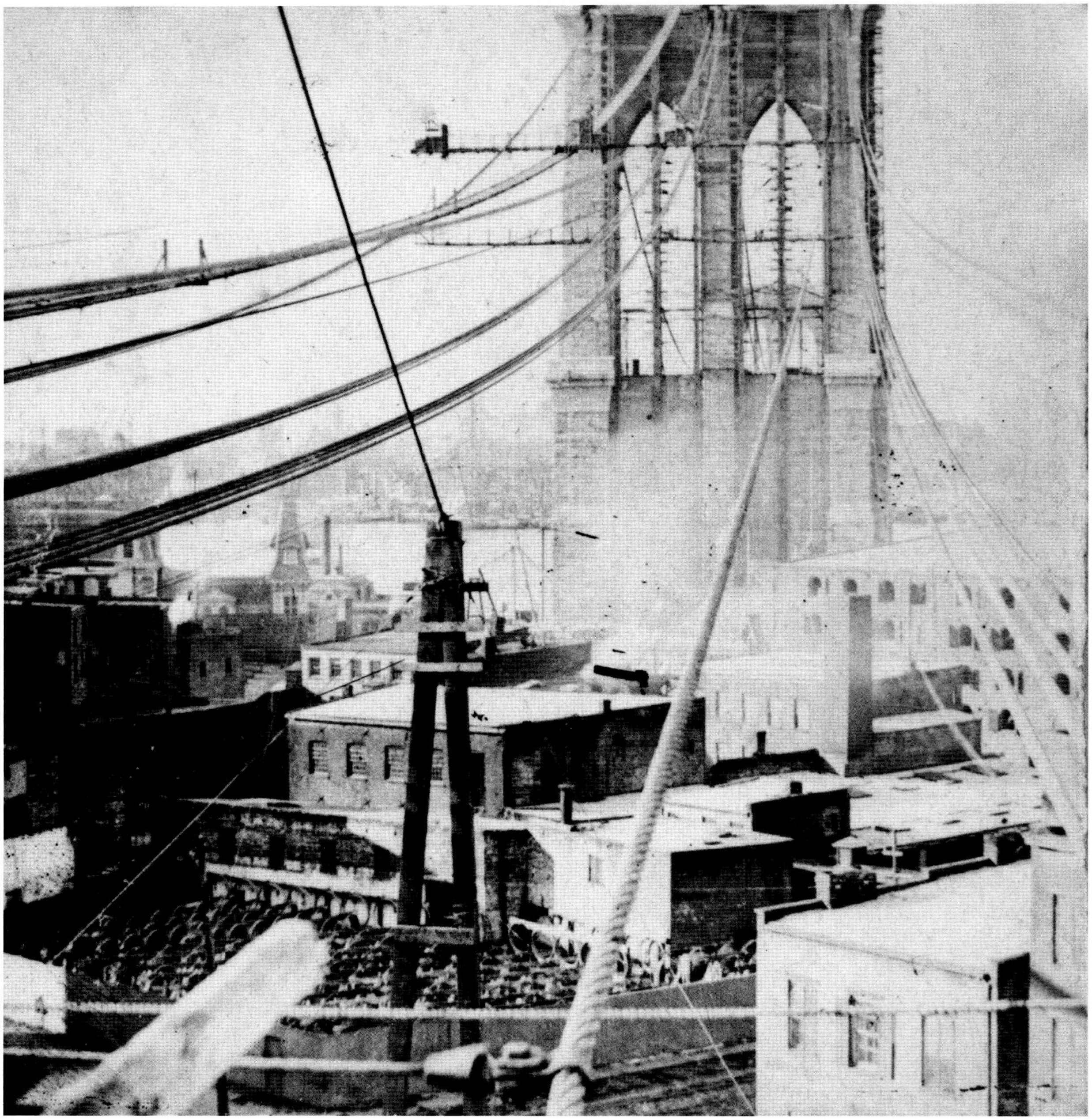

Brooklyn tower nearing completion in mid-1870s with cables and nearby storage buildings. "East River Bridge" was one of several names given to the Brooklyn Bridge.

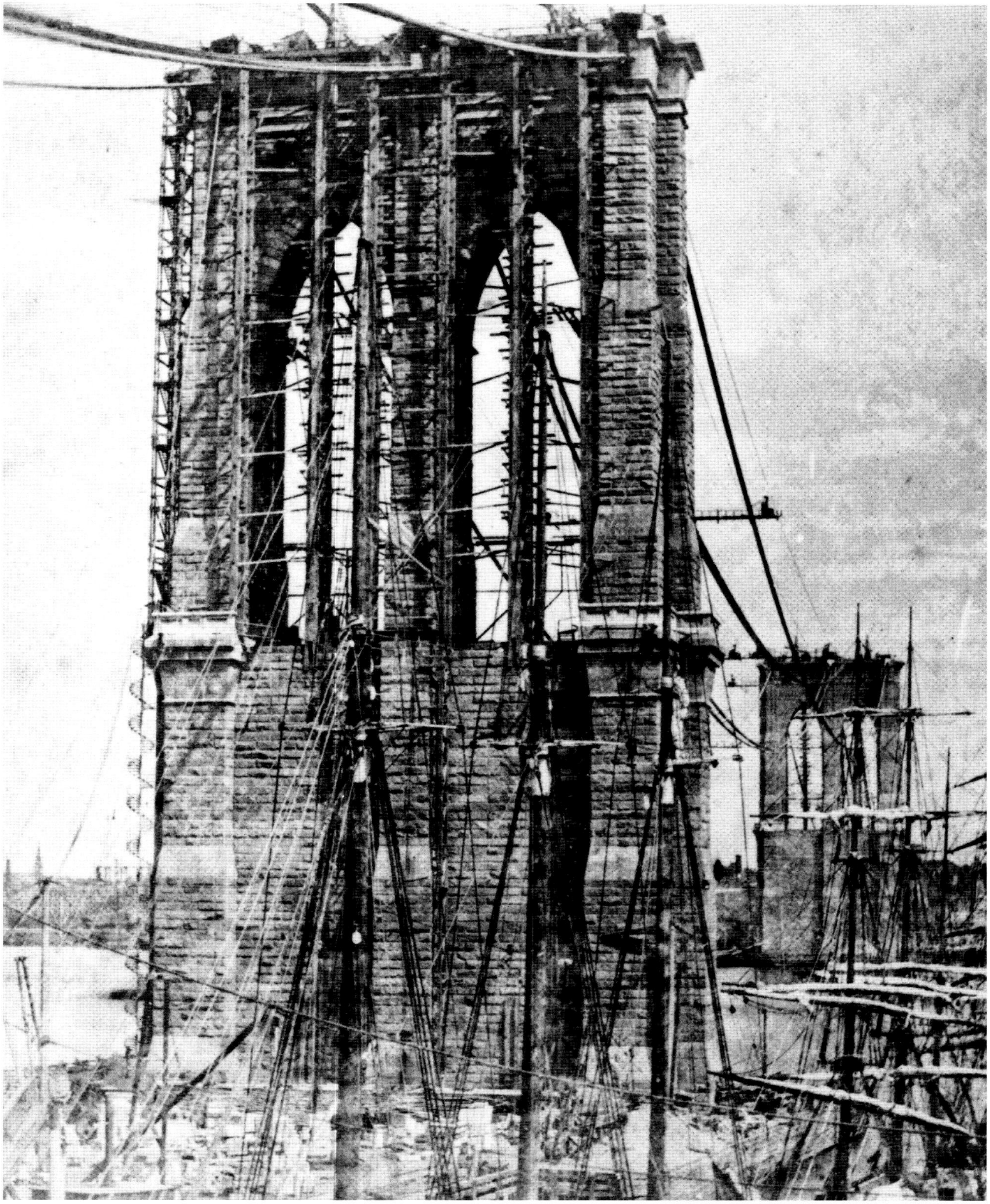

Close-up of both towers, with the New York tower near completion and cables being run between the two. At the lower right are masts of sailing ships, once the tallest structures in New York Harbor.

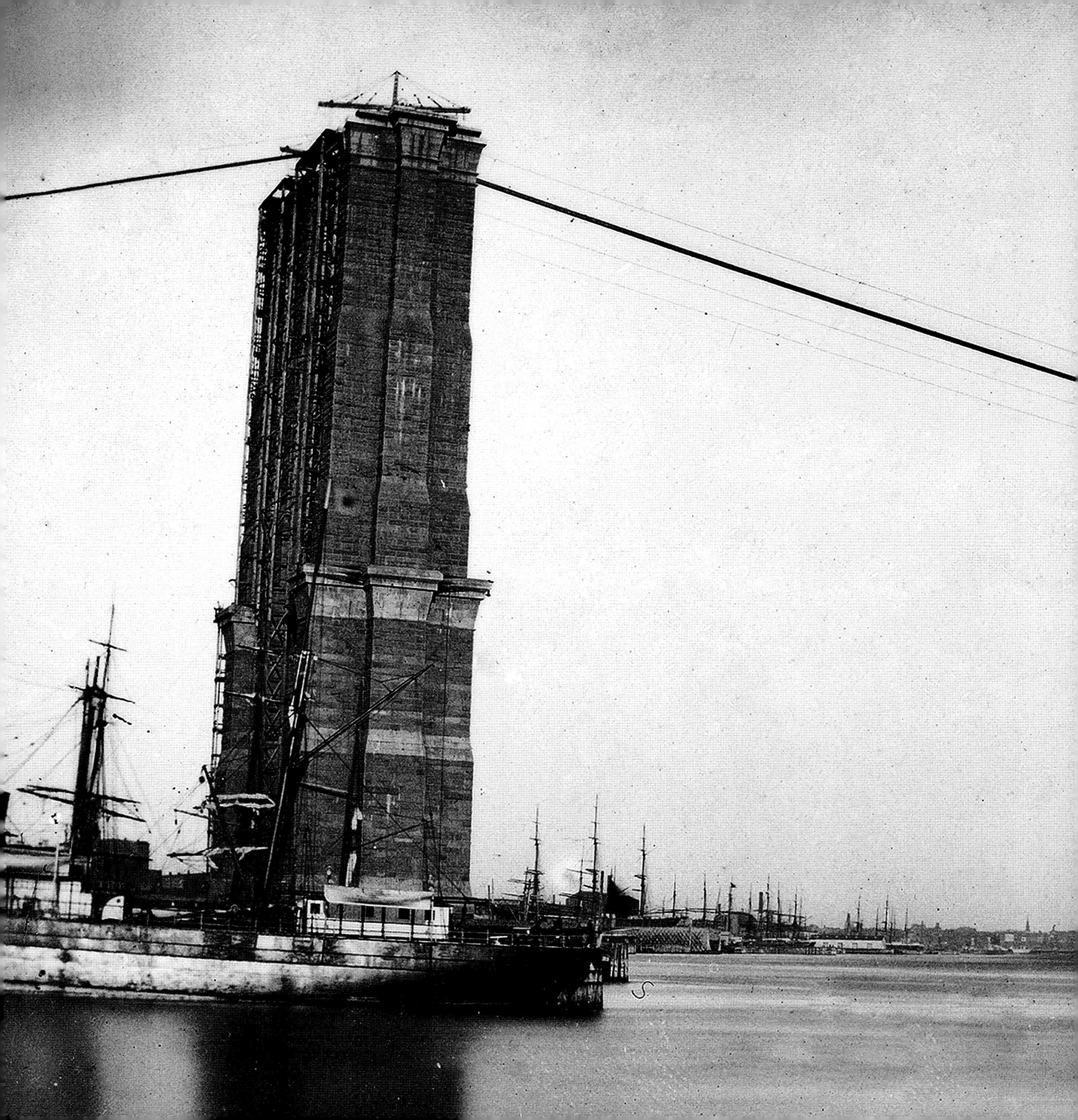

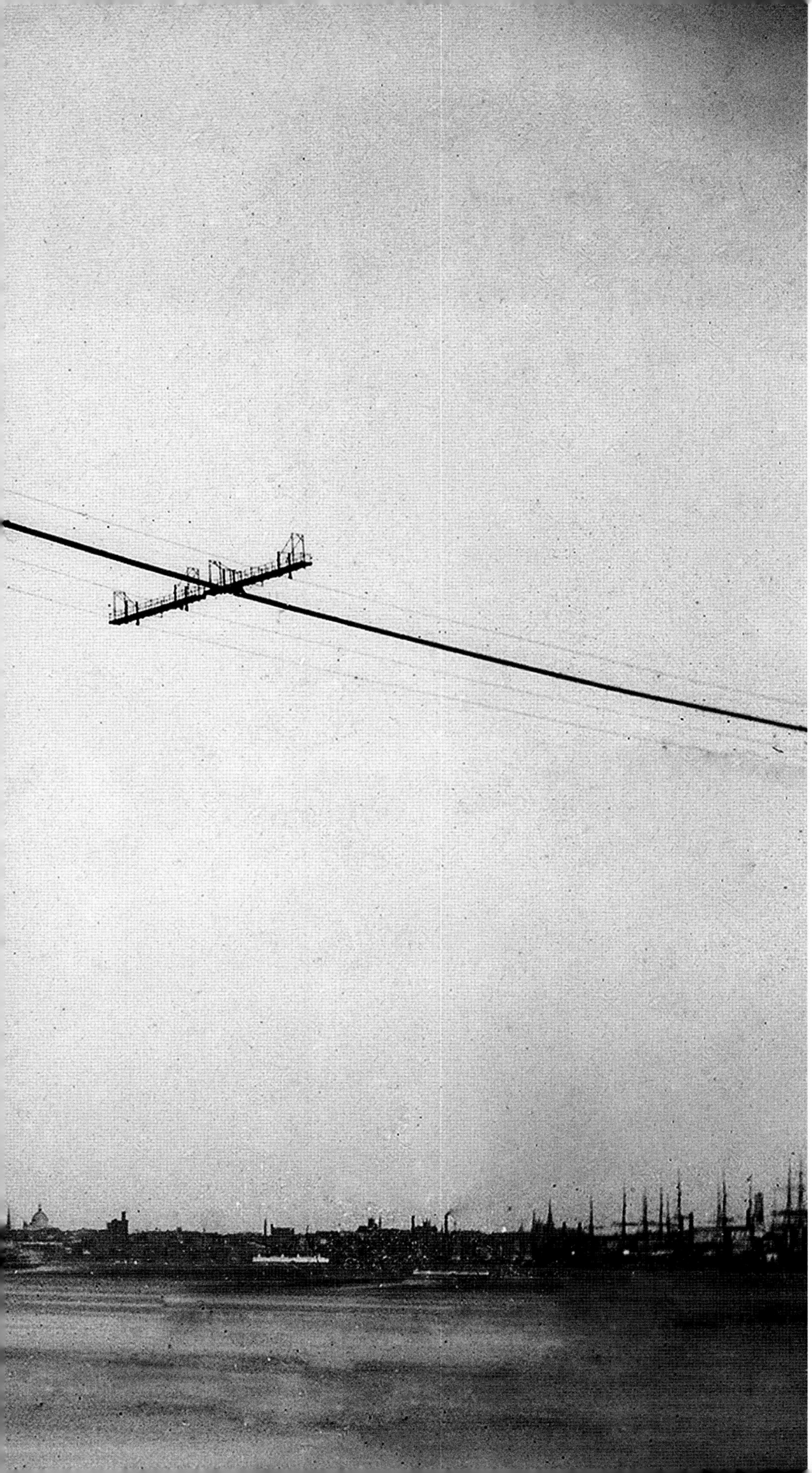

Cables from Brooklyn to Manhattan, the first bridging of the East River. This 1870s photograph by George Brainerd also shows sailing ships moored on both sides of the river.

Brooklyn Bridge construction: Viewed from Manhattan, construction of the Brooklyn Bridge shows both towers completed and cables strung from their pinnacles. The many buildings in the foreground belong to shipping-related businesses.

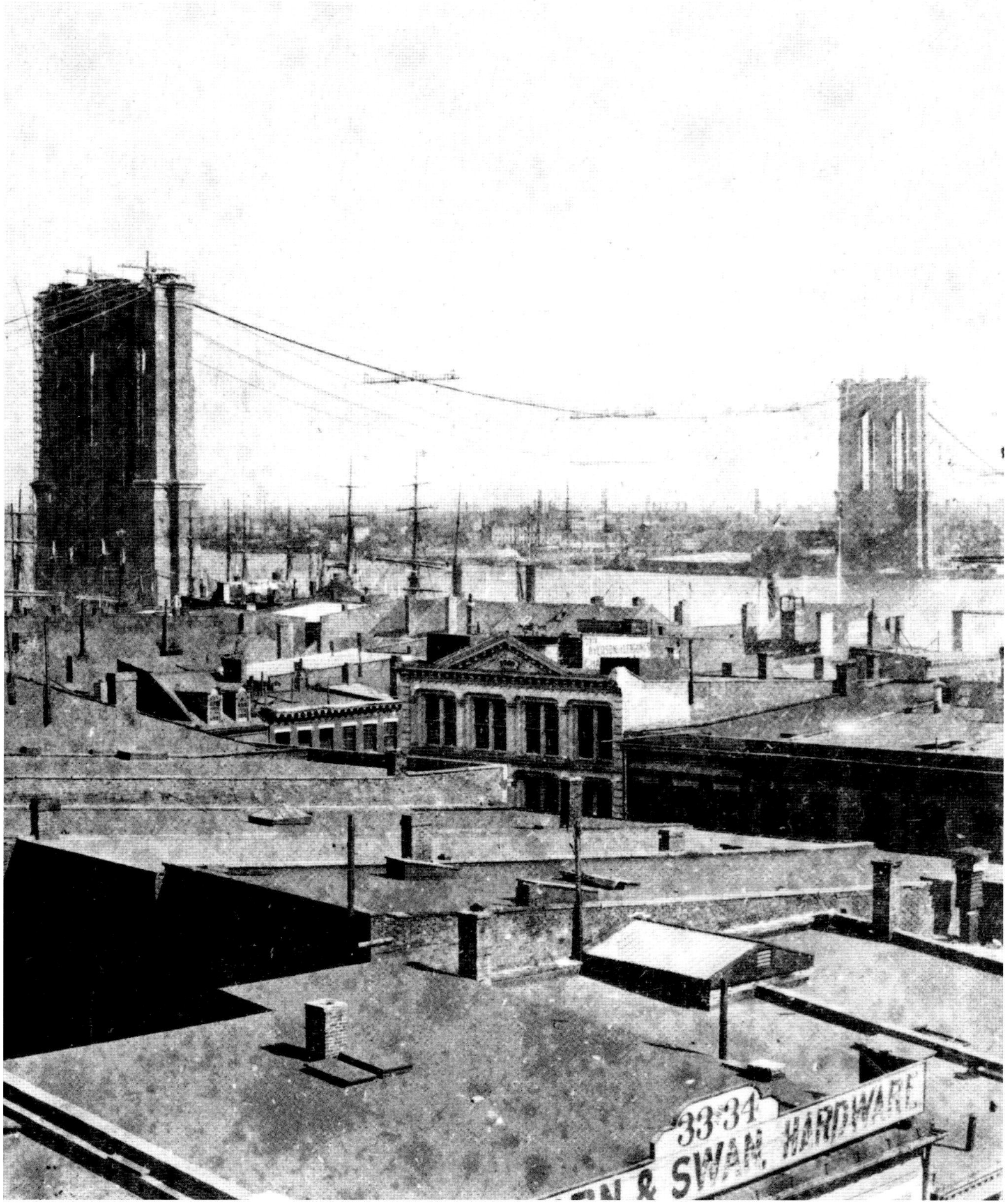

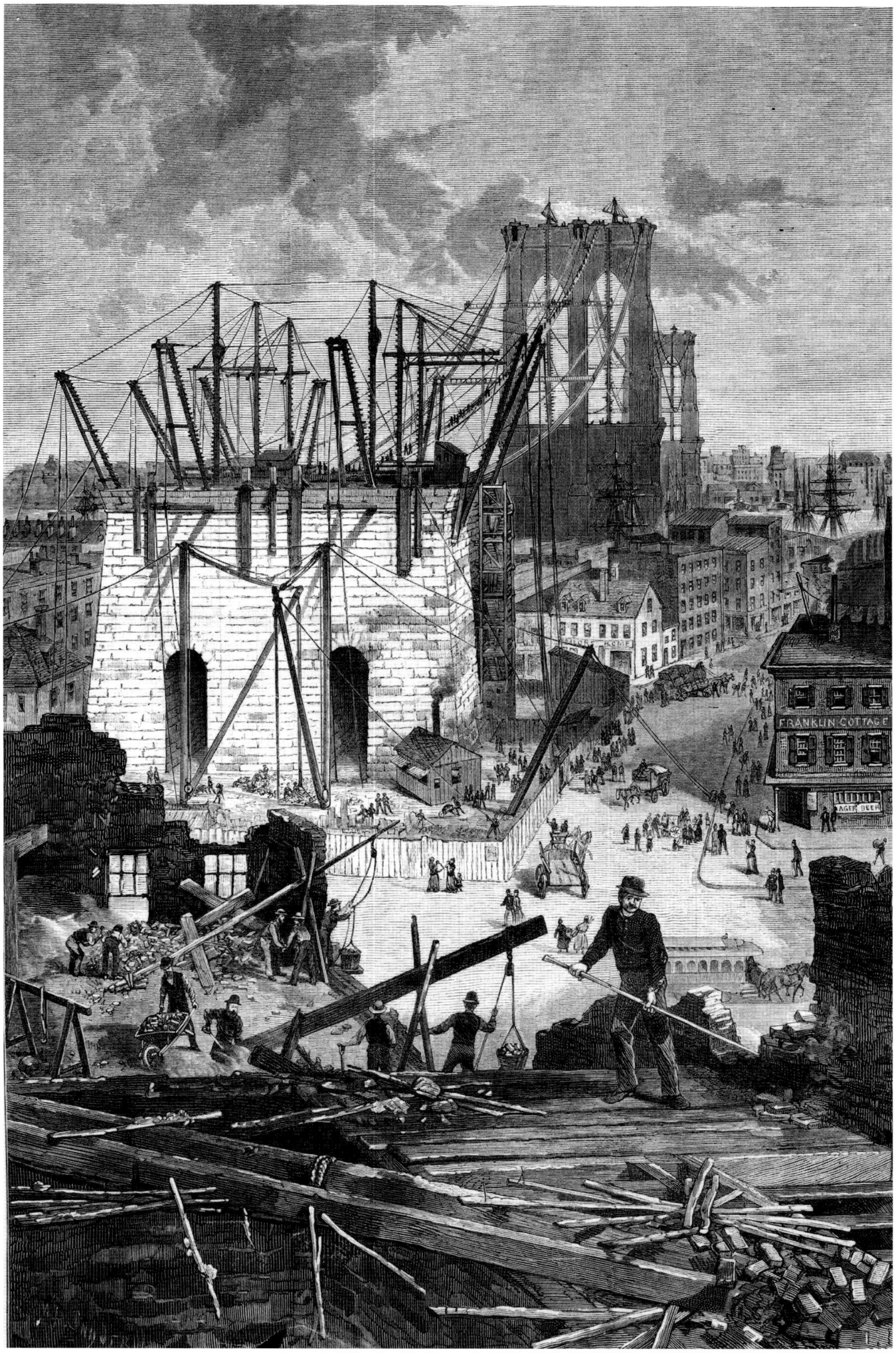

Brooklyn approach: The demolition of waterfront buildings became necessary to make room for the long approaches. This view from an engraving published in an 1877 issue of *Harper's Weekly* shows the Brooklyn anchorage, both towers, and the newly installed footbridge. Among the buildings destroyed for the Manhattan anchorage was 1 Cherry Street, home of George and Martha Washington from 1789-90 when New York was the capital. A century later, the east side neighborhood had declined into a slum.

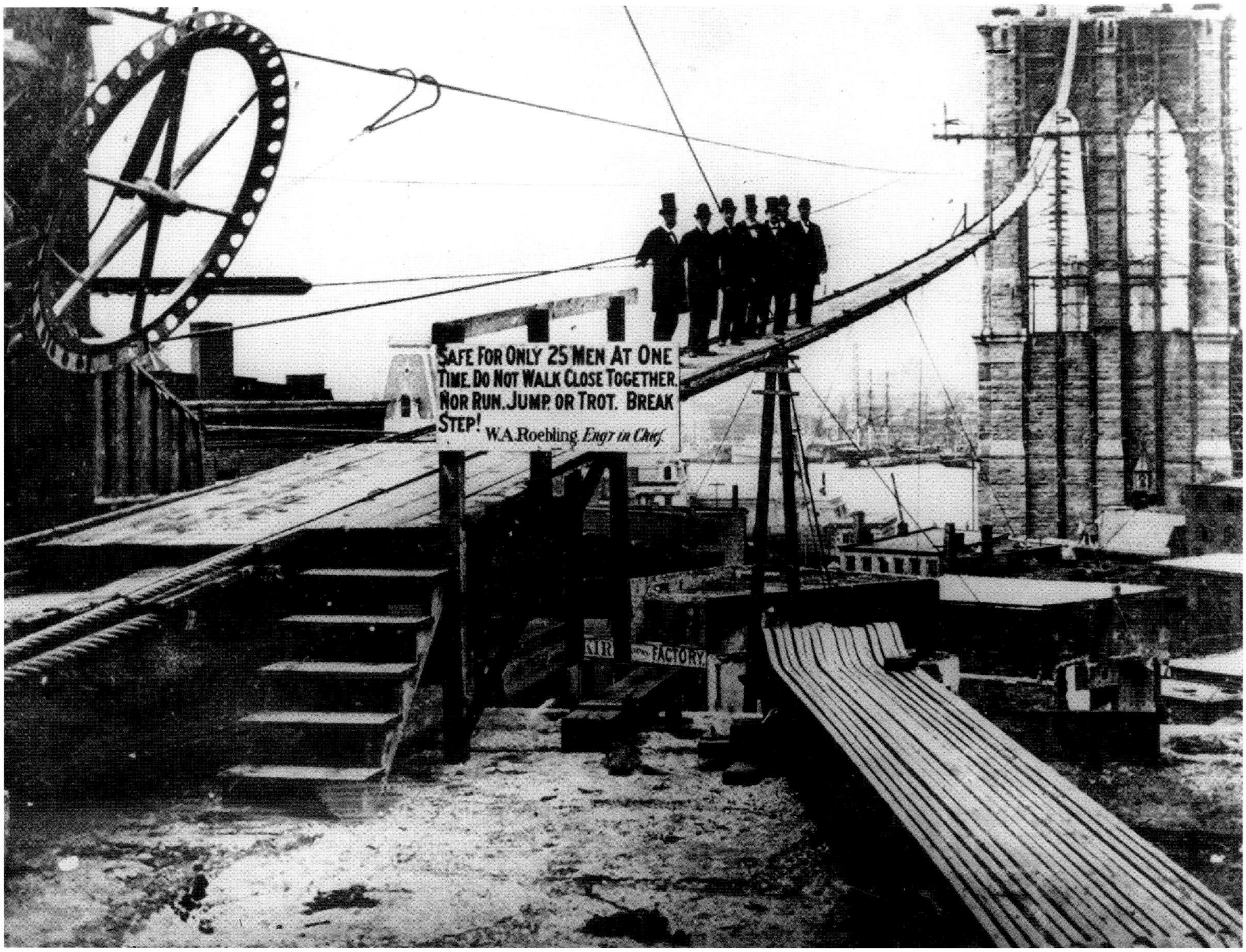

Catwalk of Brooklyn Bridge: Seven men stand on a catwalk leading from the top of the anchorage to the Brooklyn tower early in the bridge's construction. Note the sign limiting the number of people allowed on the catwalk and warning them not to walk in cadence. According to author and historian David McCullough, the visitors are bridge trustees. In the foreground are anchor bars and a wheel used as a pulley.

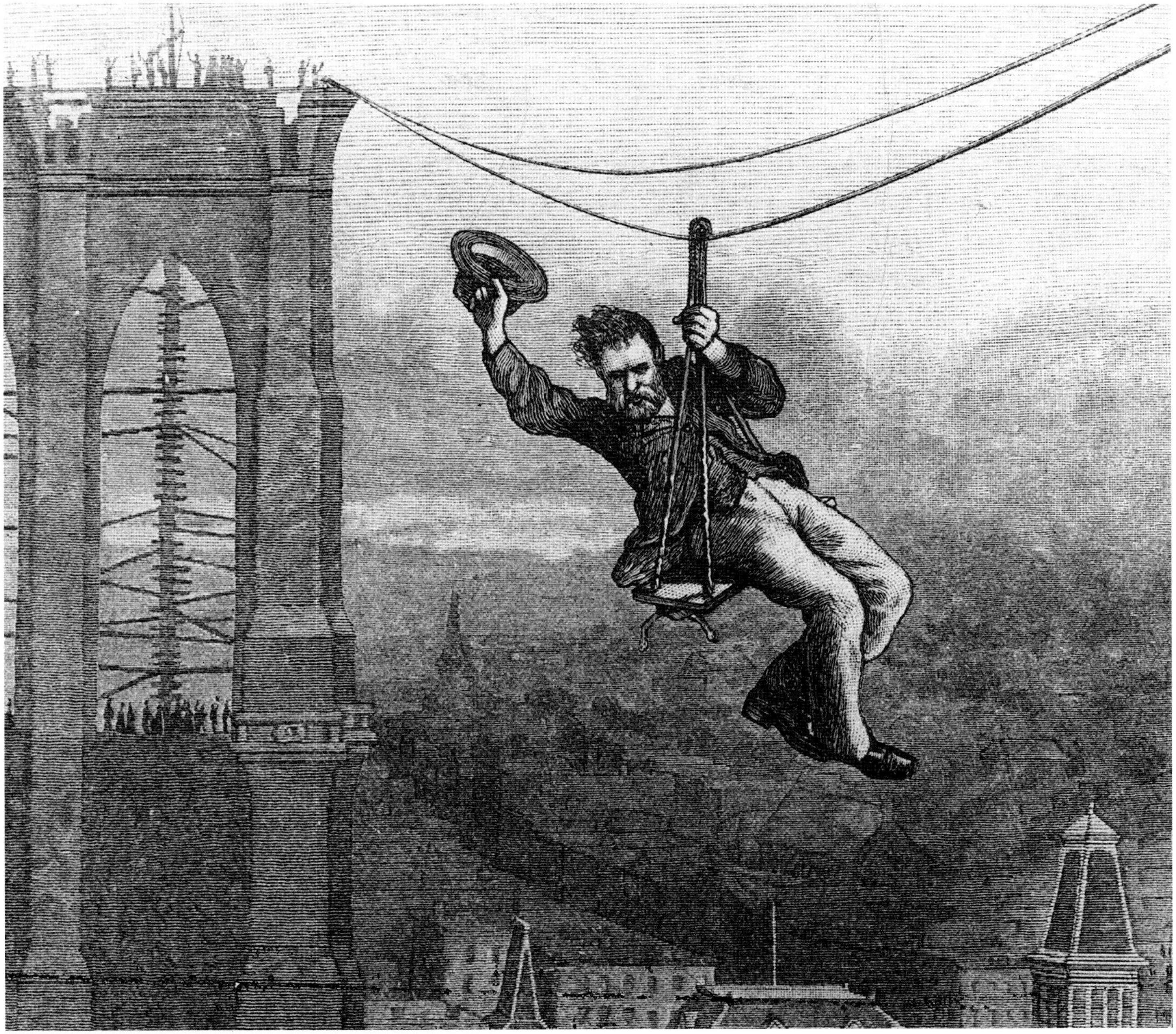

Farrington crossing East River on a "traveler": On August 25, 1876, 60-year-old E. Frank Farrington, master mechanic of the bridge, rode from the Brooklyn anchorage and then between the bridge towers on a boatswain's chair (2 ft. by 2 in. thick) attached to a wire "traveler" rope. The journey lasted 22 minutes. An estimated 10,000 people watched while he stood on the seat, doffed his hat, and blew kisses to the crowd.

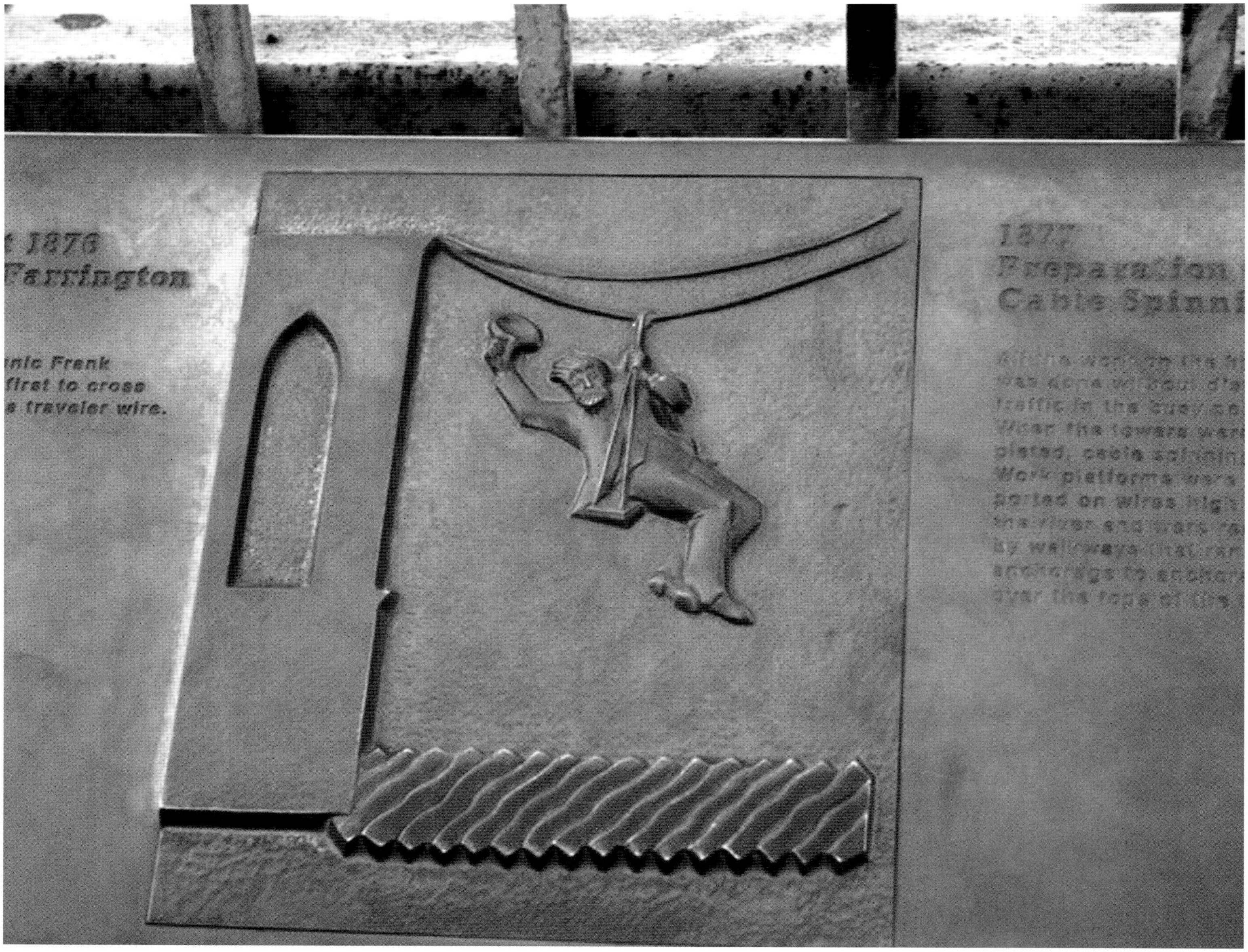

Farrington plaque: Re-created from a drawing, this image on a Brooklyn Bridge plaque is accompanied by an explanation of Farrington's exploit.

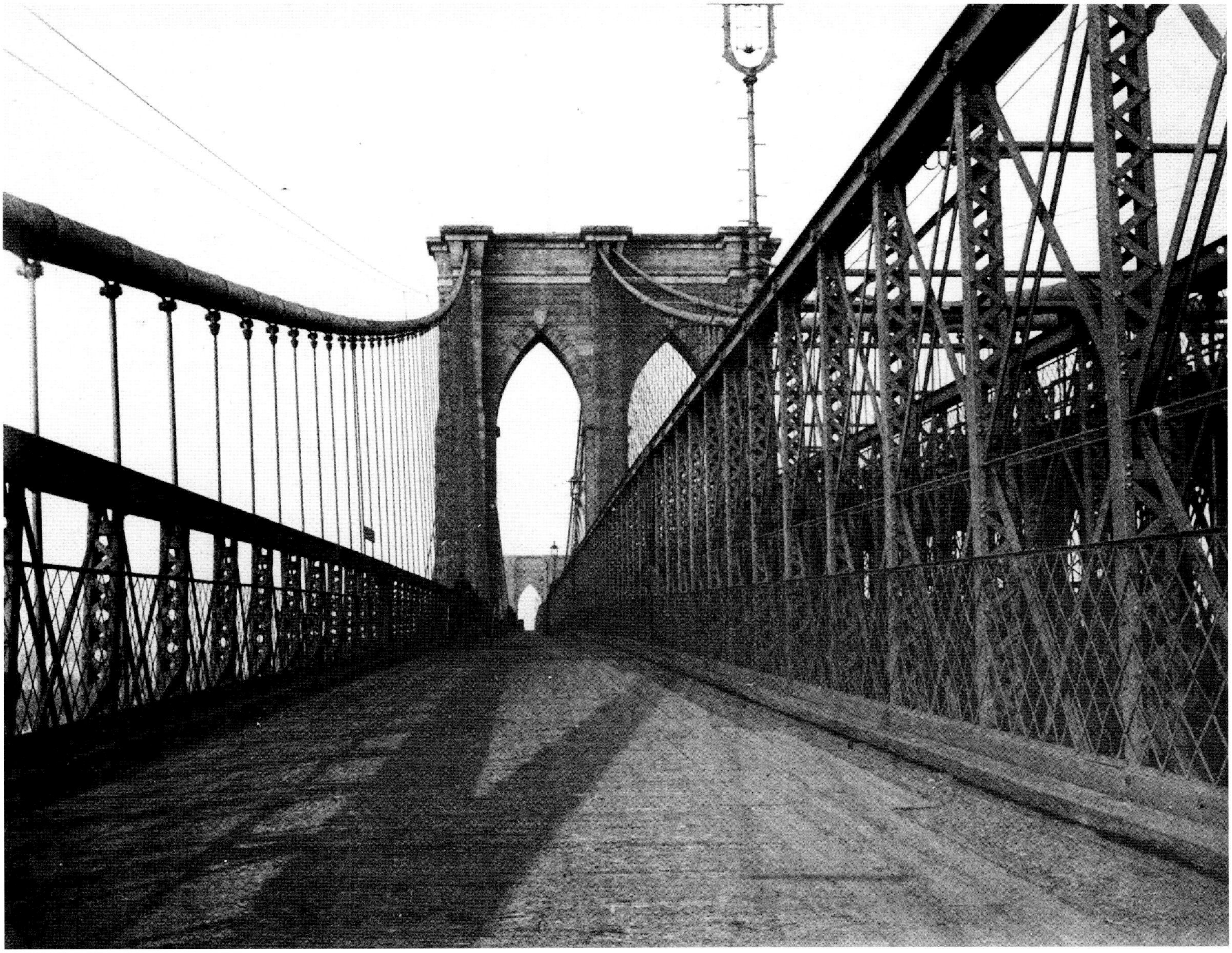

Construction on bridge roadbed: Many types of cranes were used in the bridge construction, including the gantry crane that would shift along the completed roadbed on either rails or the cable.

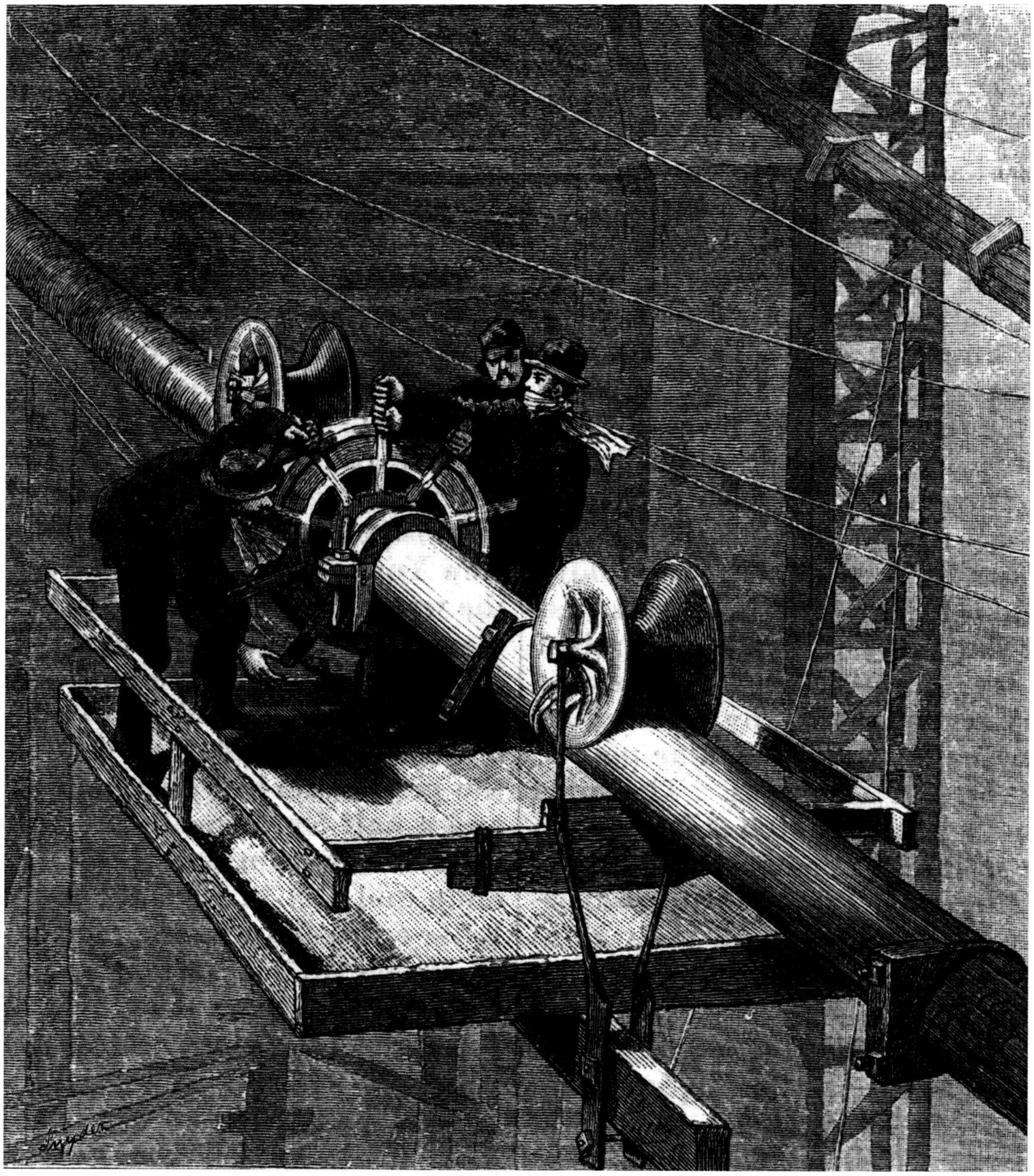

Cable-spinning machine: Workmen spinning cable with a machine of Roebling's invention. A scandal erupted when inferior quality wire was substituted by rival J. Lloyd Haigh.

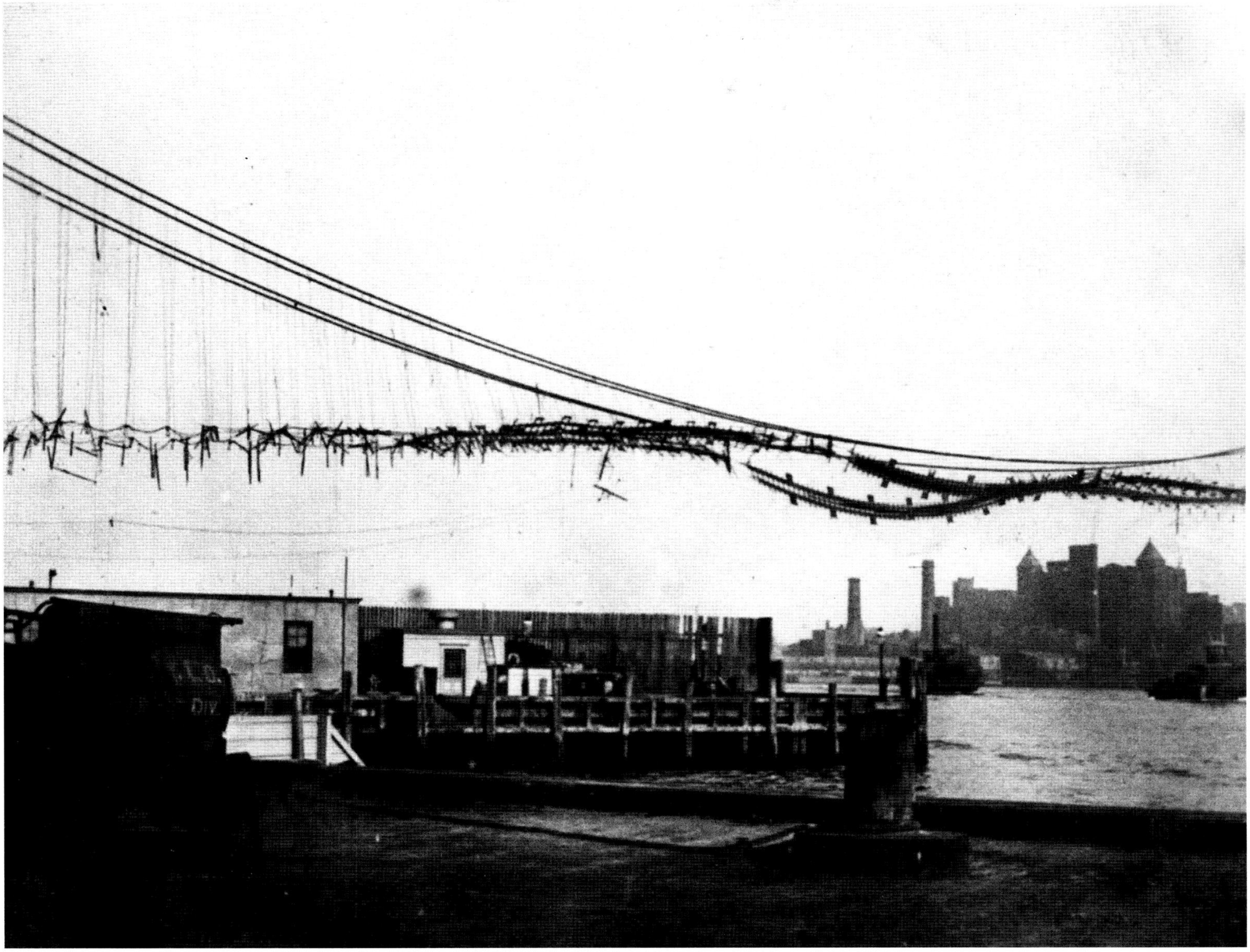

Hanging cables: As the cables were completed, workers hung suspenders from them to support the roadbed. At the bottom of the picture are waterfront buildings along Brooklyn's shore with Manhattan buildings in the background.

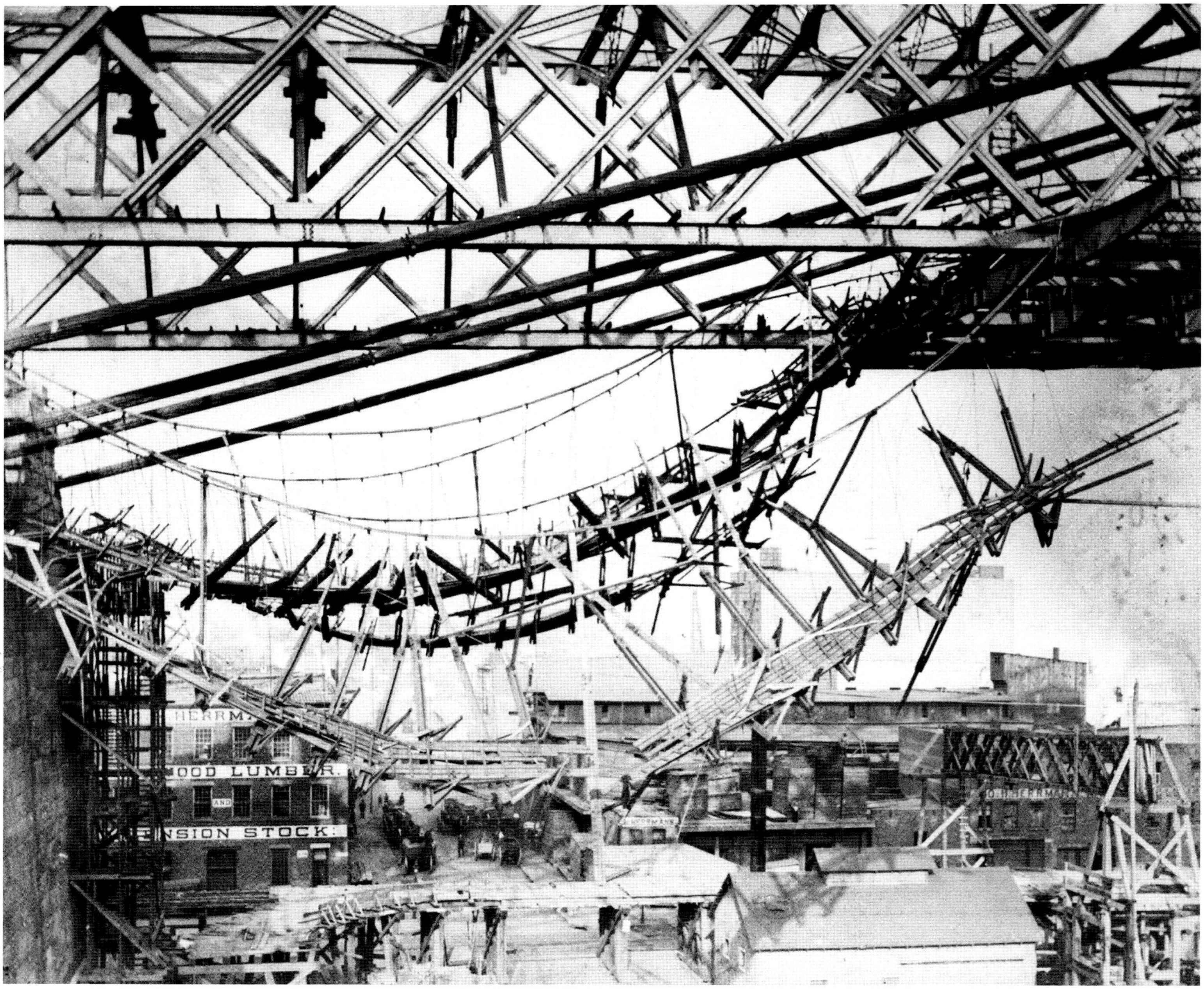

Construction progress: Cables, suspenders, walkways, and detritus from bridge construction hang from the superstructure.

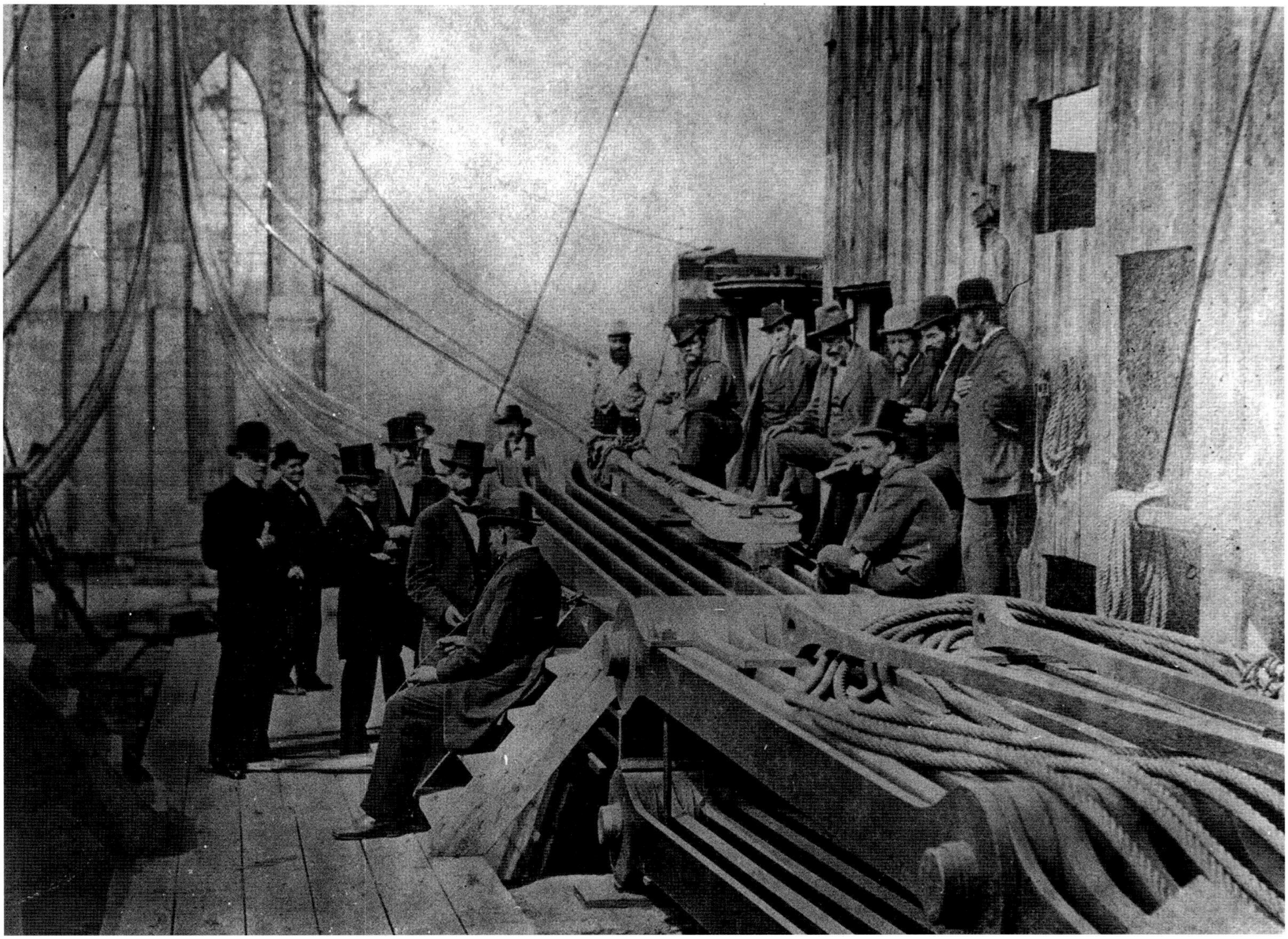

Key figures in the Brooklyn Bridge construction: Gathered on the Brooklyn anchorage in 1878 are Henry Cruse Murphy, president of the New York Bridge Company (arms folded); Wilhelm Hildenbrand, the chief draftsman (group on right, fifth from left); E. Frank Farrington, master mechanic (to Hildenbrand's left); and Charles C. Martin, first assistant engineer (next to Farrington). The cables that appeared in a previous photograph are shown here fastened to the anchorage.

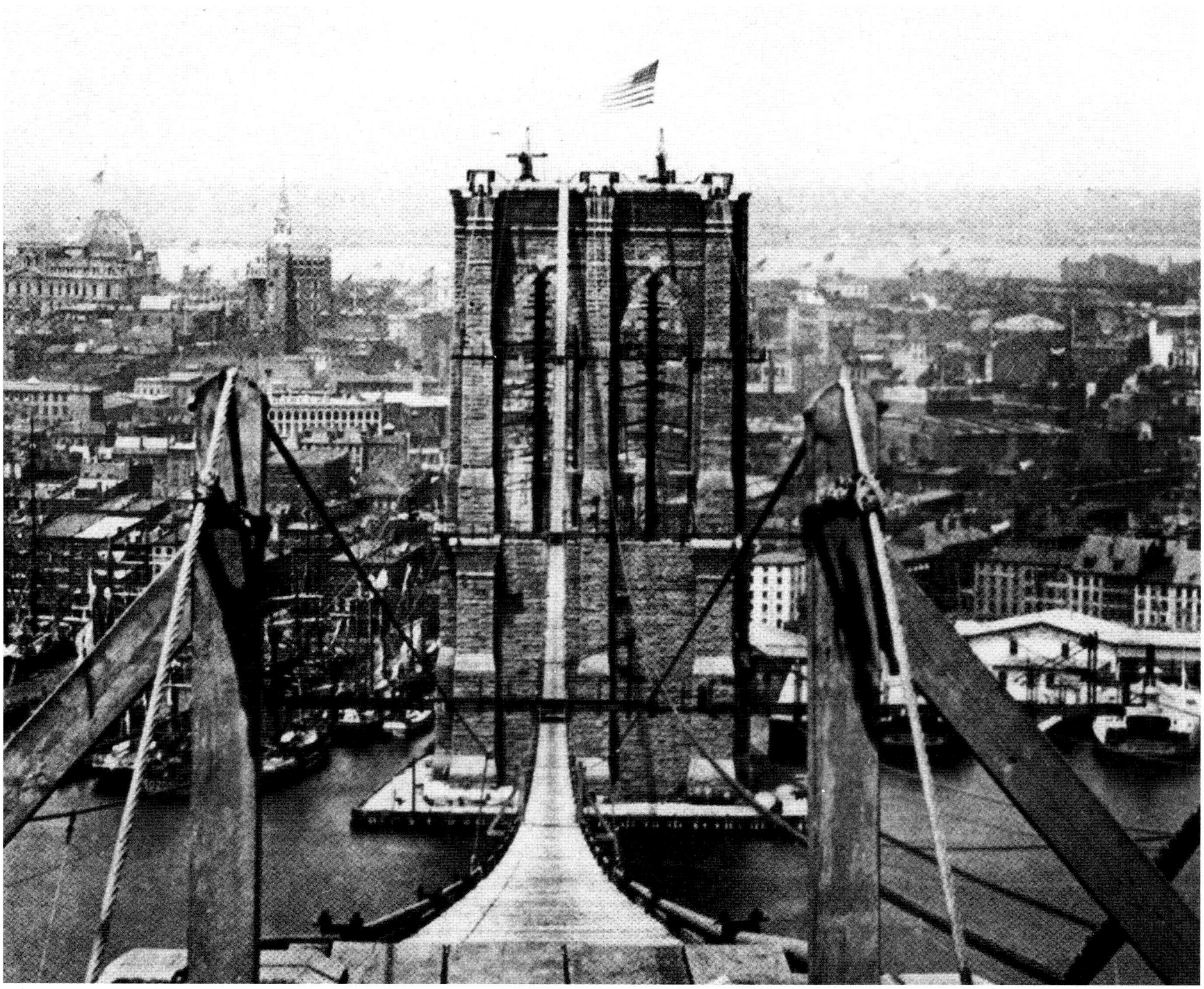

Looking toward New York tower: This view from the Brooklyn tower in 1877 is what Farrington would have seen from the boatswain's chair—without the walkway. The flag flies from the New York tower, and Lower Manhattan can be seen extending to the Hudson River.

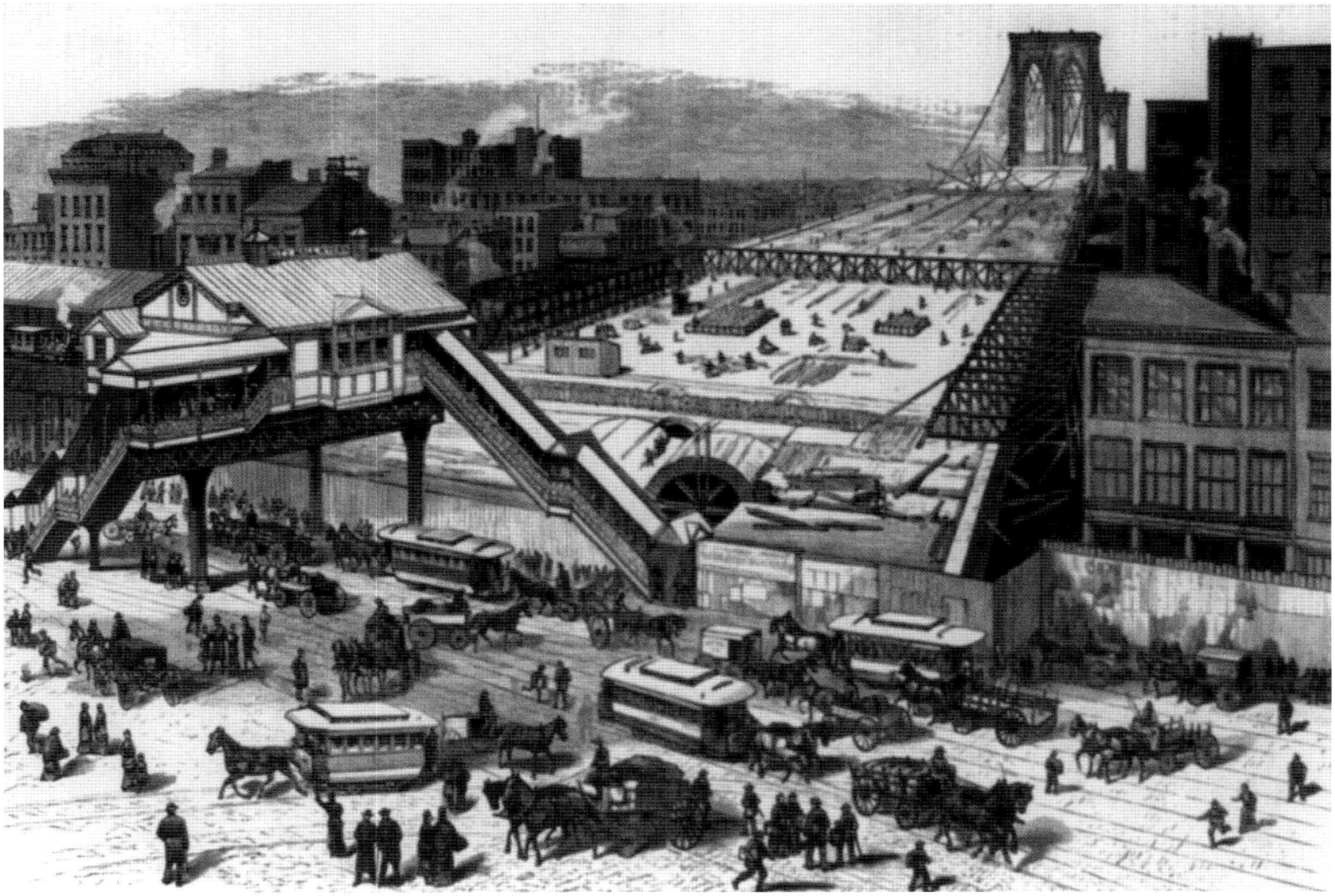

Finishing the New York approach: With completion of the New York approach, construction started on the rail terminals. This *Frank Leslie* engraving from an 1881 issue shows the view from the Hall of Records, which was near City Hall Park.

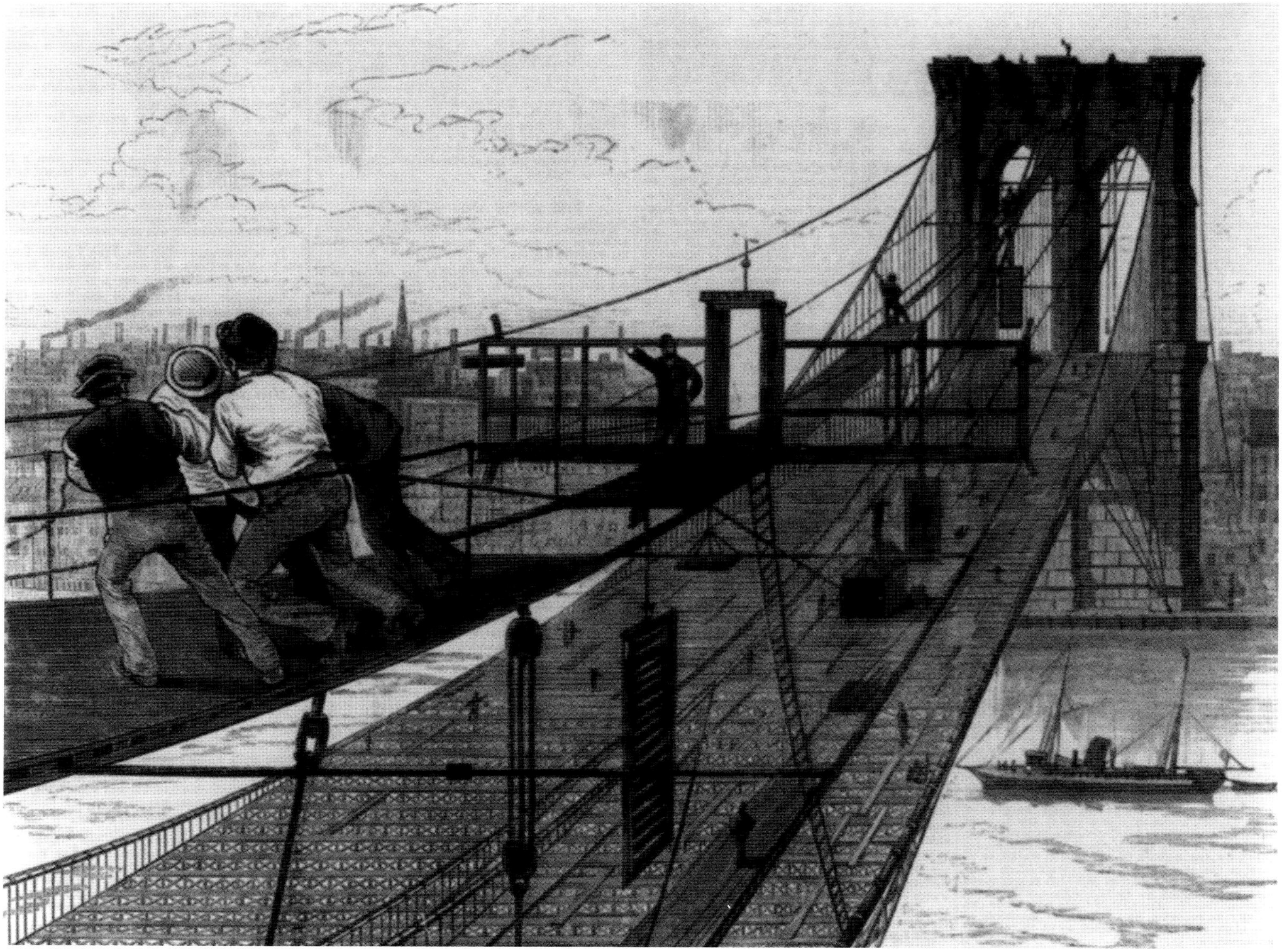

Removing the footbridge: Once the steel deck of the roadway was in place, the footbridge was no longer necessary, although its popularity had extended to the public, who could test their nerves crossing from tower to tower. This 1882 engraving comes from *Frank Leslie's Illustrated Newspaper.*

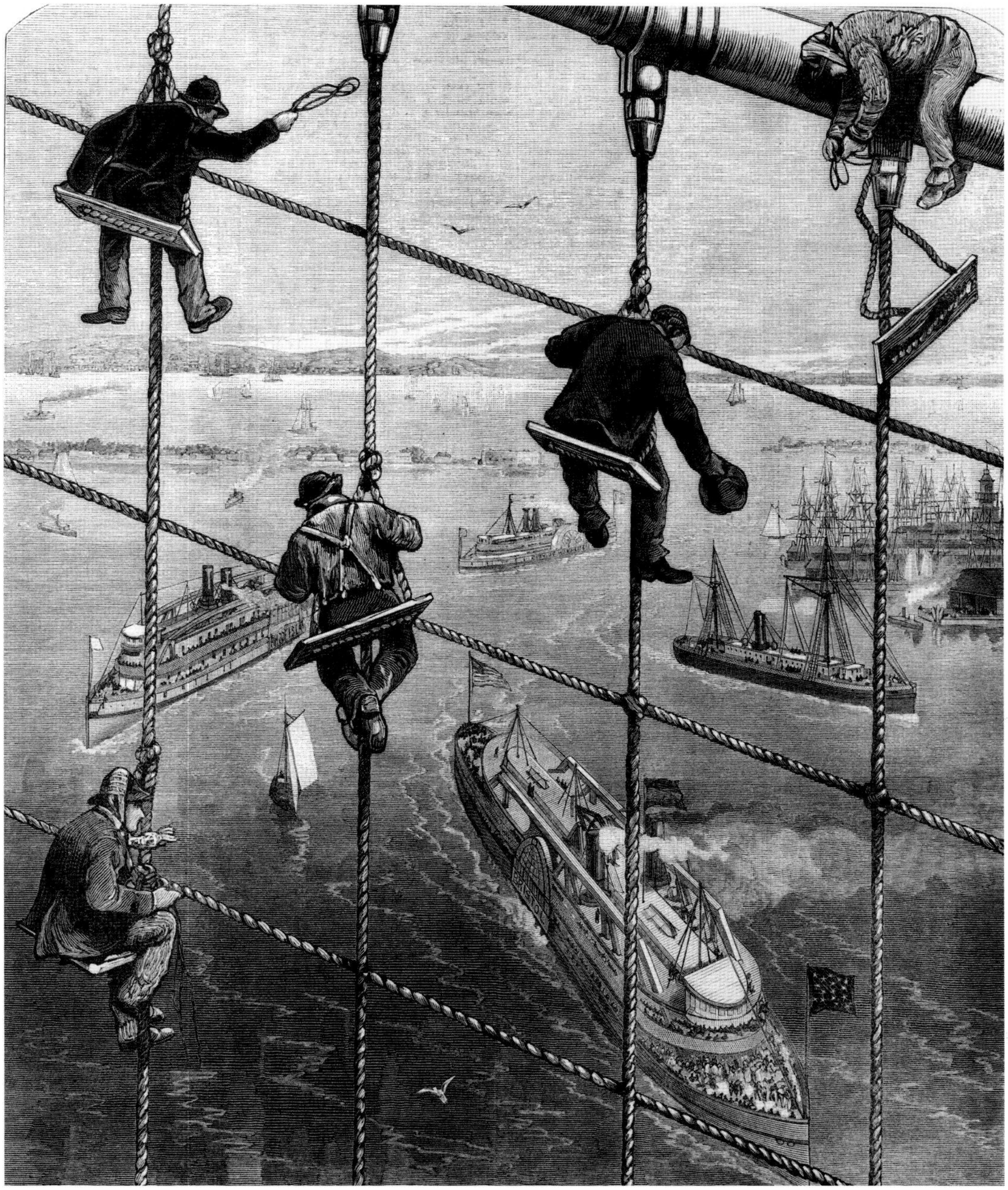

Lashing the stays: The cover of *Frank Leslie's Illustrated Newspaper* of April 23, 1883, says it all: the Brooklyn Bridge is almost finished, just a week from opening day. Workmen are attaching the suspenders which prevent the bridge from swaying too much. Below steamships move by as well as sailing vessels.

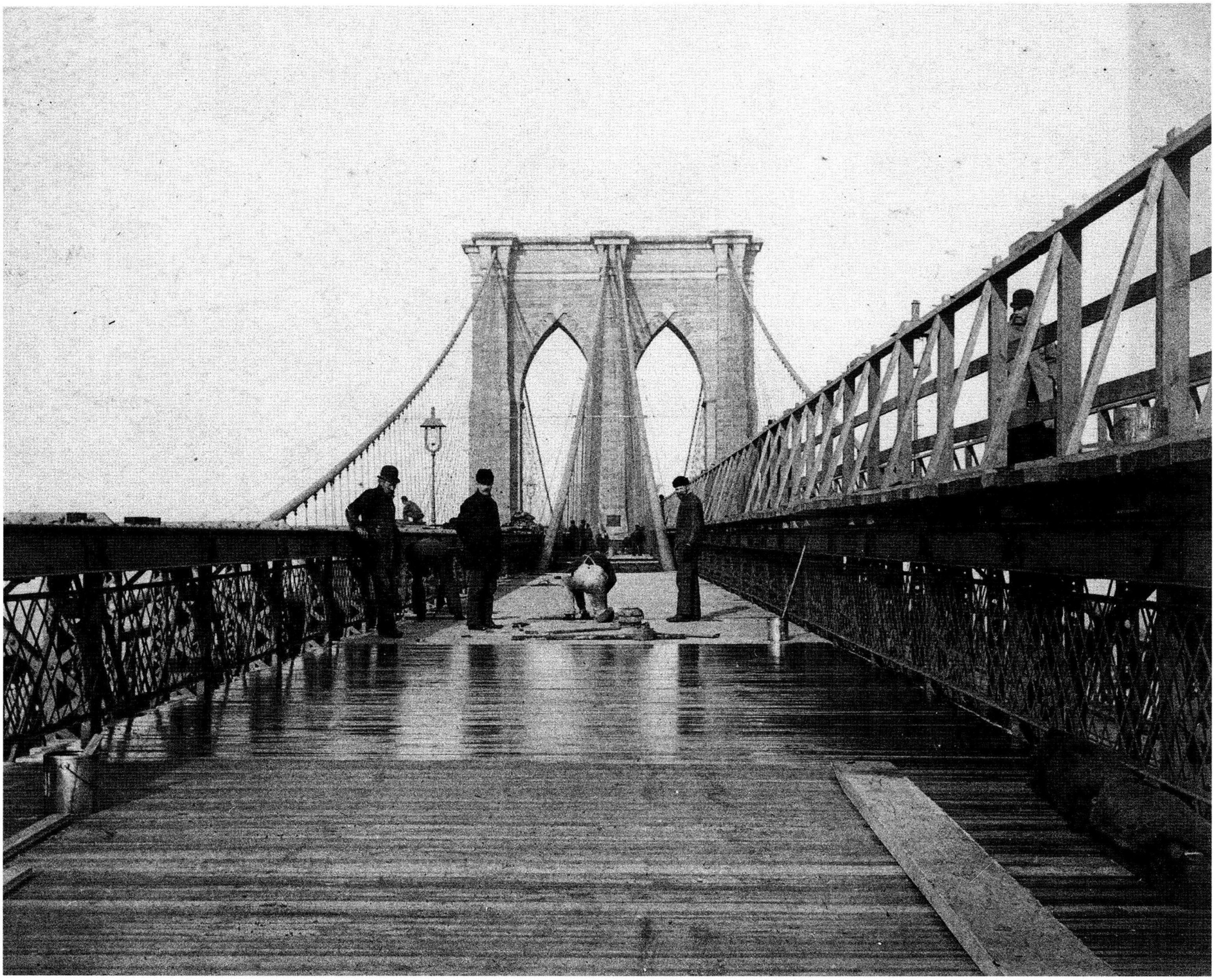

Completing the wooden walkway: Called the Promenade, the walkway was last to be completed. Here a worker with tar and paint is observed by three other men, possibly supervisors. The length of the Promenade is 1,525 feet.

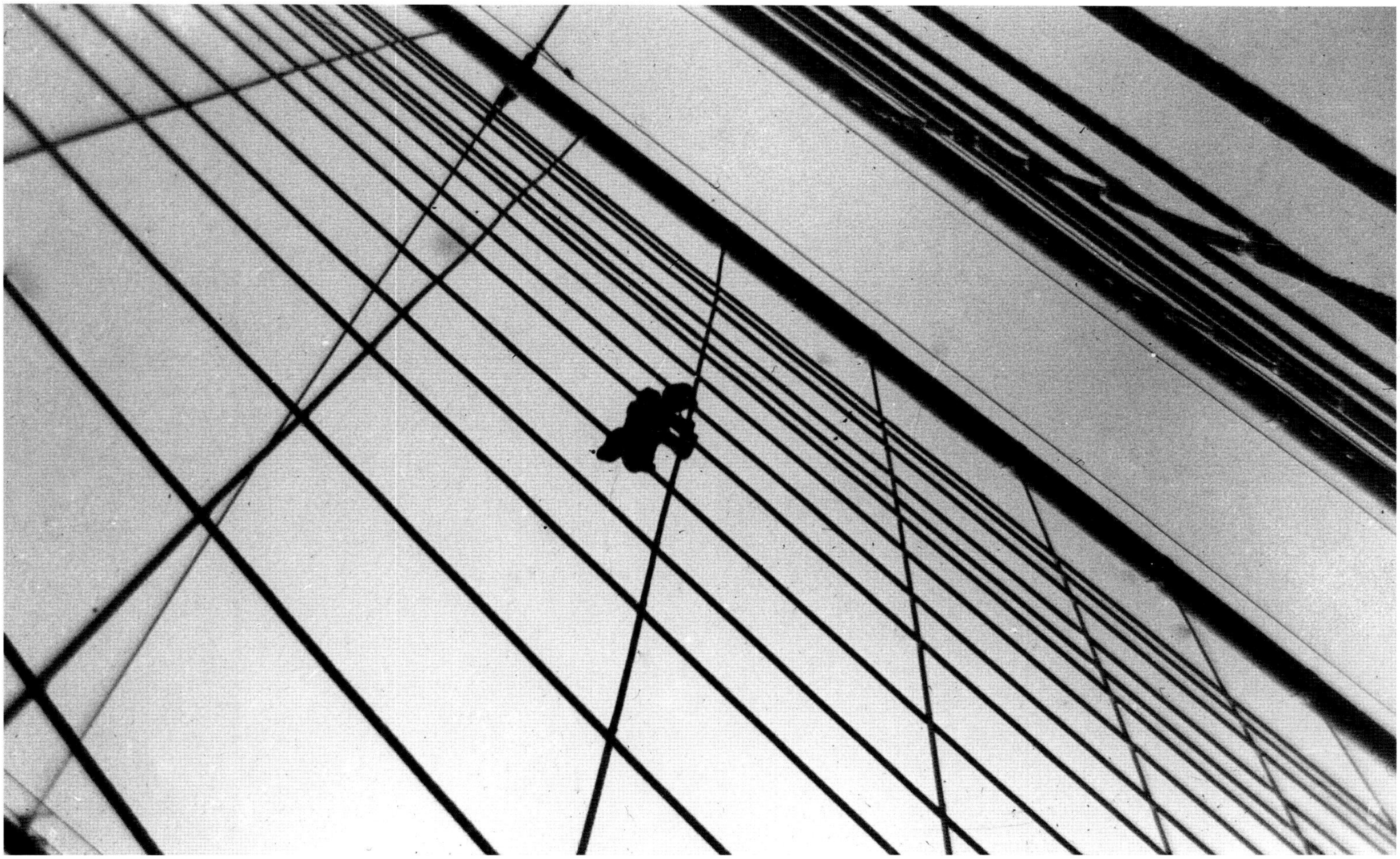

Painting the bridge: The first coat of paint was applied in the 1870s, but the bridge still needs a coat every five years. About 20 workers were reported to have been killed in construction of the bridge, but no record exists as to whether any of these men were painters.

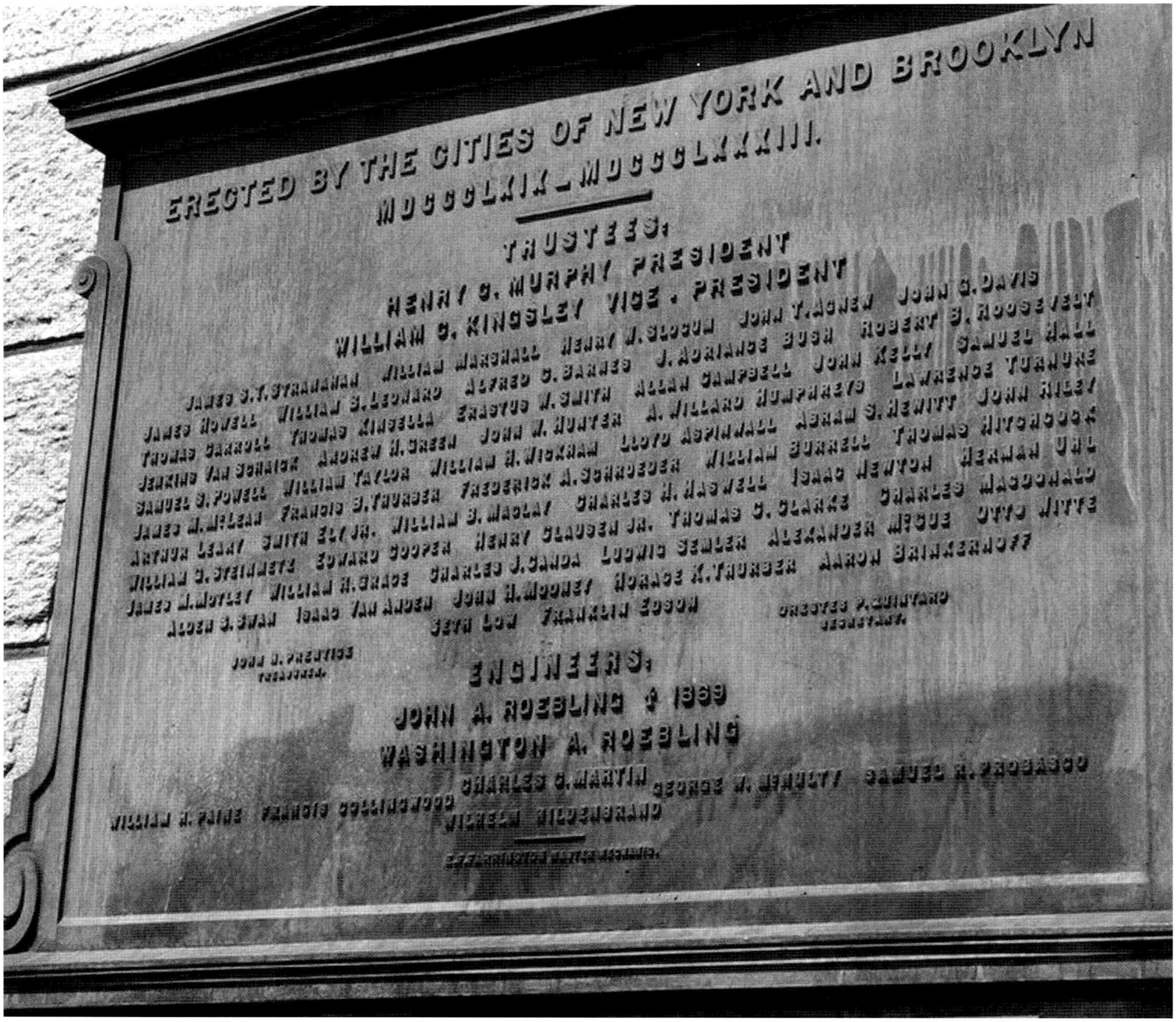

Plaque: A plaque naming the trustees of the New York Bridge Company was mounted on the Brooklyn Bridge.

Triumph: The Eighth Wonder of the Modern World

(1883)

The Brooklyn Bridge's formal ceremonies opened on the windy but sunny May 24, 1883. Declared an official holiday in Brooklyn, the celebration was marked by vast crowds around the bridge and flags in each window of Brooklyn City Hall. Mayor Seth Low, who had tried to have Roebling fired, proclaimed it "the People's Day," calling it "an American triumph."

Beneath the bridge, ferries, warships, yachts, and both public and private vessels lingered and blew whistles. The bridge, boosters proclaimed, put Brooklyn on the map; it was the "Eighth Wonder of the Modern World." Clairvoyants proclaimed that the East River would boast of 100 bridges by 1983. After 14 years of construction, vaudevillians declared, "All that trouble to build a bridge, just to get to Brooklyn." And Irish protesters objected to the opening on Queen Victoria's birthday.

In a procession of 25 carriages, Chester Arthur, who had succeeded to the presidency after Garfield's assassination in 1881, rode to the bridge from Madison Park in Manhattan with New York mayor Frank Edson and the governors of New York, New Jersey, and Rhode Island. He then walked across the bridge to the Brooklyn tower to shake hands with Mayor Low.

At 12:40 P.M., three hours of speeches in the Brooklyn train terminal endorsed Peace, Courage, Science, and Commerce. Chairman of the bridge board Abram Hewitt addressed the crowd, but neither the president nor New York governor Grover Cleveland spoke. Jules Levy, a popular trumpeter, was not introduced despite being on the program, so he stepped forward and played "The Star-Spangled Banner" with "Hail, Columbia" as an encore and "Yankee Doodle" for his exit music. At 8:00 P.M., 14 tons of fireworks exploded from the towers and center span, balloons were set adrift, and 500 rockets were fired in unison. At 11:20 P.M., 22 tollbooths opened on the bridge, and at midnight, the public was invited to cross the Promenade at a cost of one penny per person.

A public celebration was held at the nearby newspaper offices of Richard Fox's *Police Gazette.* And private receptions were held at the Brooklyn homes of Seth Low and Washington Roebling. Emily Roebling welcomed President Arthur, the mayors of Brooklyn and New York, and three governors into her home to meet her 46-year-old husband, who never set foot on the completed bridge. Emily was the first person to cross the completed bridge, riding in a carriage and carrying a white rooster in her lap as a symbol of victory.

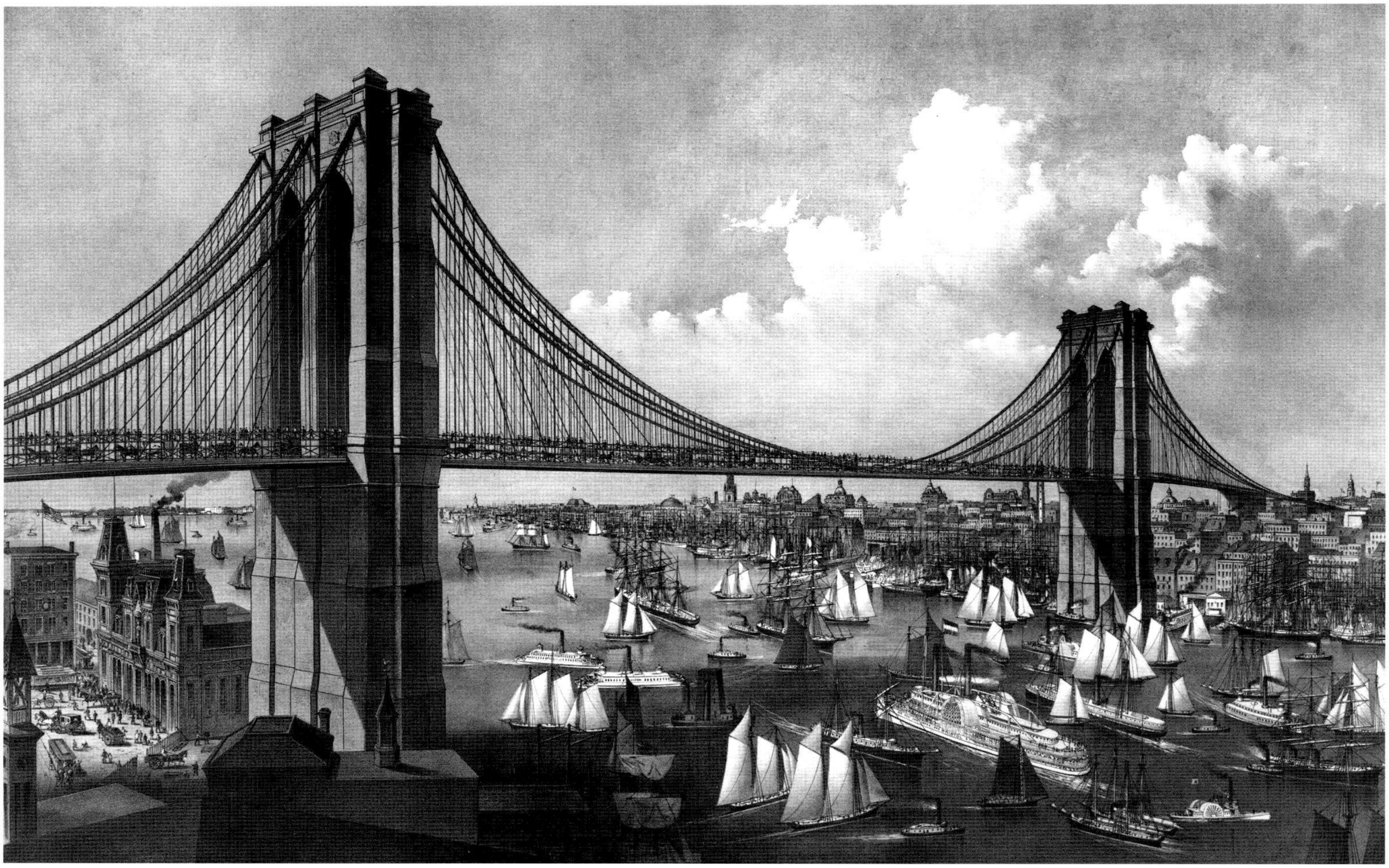

The Great East River Suspension Bridge: In 1883, the Brooklyn Bridge was ready to debut. As the day for its unveiling approached, dire predictions circulated that it wouldn't be completed in time for its own premiere. But the curious came and gawked. A flotilla of boats and bevies of women in carriages joined the crowds that hurried over to the Great Bridge, as depicted in this *Currier and Ives* lithograph. It was the biggest occasion since the opening of the Erie Canal.

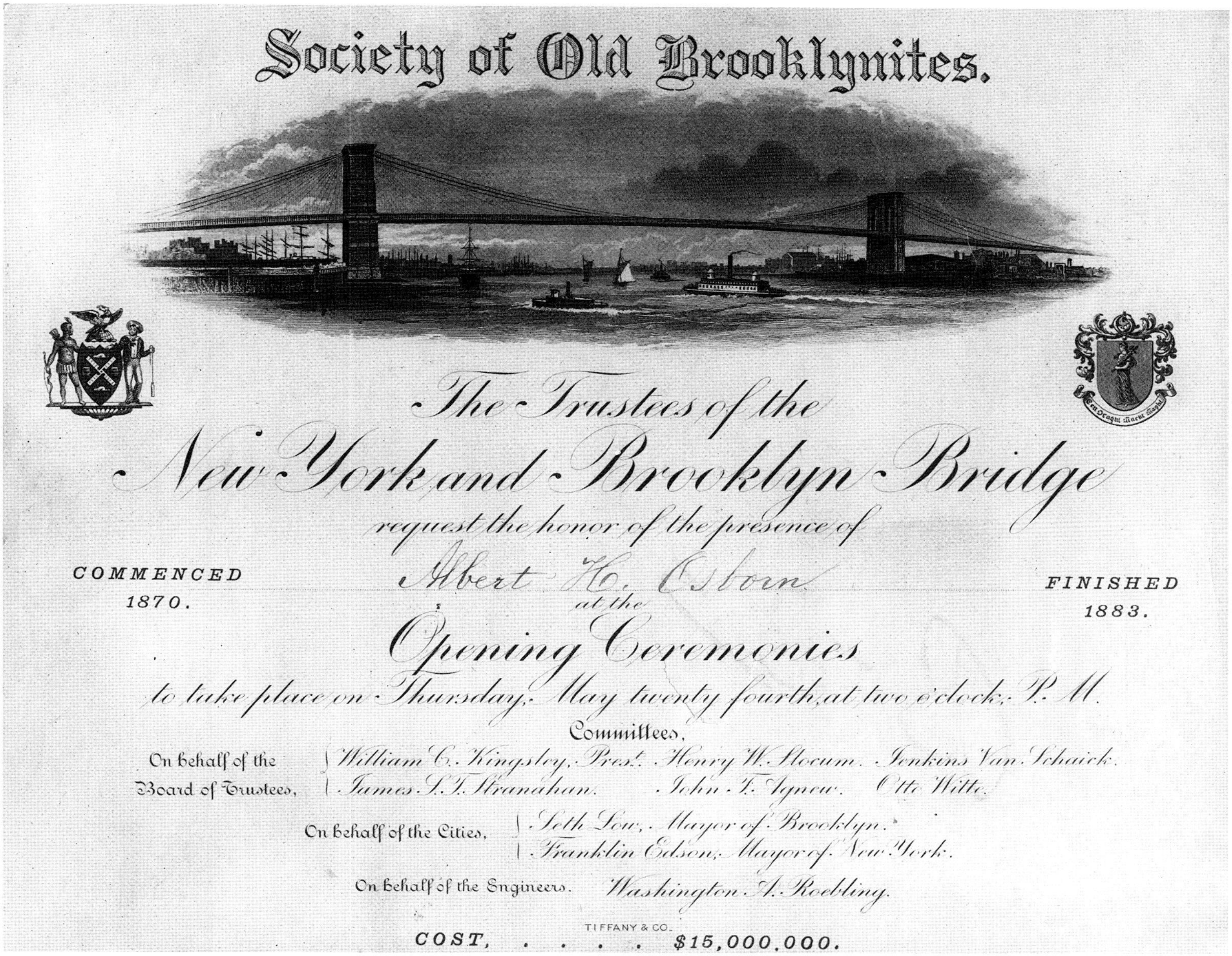

Society of Old Brooklynites.

The Trustees of the

New York and Brooklyn Bridge

request the honor of the presence of

Albert H. Osborn

COMMENCED 1870.

FINISHED 1883.

at the

Opening Ceremonies

to take place on Thursday, May twenty fourth, at two o'clock, P.M.

Committees,

On behalf of the Board of Trustees, William C. Kingsley, Prest. Henry W. Slocum. Jenkins Van Schaick. James S. T. Stranahan. John T. Agnew. Otto Witte.

On behalf of the Cities, Seth Low, Mayor of Brooklyn. Franklin Edson, Mayor of New York.

On behalf of the Engineers. Washington A. Roebling.

TIFFANY & CO.

COST, ... $15,000,000.

Invitation to the opening ceremonies: This general invitation to the grand event was sent to the Society of Old Brooklynites. Coming from the Trustees of the bridge, the Tiffany & Co. invitation carefully explained the 1870 starting and 1883 completion dates as well as the $15 million cost. Invitations from the Roeblings were less ostentatious.

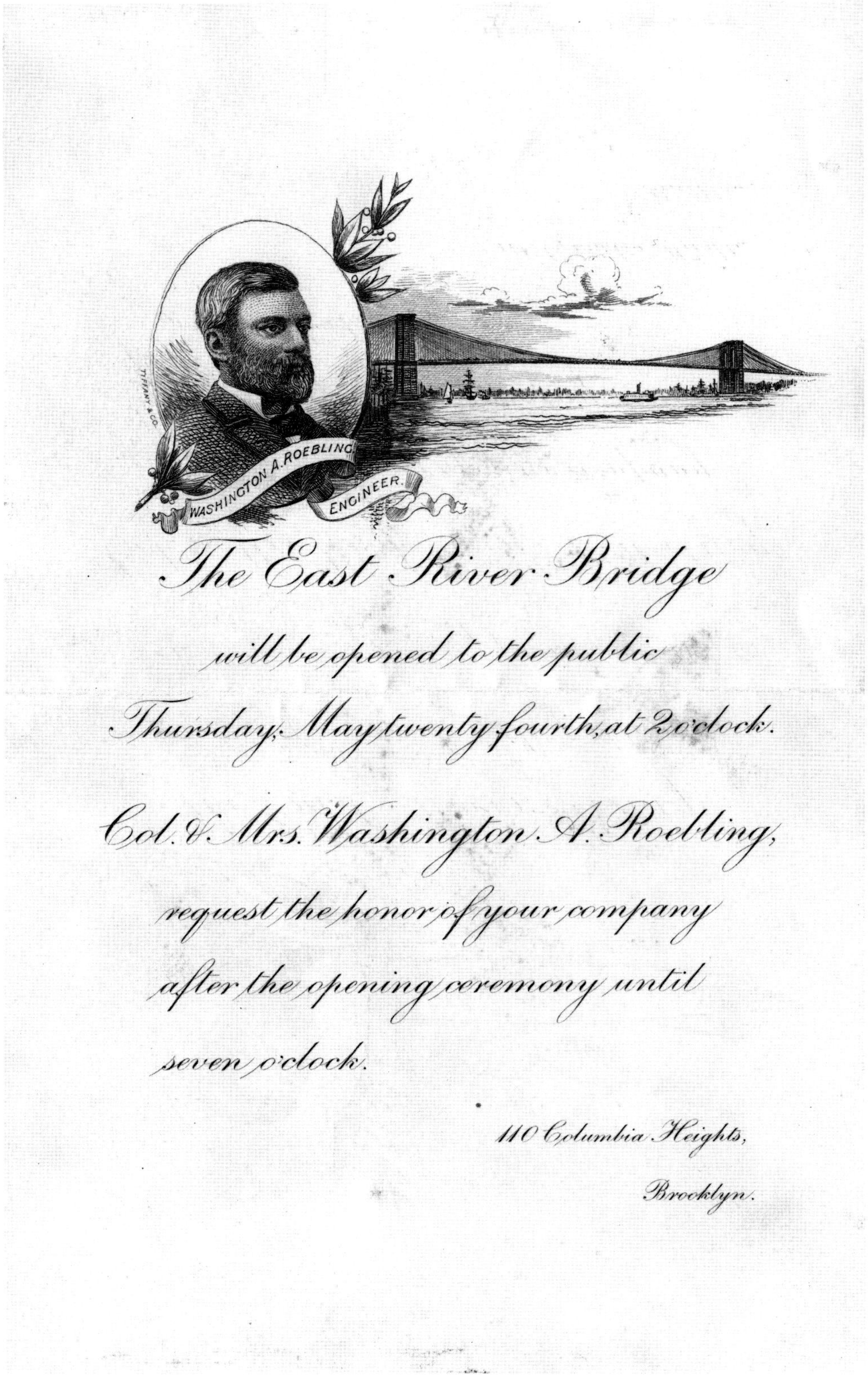

WASHINGTON A. ROEBLING.
ENGINEER.
TIFFANY & CO.

The East River Bridge
will be opened to the public
Thursday, May twenty fourth, at 2 o'clock.

Col. & Mrs. Washington A. Roebling,
request the honor of your company
after the opening ceremony until
seven o'clock.

110 Columbia Heights,
Brooklyn.

Roebling invitation: A formal invitation requests a visit to the Roebling home to celebrate the bridge opening after the official festivities.

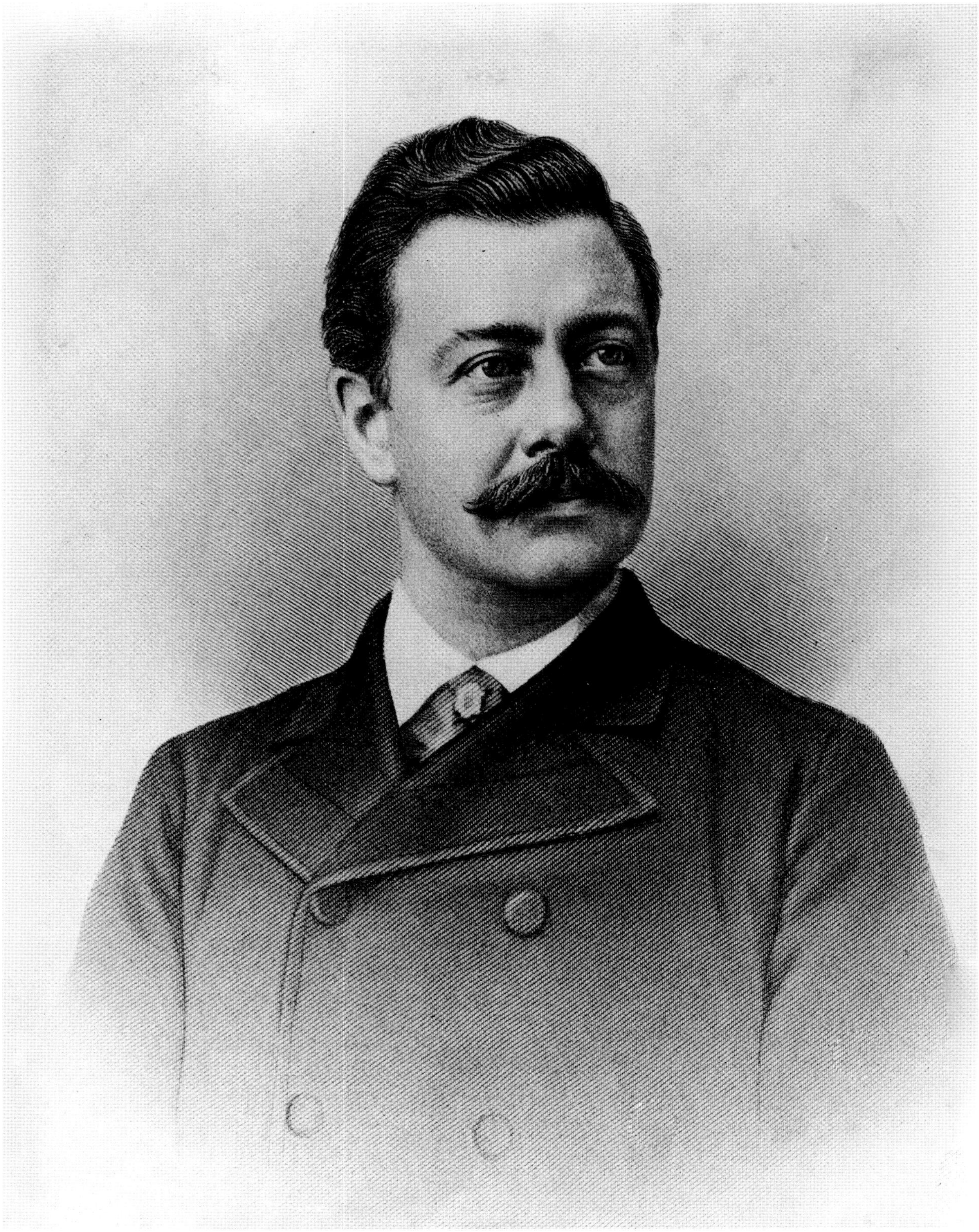

Seth Low: A member of a prominent Brooklyn family with roots in the Brooklyn Heights community, Low was elected mayor of Brooklyn in 1881 and won re-election in 1883. At the bridge's opening, Low behaved as if his attempt to remove Roebling had never occurred, declaring the day an official Brooklyn holiday. Seth Low went on to serve as president of Columbia College before becoming the second mayor of the consolidated New York City.

The president crossing the suspended highway: On "the People's Day," May 24, 1883, President Chester Arthur crossed the Brooklyn Bridge in a ceremonial march with New York mayor Frank Edson and Governor Grover Cleveland to greet Brooklyn mayor Seth Low. Celebrities and the 201 full-time bridge employees attended the procession as guns boomed from the Navy Yard below. A lone photographer recorded the event from the New York tower, and artists Schell and Hogan drew this *Harper's Weekly* view.

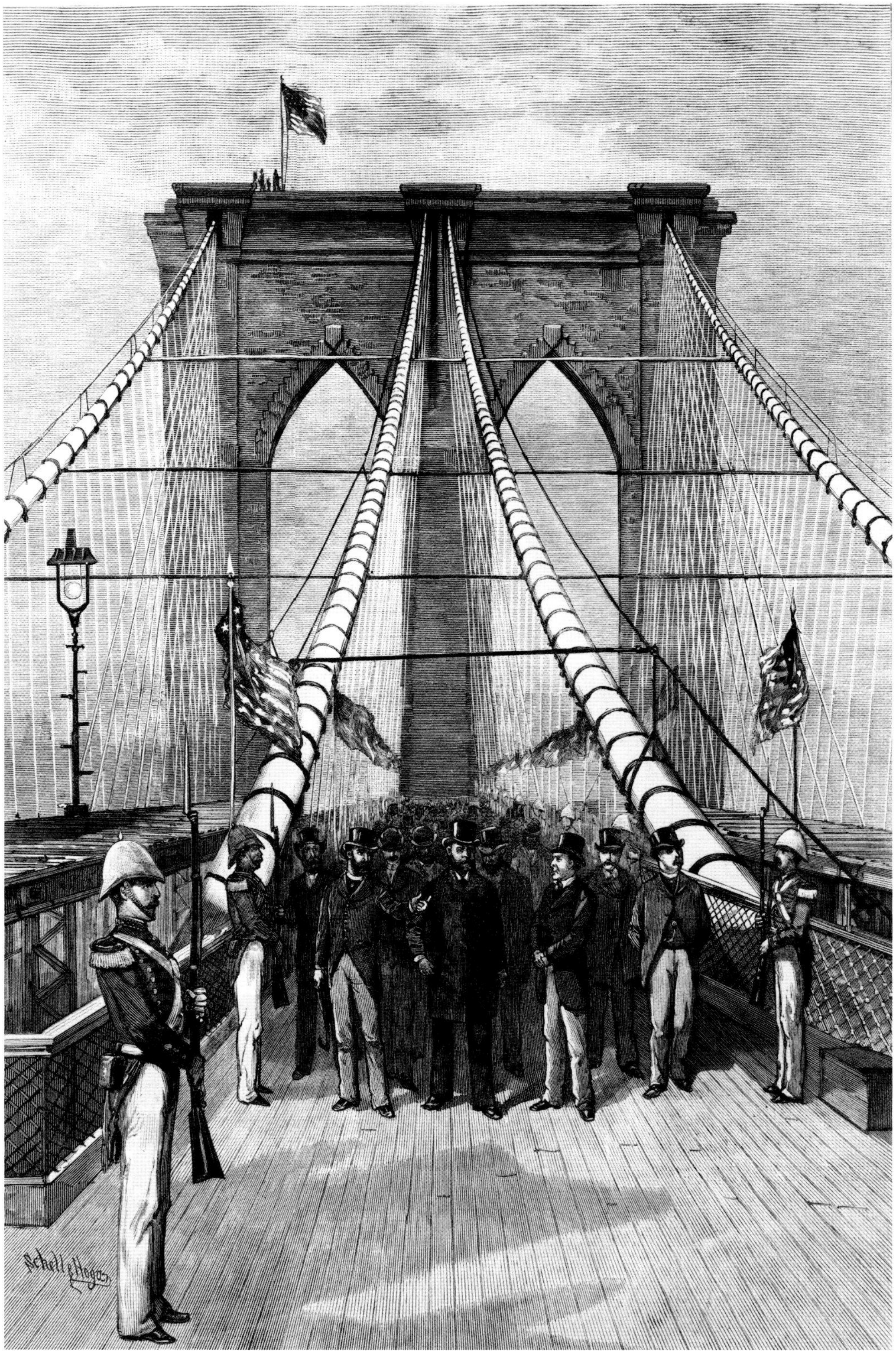

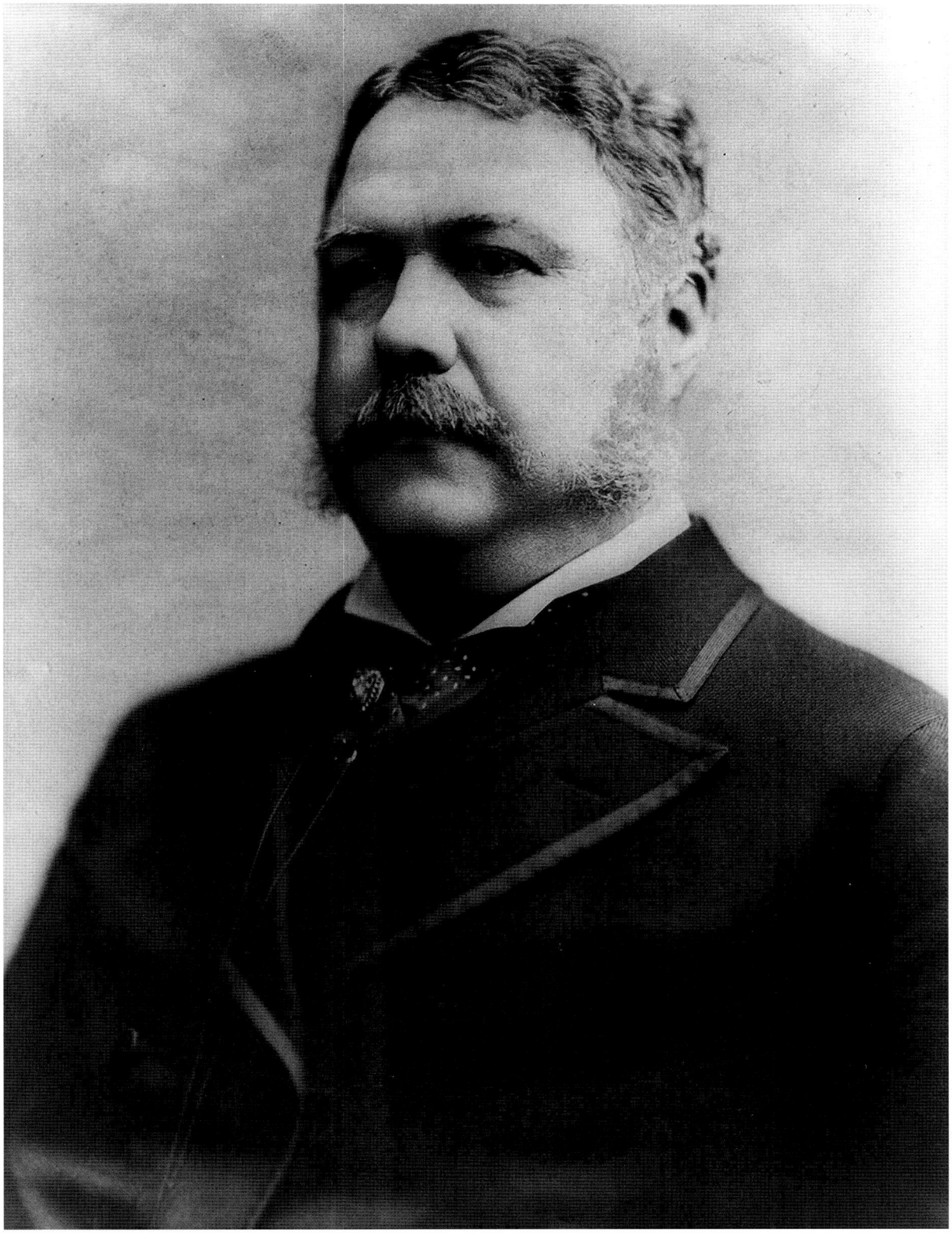

President Chester Arthur: Arthur, vice-president under James Garfield beginning in 1880, succeeded to the presidency after Garfield was assassinated in 1881. Considered handsome, he was known as "the American Apollo" and "Elegant Arthur." While proceeding to the bridge ceremony with 24 other carriages, the President's horses bolted but were quickly restrained. Arthur then walked from Manhattan, meeting Mayor Low at the Brooklyn tower. By 1883, Arthur had been diagnosed with Bright's Disease, which soon began to affect his health. Two years later, Arthur was followed in the presidency by Grover Cleveland, New York's governor.

Abram Hewitt: As president of the Brooklyn Bridge board of trustees and also holder of the mortgage, Hewitt exerted considerable influence. On his recommendation, wire from J. Lloyd Haigh was substituted for Roebling's wire—until inspectors found it to be faulty. Hewitt gave the key speech at the bridge's opening celebration and went on to become mayor of New York City.

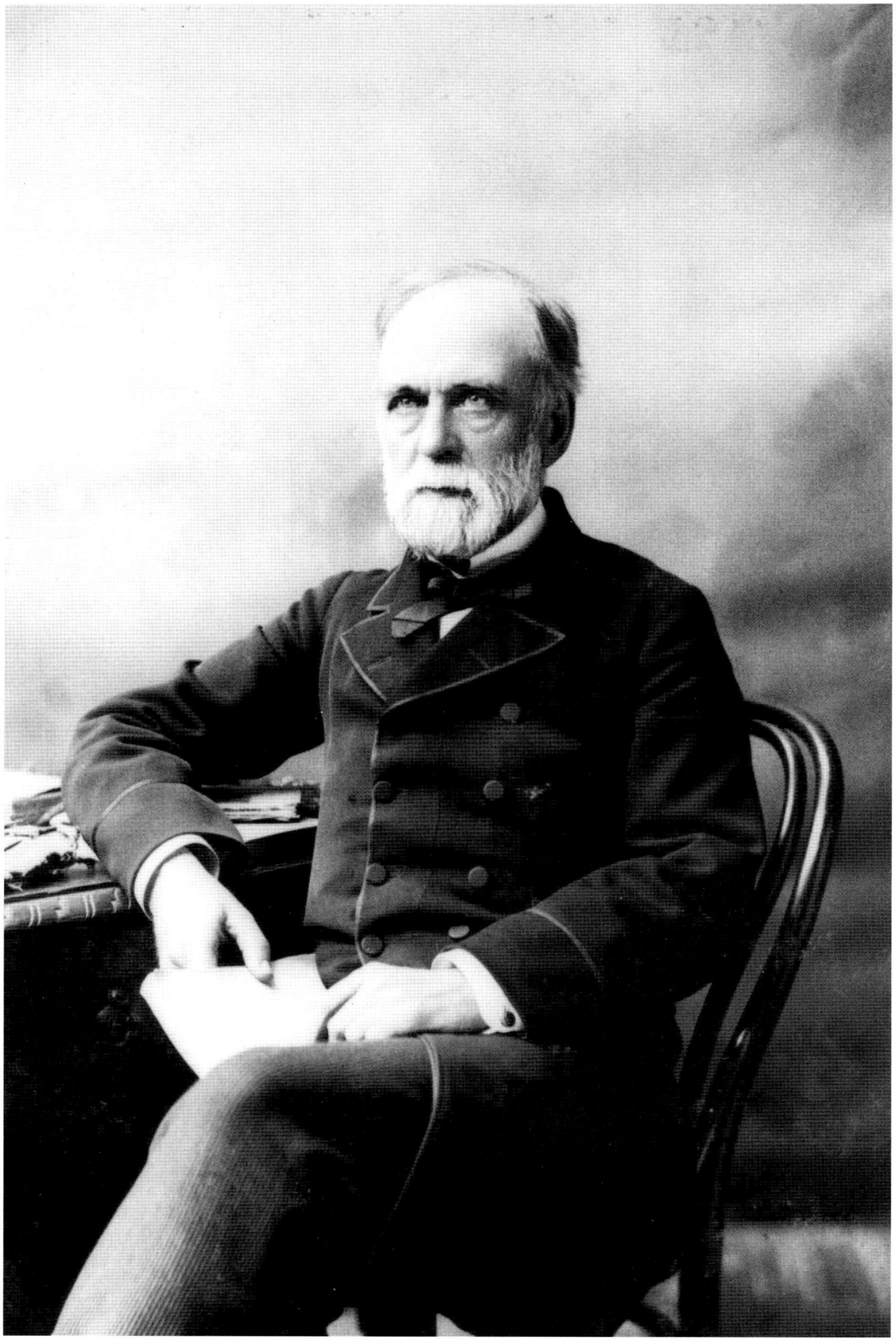

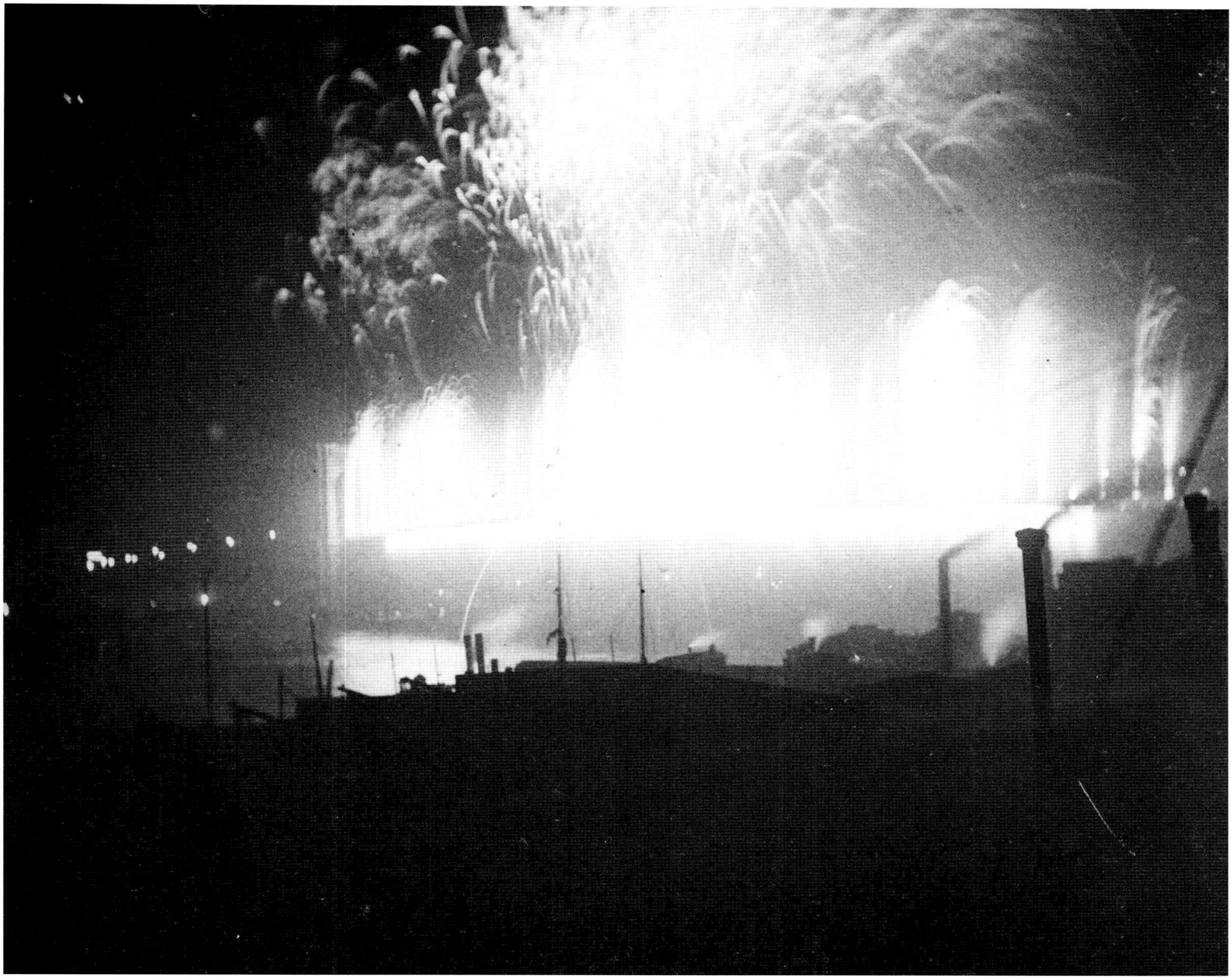

Fireworks on the Brooklyn Bridge: As night approached, the bridge was sealed off, and at 8:00 P.M., an hour of fireworks commenced from the towers and the center span. Over 10,000 fireworks and 500 rockets exploded in celebration of the event. According to author and historian David McCullough, the bridge dominated the imagination. The inscription on this photo appropriately reads "Niagara Falls and things generally."

Bridge toll tickets: A sampling of tickets required to cross the Brooklyn Bridge after its opening. Pedestrians, passengers, bicyclists, wagoneers, and even animals paid tolls for the privilege of crossing, even though the bridge was publicly financed. According to David McCullough's book *The Great Bridge: The Epic Story of the Building of the Brooklyn Bridge,* H. D. VanKeuren was first to pay a toll; Mrs. C. G. Peck was the first woman to cross (after Emily Roebling); and the first wagon to cross was driven by Charles Overton of Coney Island. The *New York Times* recorded the first beggar, the first drunk, the first (empty) hearse, the first black person, the first policeman, the first musician (a bagpiper), and the first dude.

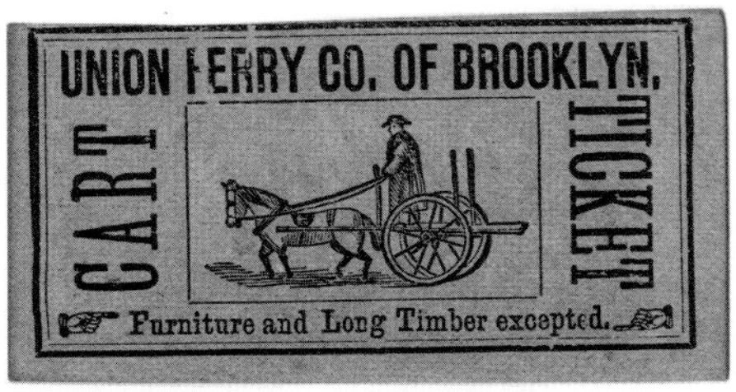

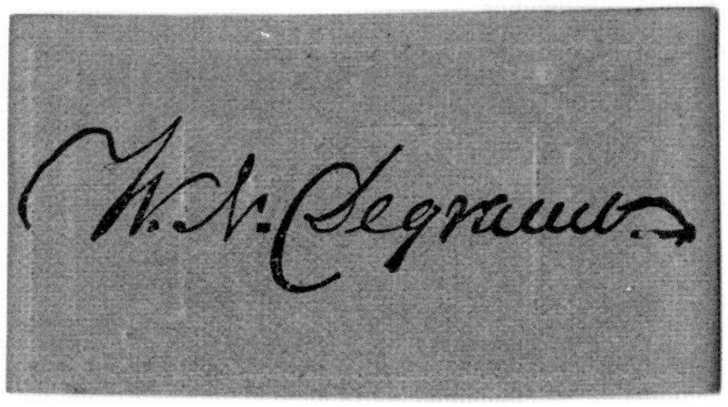

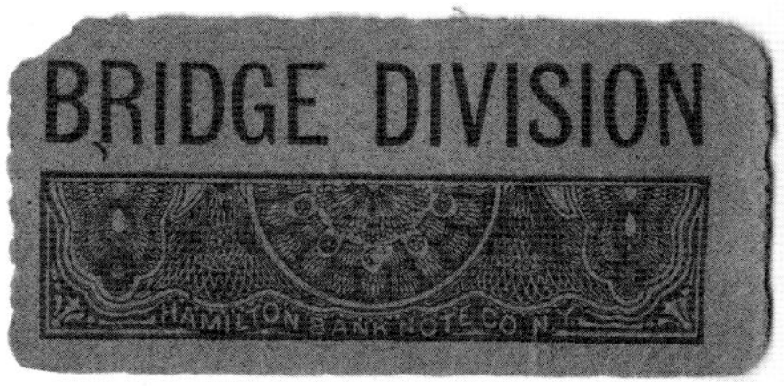

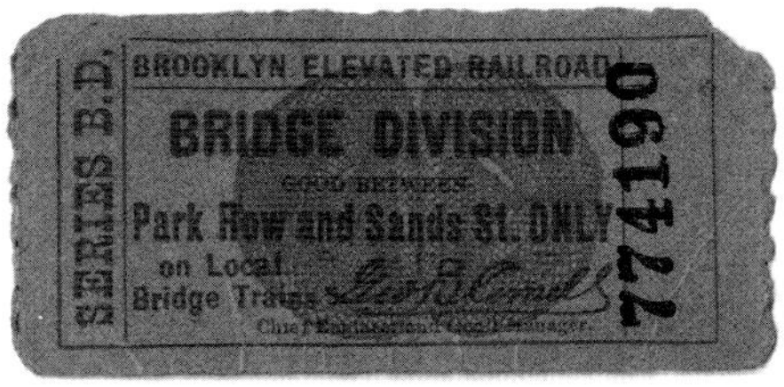

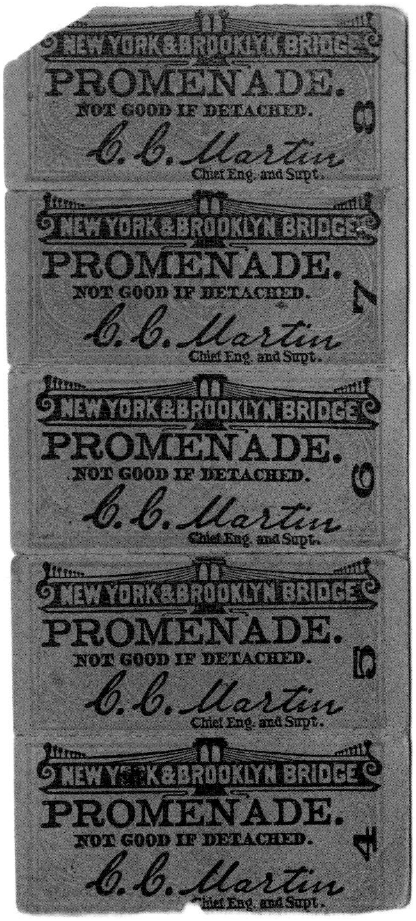

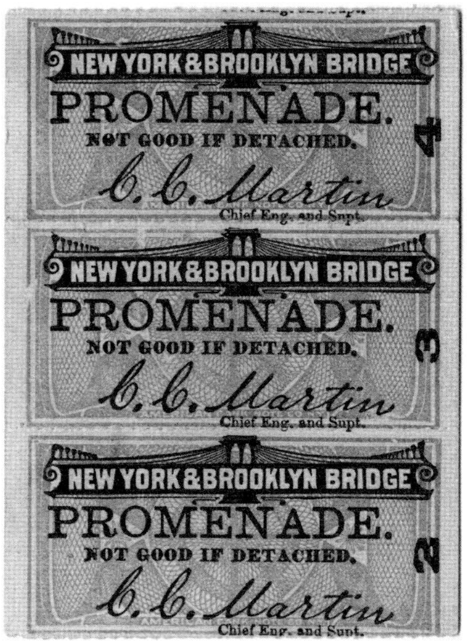

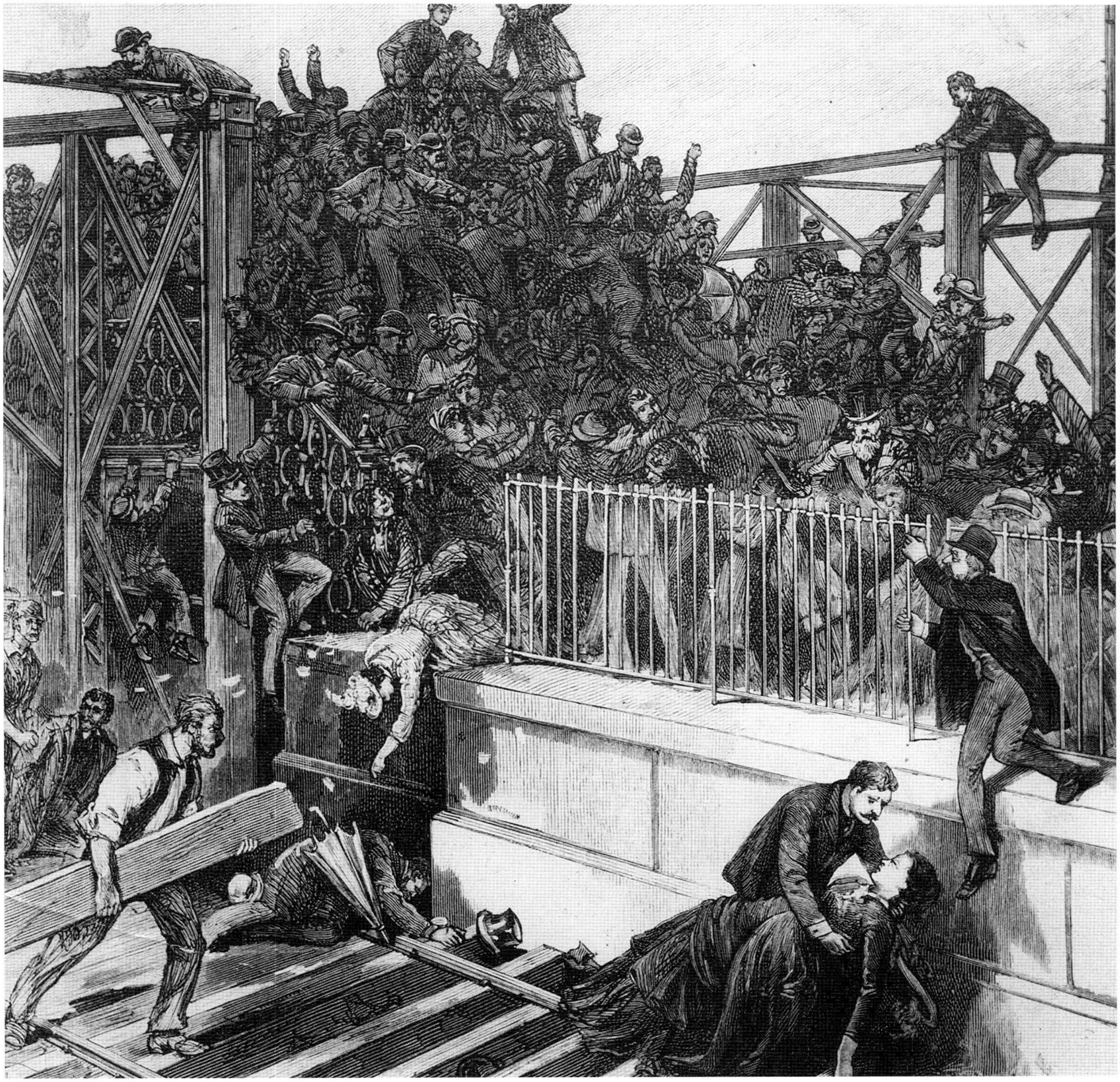

Tragedy on the bridge: A week after the glorious opening of the bridge, on the Sunday of the Memorial Day weekend, an unknown incident led the 20,000 people on the Promenade to believe the bridge was collapsing, causing a mass riot on the bridge and stairs. Twelve people died from the trampling mob. *Harper's Weekly* demanded a stronger police presence at the bridge, but circus owner P. T. Barnum marched 22 elephants across the bridge to prove it safe.

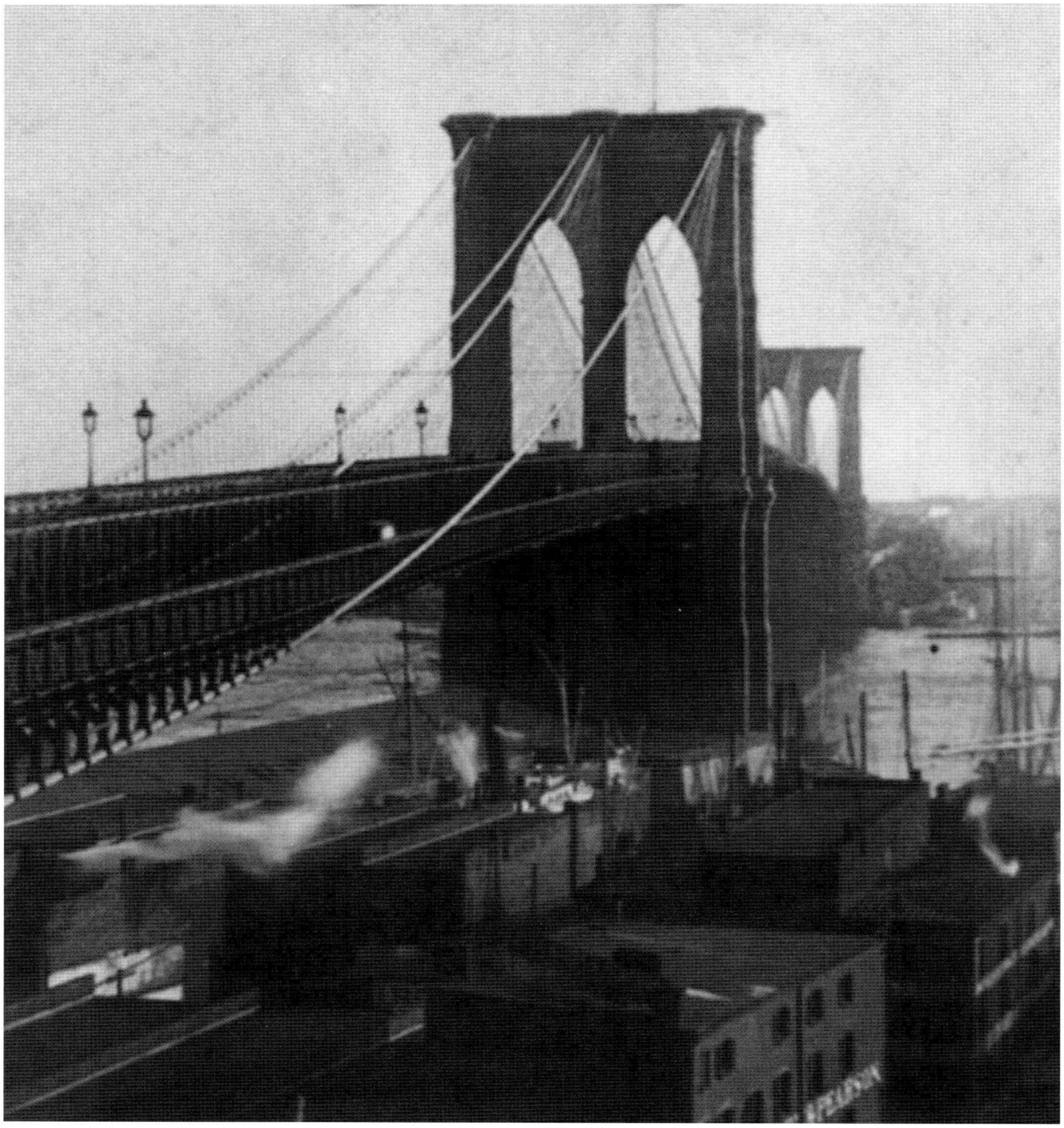

After the excitement, pedestrians could cross the Brooklyn Bridge, which dominated both Brooklyn and New York as no creation had before. As shown here, the bridge elegantly hovers over warehouses and ships' masts.

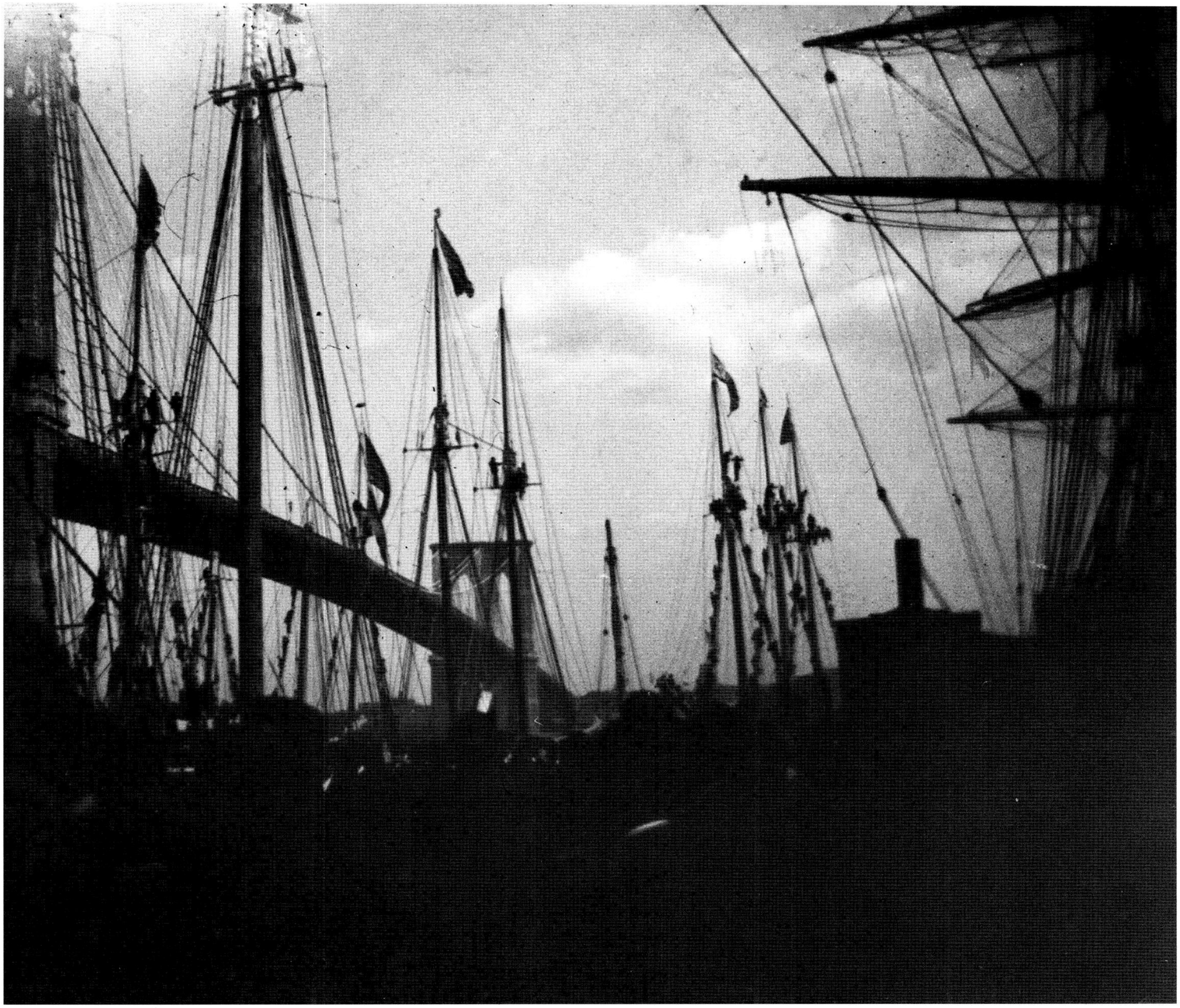

Bridge and masts: Next to the new bridge, masts from sailing vessels still pointed their pennants toward the sky. The Brooklyn Bridge would outlive them all.

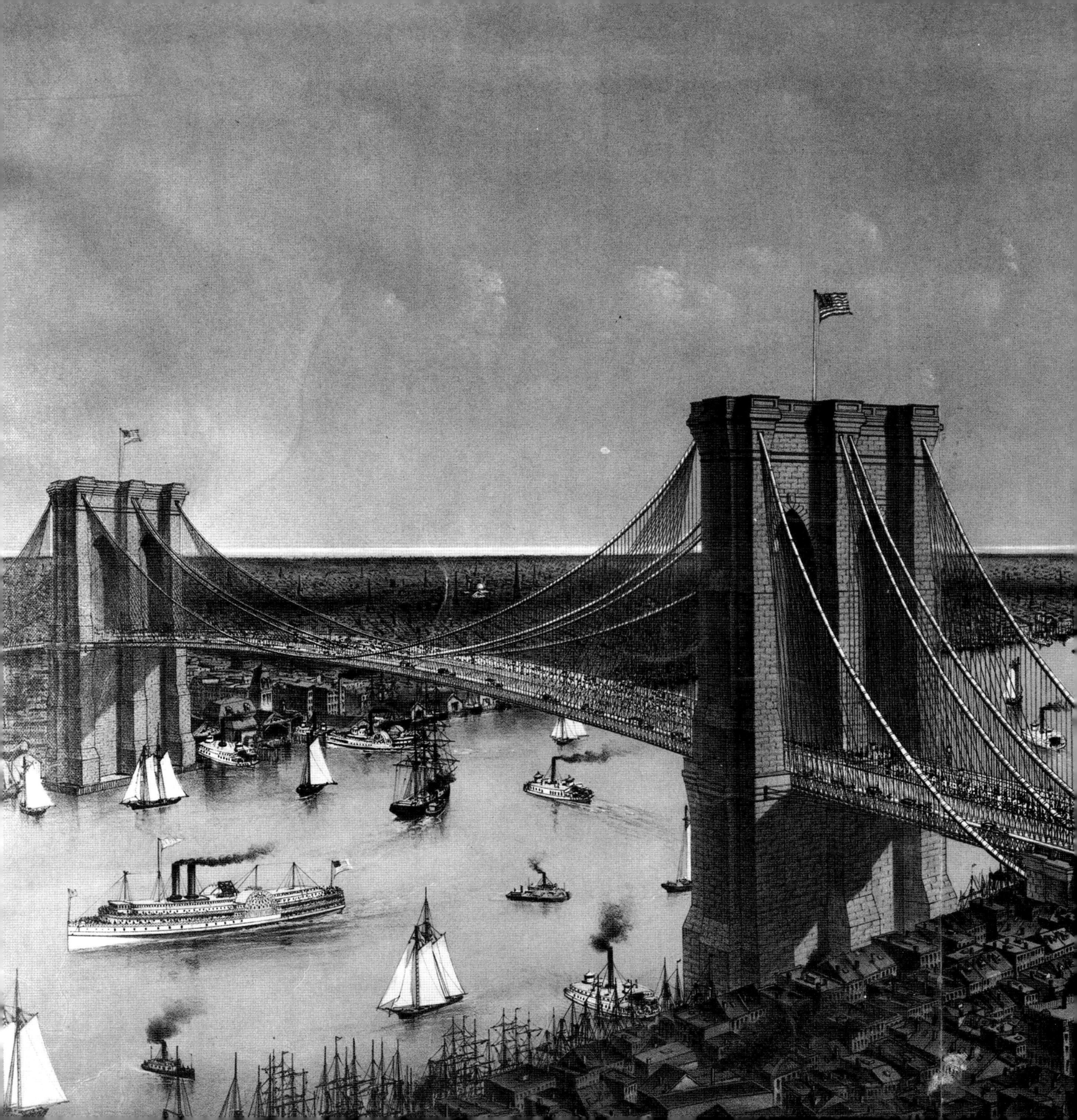

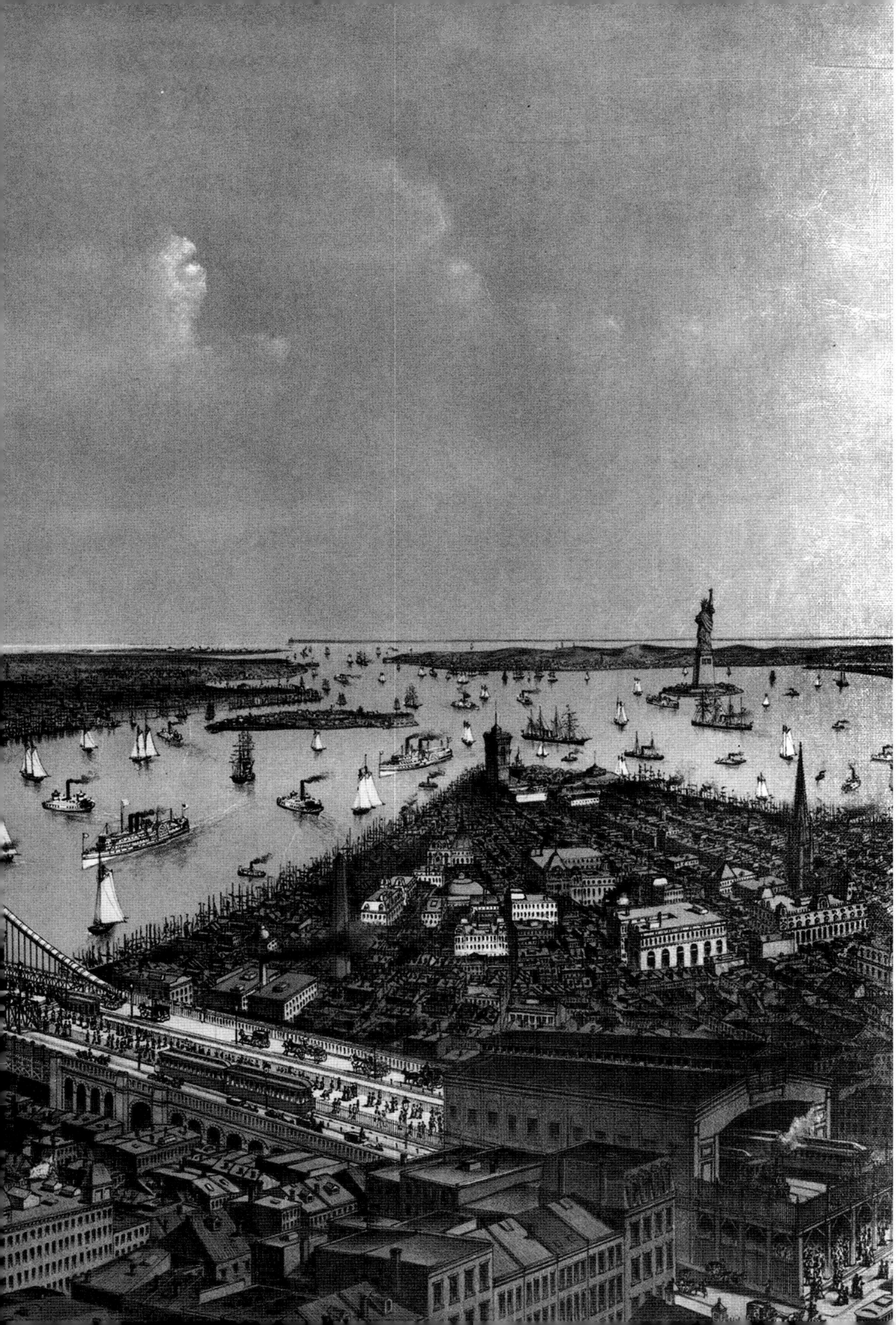

Bird's-eye view: An 1885 *Currier and Ives* chromolithograph panorama of New York Harbor with its new Brooklyn Bridge. The oversized terminal juts into Manhattan, with both the bridge and the river jammed with vehicles. The new Statue of Liberty (dedicated in 1886) stands to the right, and below the statue, Trinity Church and Federal Hall.

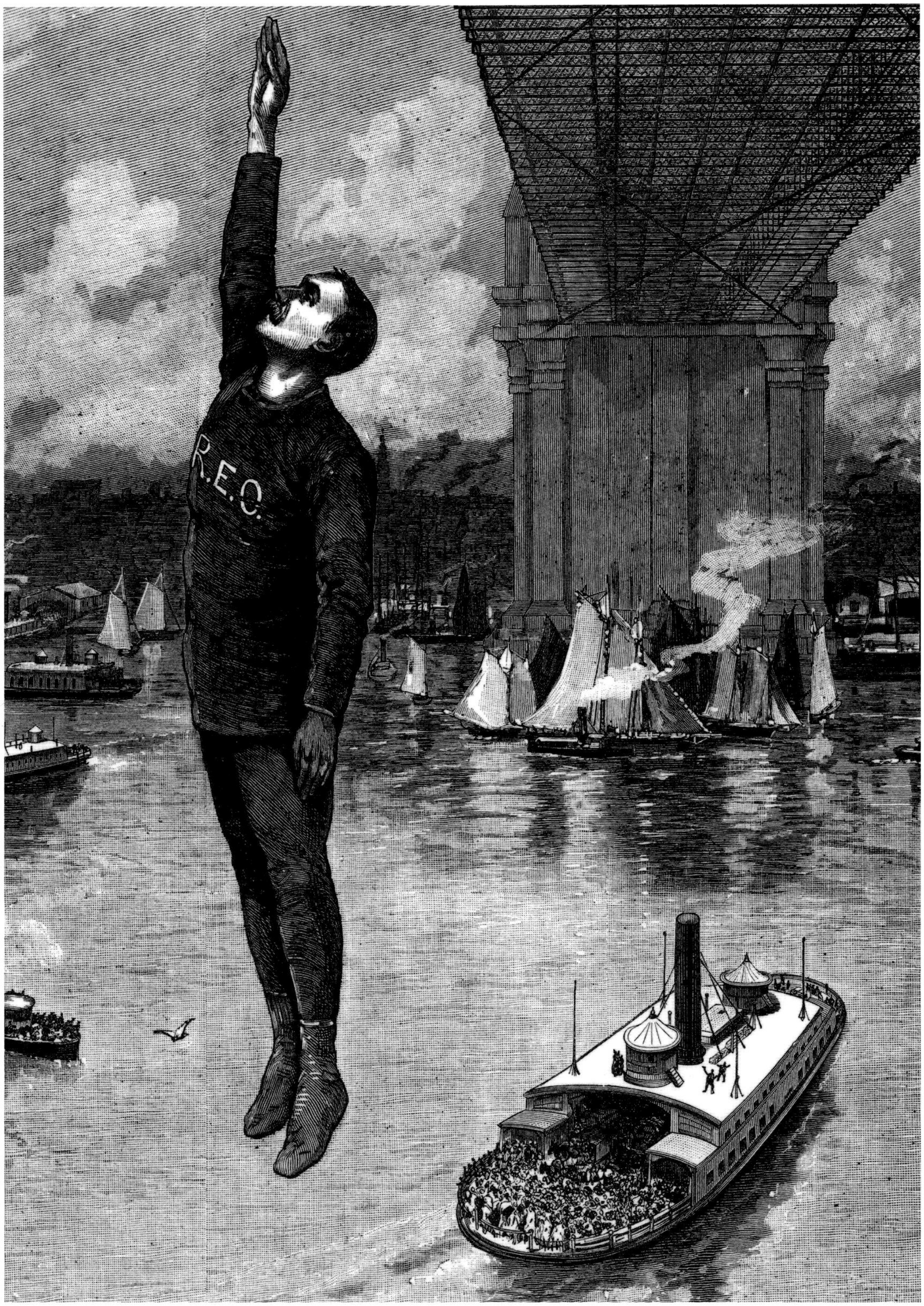

Robert Odlum's jump: In 1885, Odlum, the first to jump off the Brooklyn Bridge, started a trend that increased until the end of the nineteenth century. Steve Brodie claimed he jumped in 1886 but never proved it. In 1895, Clara McArthur, the first woman, jumped and lived. Robert Donaldson threatened to jump very publicly in 1892 but never did. Johnny Weissmuller did so in the movies as Tarzan. And Superman saved a soul from jumping. Paul Boyton, founder of Sea Lion Park in Coney Island, attempted to save Odlum, who died soon after from injuries sustained upon hitting the water.

View from the Bridge: Sky, Land, and Sea

(1884–1920)

Since the day the public discovered the Brooklyn Bridge, when 150,000 people and 1,800 vehicles crowded onto its span, it has been all things to all people: transportation to and from Brooklyn and Manhattan; a perfect point to view the city, its harbor, the Statue of Liberty, and Governor's Island; and a height from which pedestrians can marvel at the rich assortment of boats, including the museum pieces below at South Street Seaport.

Before the bridge spanned the river, the development of industry increased the importance of the shoreline for factories and use of the water for transportation of goods. Shipping gathered in the Atlantic Basin and up the Gowanus Canal, and oceangoing skiffs and schooners docked and unloaded passengers and cargo at the shores. Ferries filled the waters, transporting passengers between Brooklyn and Manhattan. However, with the addition of the bridge, New York's reliance on ferries began diminishing, and fealty transferred to the magnificent new bridge overhead. While pedestrians and passengers crossed its span, steamships and tugs crowded the waters underneath. Other bridges, more utilitarian, had joined Brooklyn's landscape, supplementing but never replacing the Brooklyn Bridge.

Soon the cable cars became trains and trolleys, and wagons became automobiles. However, the people persisted. During a period of economic stress, routine maintenance had been neglected, but even this was rectified. Washington Roebling declared its minimal sway and sag a "natural oscillation." Then the cities of Brooklyn and New York drifted toward a union and wed.

And still they came: passengers, strollers, tourists, business people, and bicyclists. They stared from train windows and through automobile glass, from the bridge's benches, from the railings, and through the cables at the skyline, at maps, at landmarks (albeit they stood on one), at the water below.

As Seth Low commented, "No one who has ever been upon it can forget it."

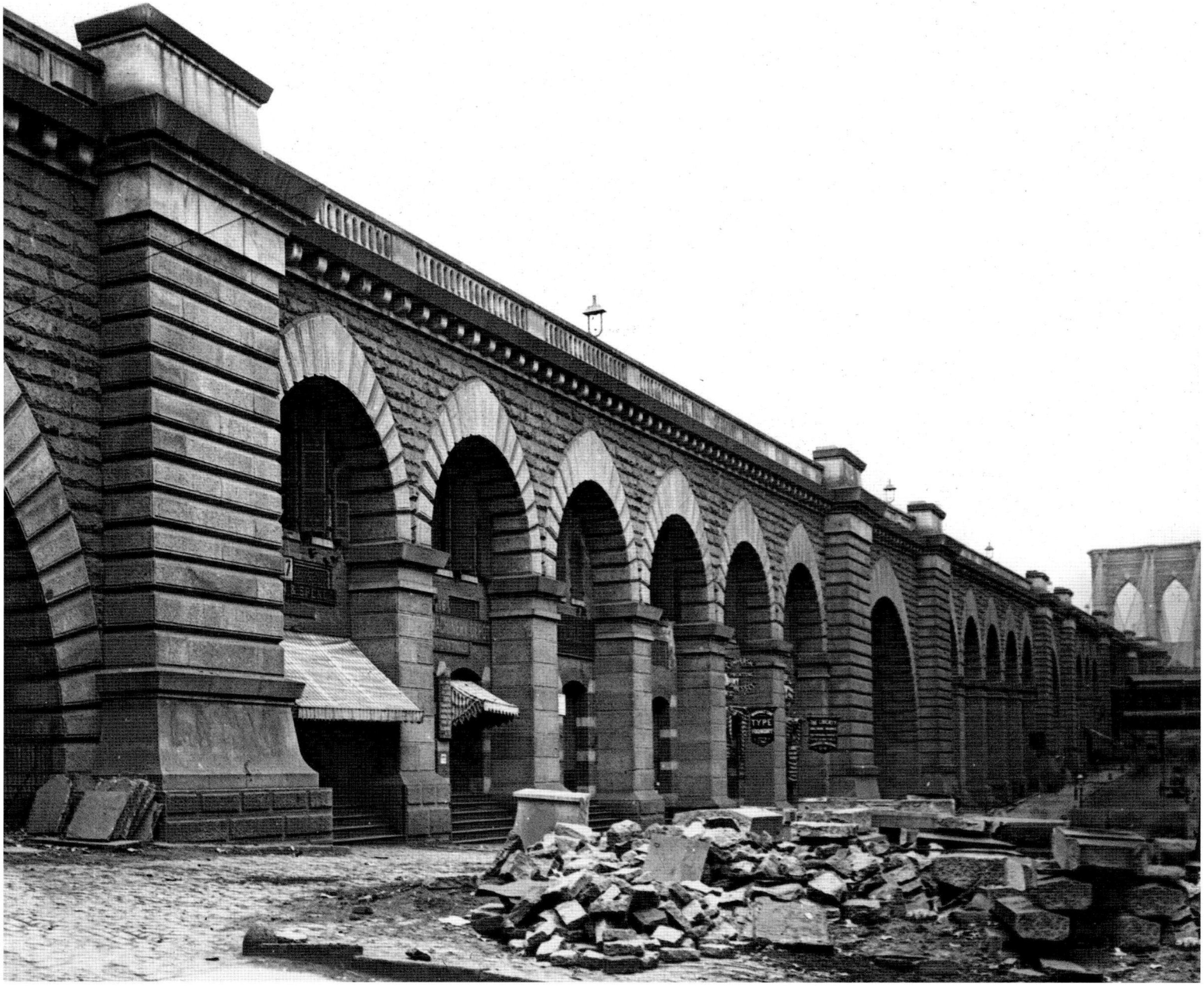

Brooklyn Bridge arches: The arches underneath the bridge approaches contain storage warehouses as well as rental space for commercial businesses. In the 1930s, a luncheonette occupied one of these spaces. The rubble in the right foreground indicates that this is an early photograph.

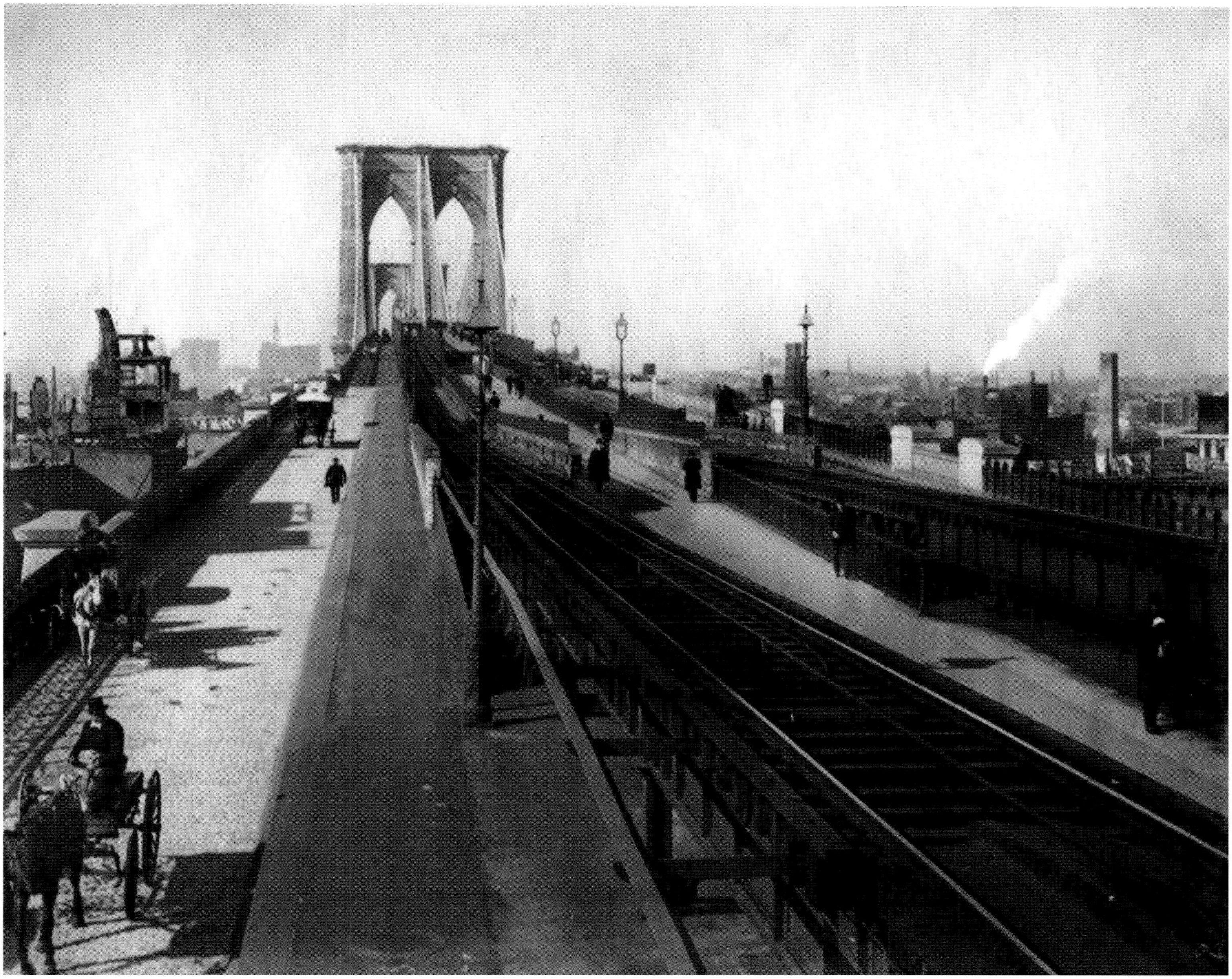

Busy times: A photograph of the Brooklyn Bridge in action from the Manhattan terminal at Park Row. This photo provides a good example of the variety of traffic that crossed the bridge: trolleys, carts, wagons and carriages, an elevated train, and pedestrians.

New York terminal to Brooklyn Bridge: Superimposed on the New York terminal in this *Harper's Weekly* cover is the face of John Roebling, who originally conceived of trains crossing the bridge. Not only was the bridge strong enough to carry trains, trolleys, and other vehicles, but its strength was multiplied after a wire fraud was exposed. George McNulty designed the two-story terminal buildings, with the New York station almost twice the size of the one in Brooklyn.

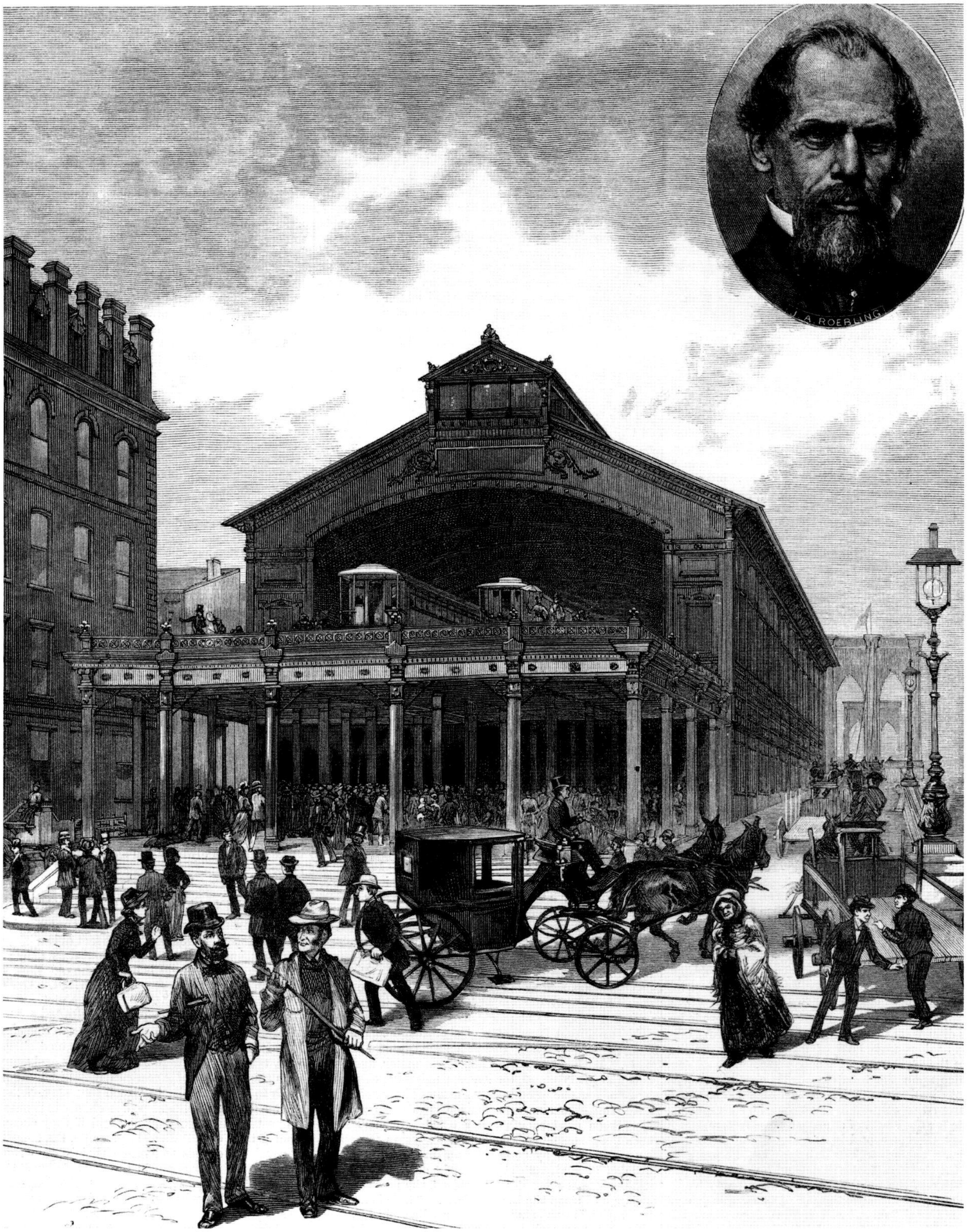

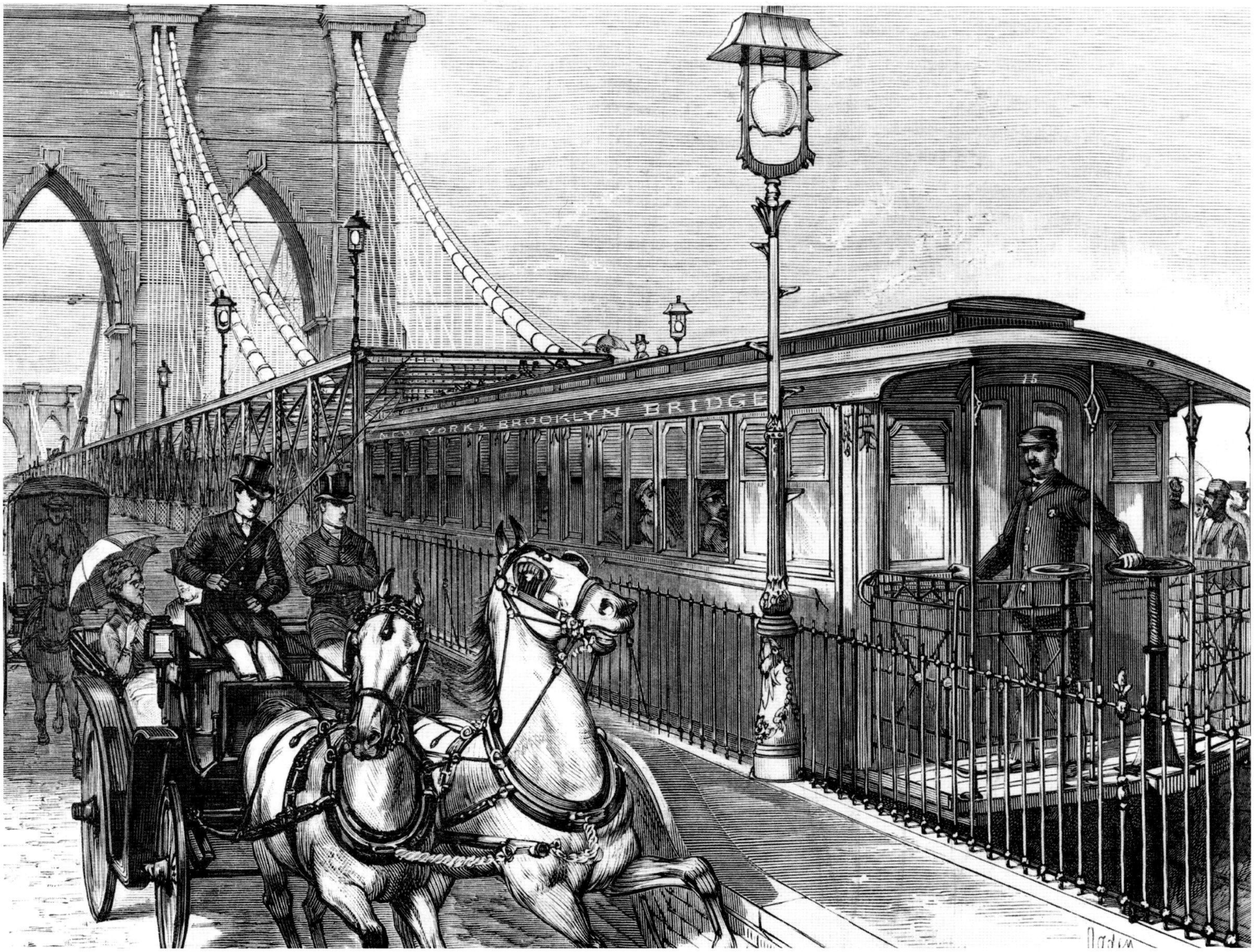

Cable car on bridge: Cable cars, operated with a continuously running steel cable driven by two steam engines, took five minutes to cross the bridge and cost five cents for a one-way ticket. Trains ran on center tracks, obviously scaring the horses in this interpretation from *Harper's Weekly.* At the end of the run, the cable car reversed course.

Tracks for cable car: The cable car route ran down the center of the bridge. The engines powering the machinery were located beneath the bridge, and a boiler house and smokestack stood at Washington and Prospect streets, now in DUMBO

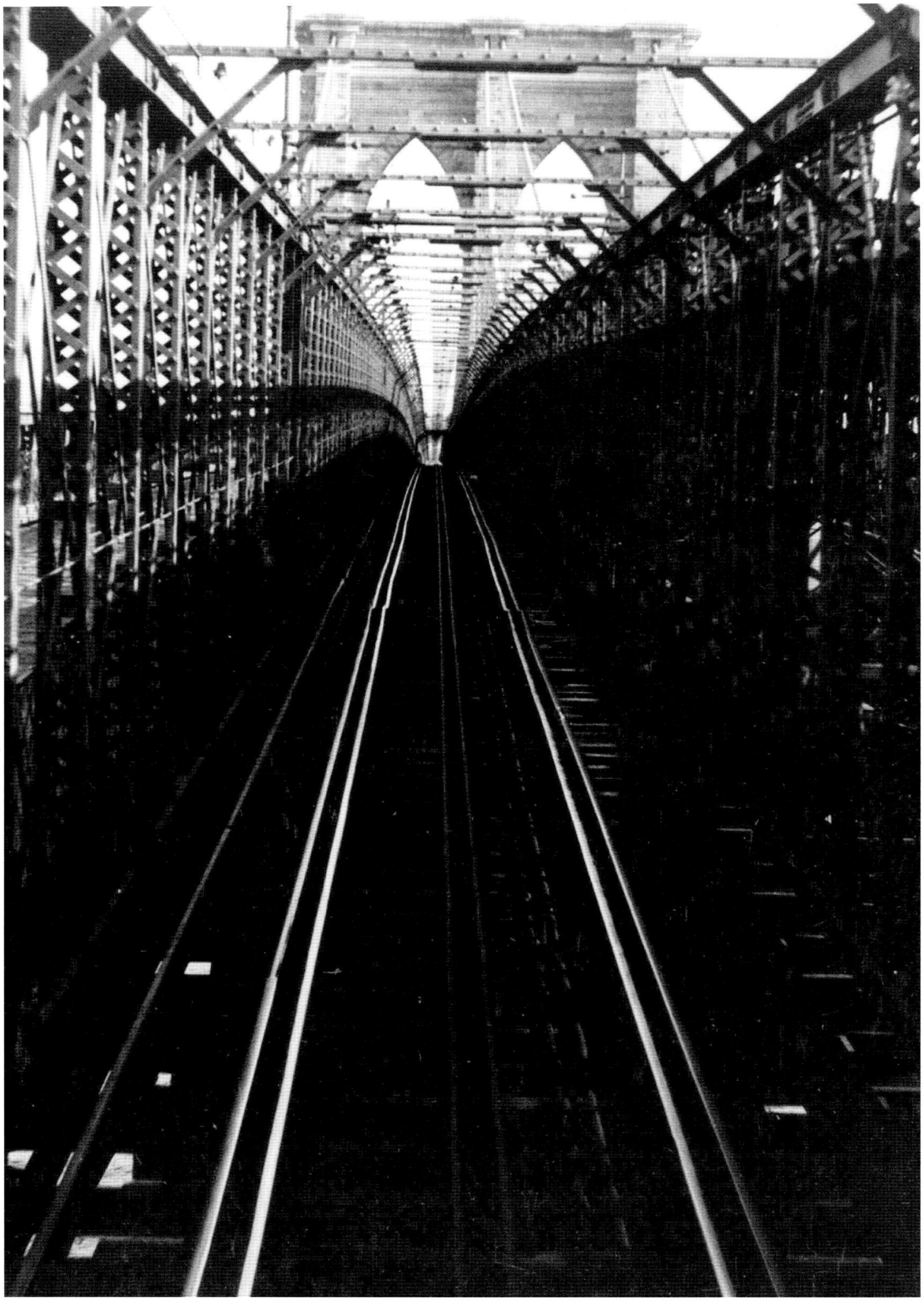

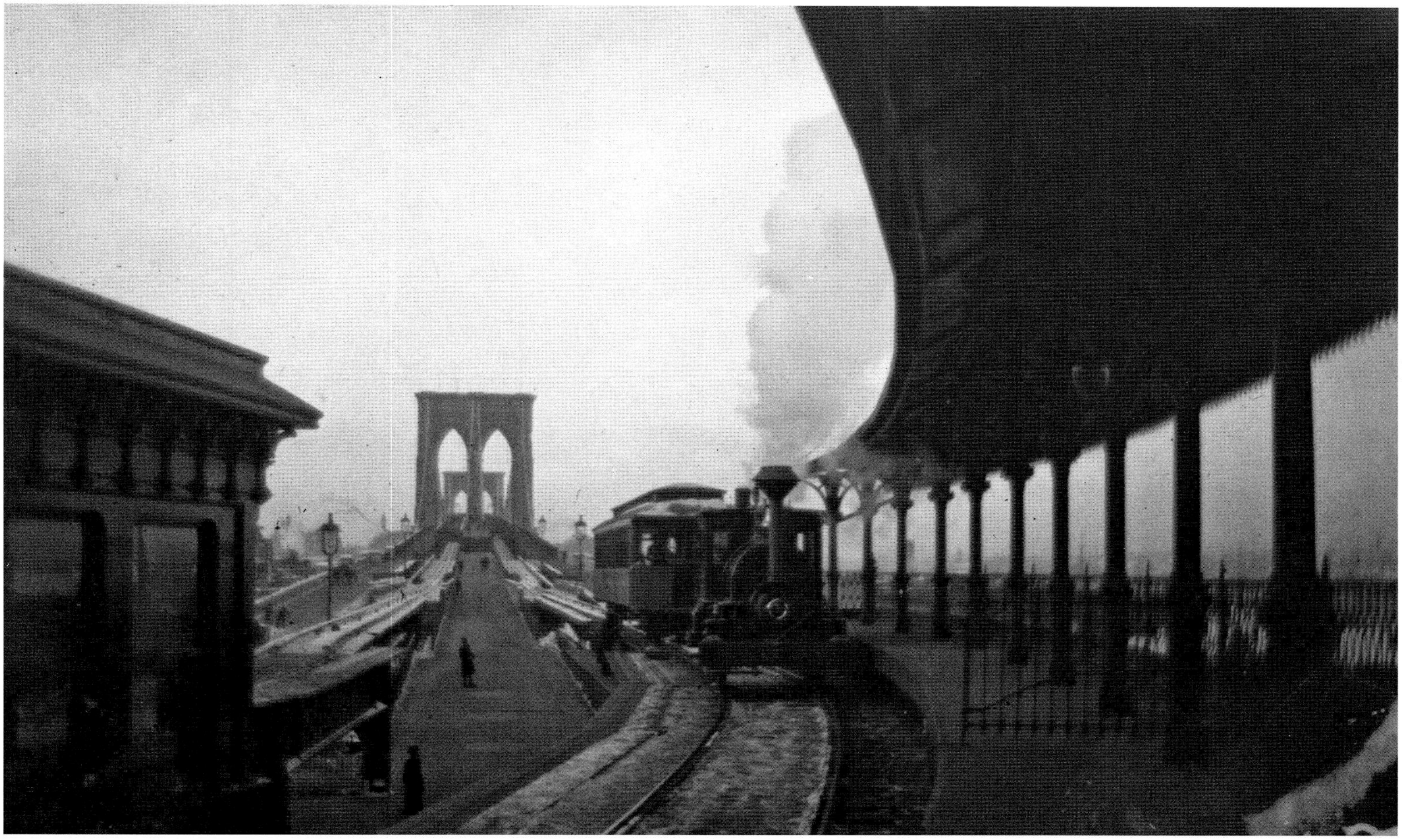

Steam engine crossing Brooklyn Bridge: This train is approaching the terminal. Steam engines alternated with cable cars on the bridge and often shunted trains at the terminal. Service started in September 1883 following the opening of the bridge.

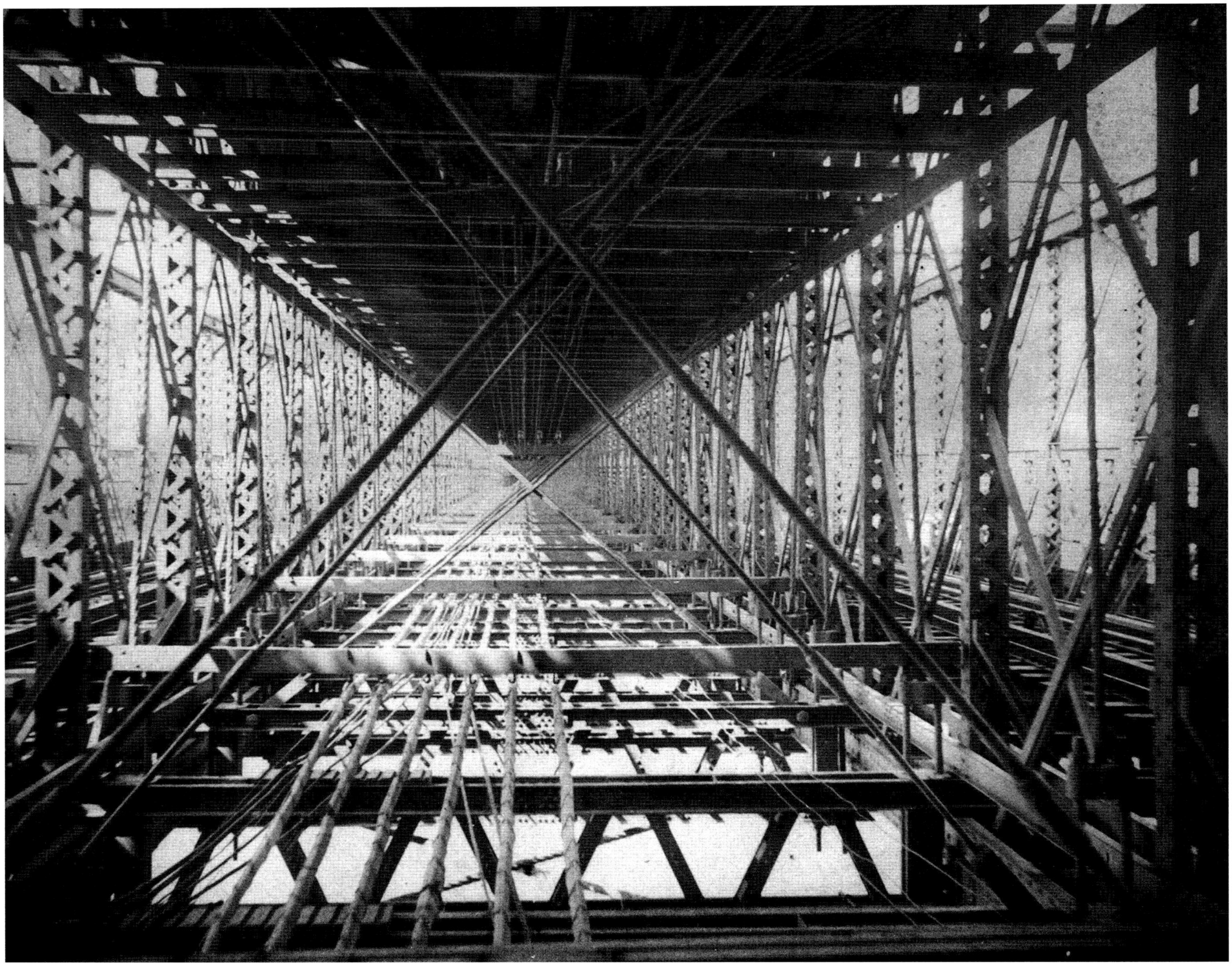

Supports under Brooklyn Bridge: Girders, cables, and other supports crisscross beneath the walking path. To the left and right run trolley tracks.

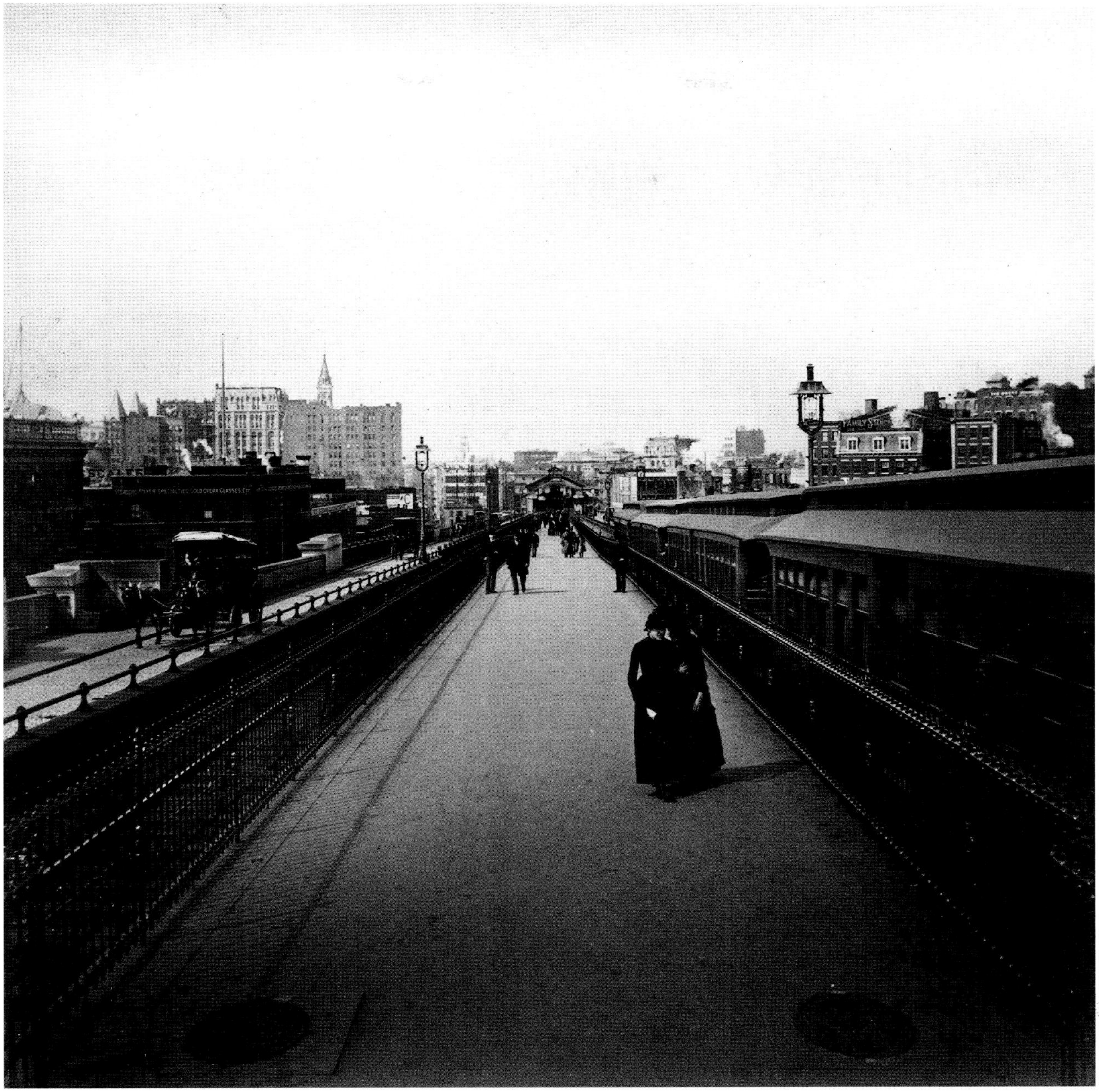

Entrance to pedestrian path: From opening day when 150,300 people crossed the bridge, the Promenade has attracted crowds. With a view toward Manhattan, this 1888 scene photographed by Breading Way shows pedestrians walking between other traffic: horses and wagons on the left and a train on the right. The unique bridge lamps became the last fixtures added to the bridge.

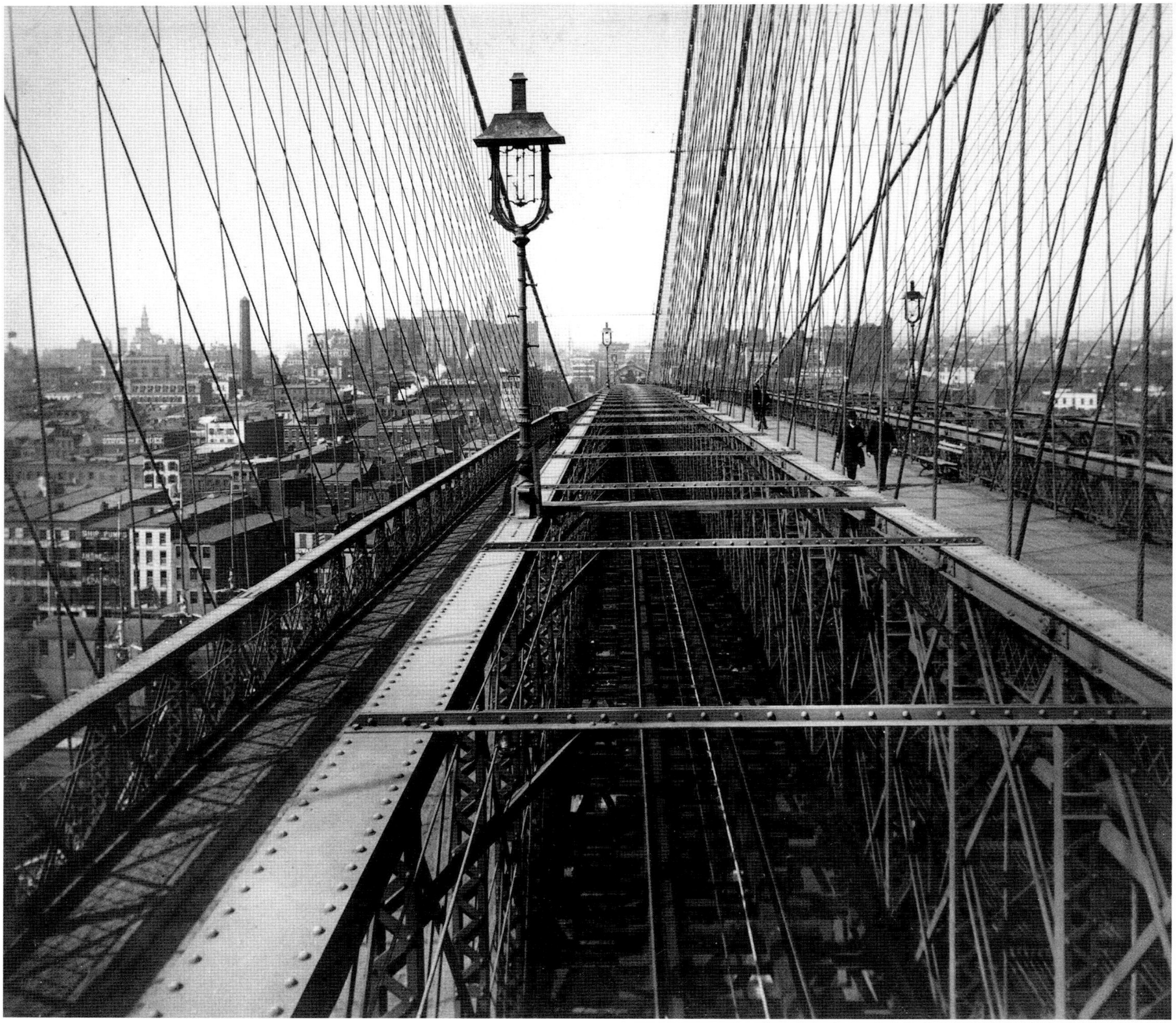

Parallel routes to Brooklyn: With the walkway on the right and train tracks below, girders form symmetrical lines to Brooklyn. The light to the left was one of 70 arc lights that Roebling ordered from the United States Illuminating Company with dynamos located in the Brooklyn terminal. As seen in this 1888 photo, the light posts are mounted on the steel truss work at intervals of 100 feet. This was the first use of electric lights over a river.

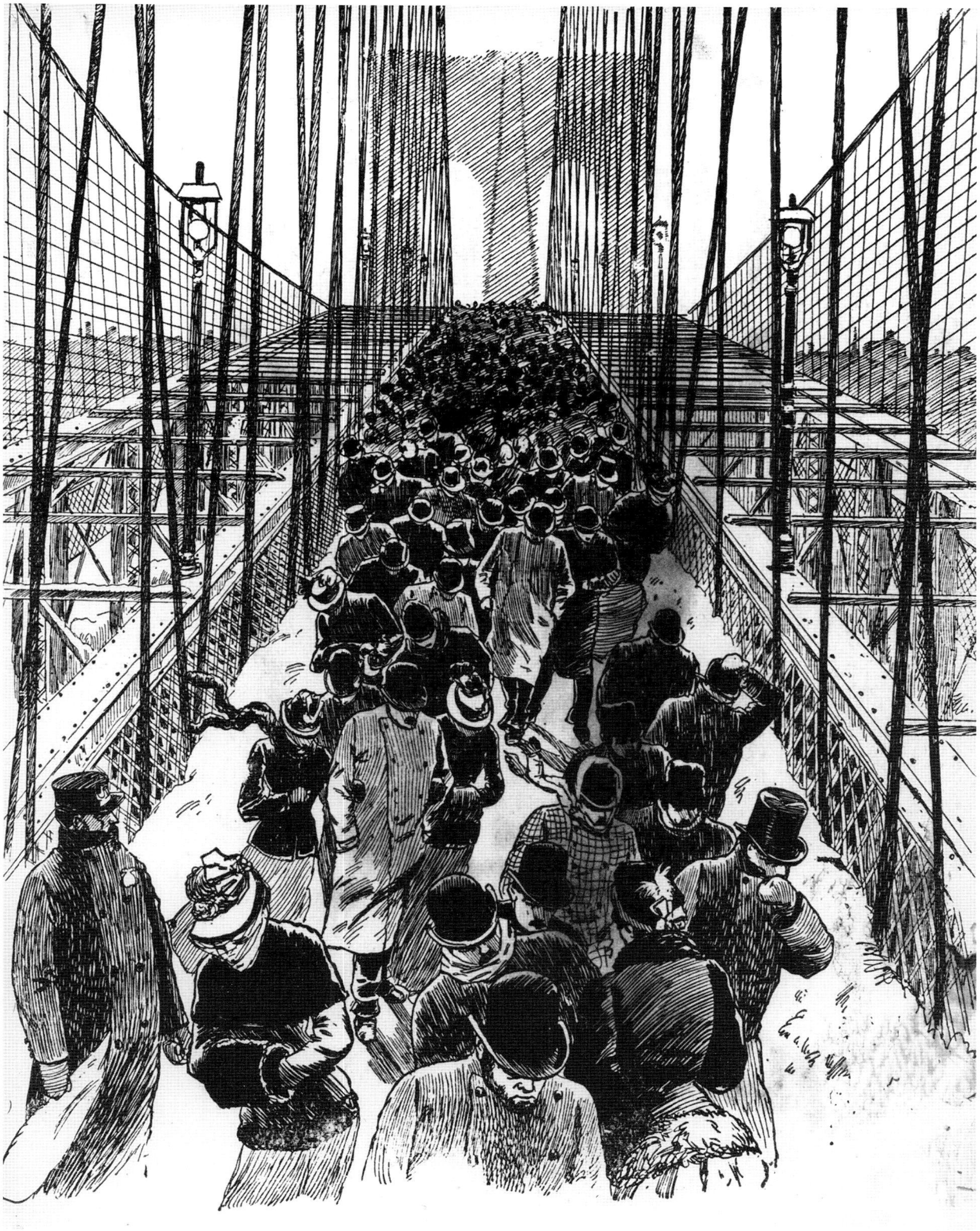

Winter crossing: Crowds walk across the bridge in 1892 during a winter snowstorm.

Following Spread: Crossing over the bridge: Within the first 30 years of the existence of the Brooklyn Bridge, more varied vehicles crossed it than crossed any other bridge: pedestrians, cable cars, steam trains, trolley cars, horses and wagons, and, eventually, automobiles. In this 1894 picture taken by J. S. Johnston, pedestrians and trains share the bridge.

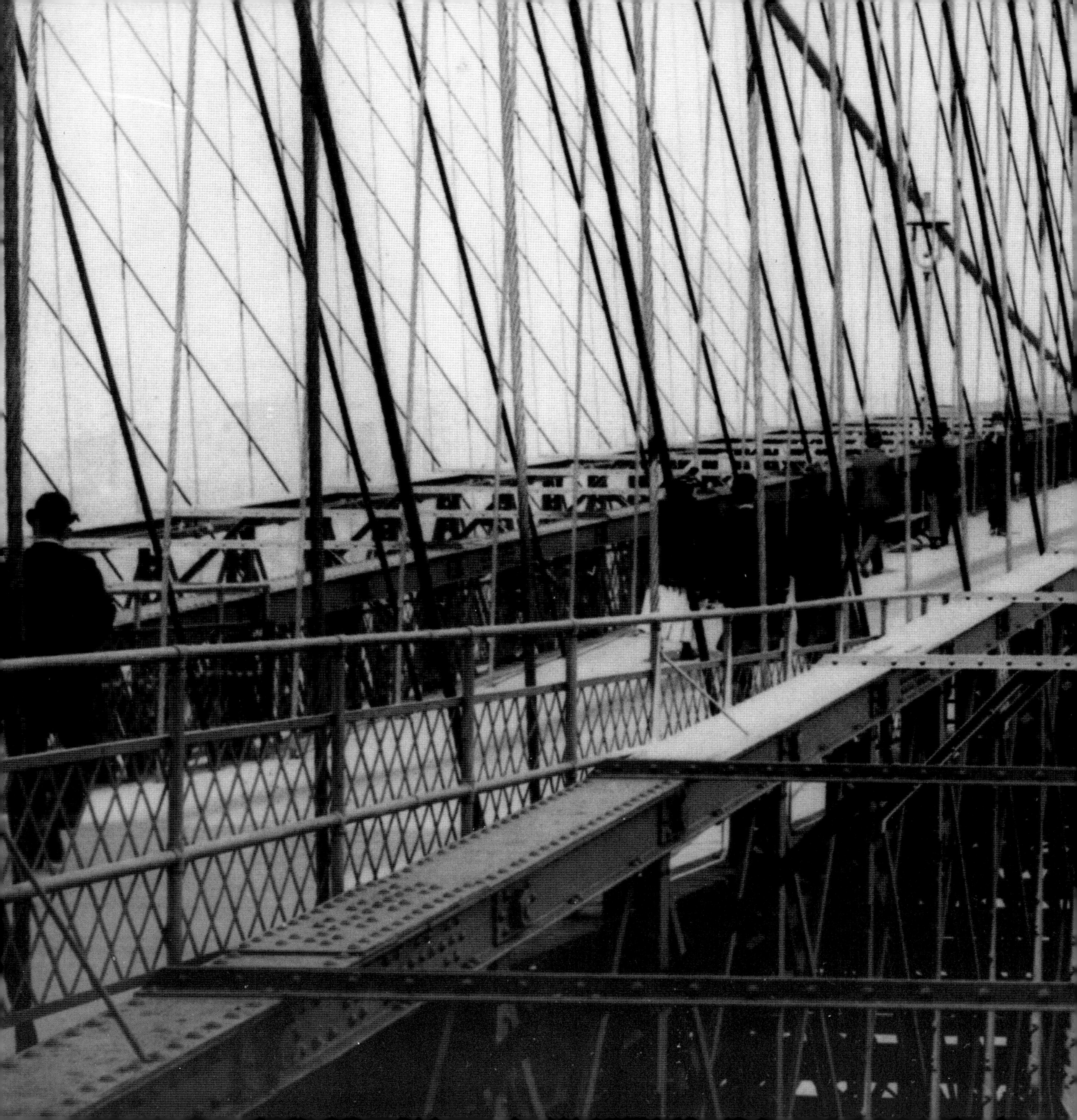

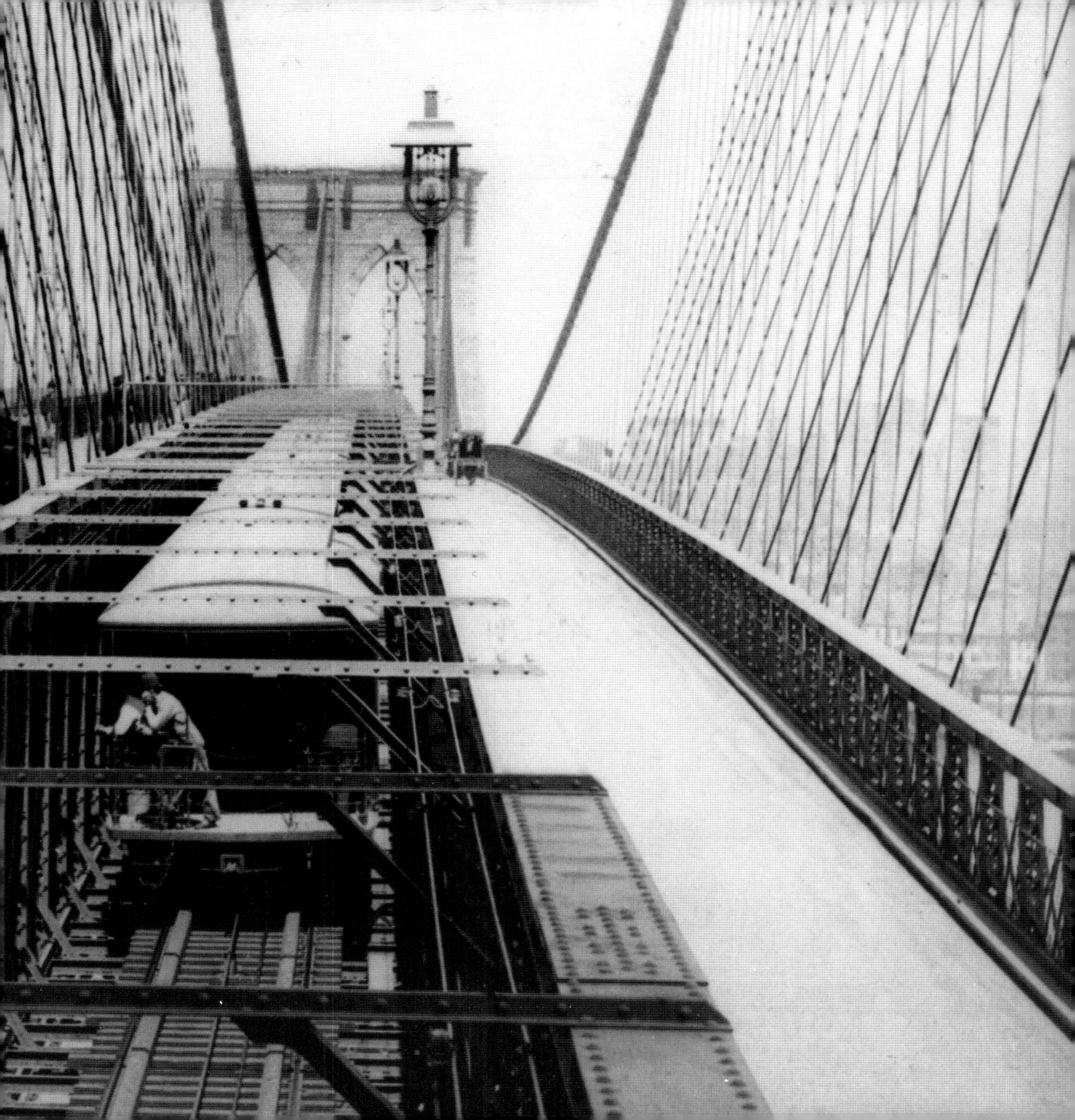

As soon as electric trolley cars were invented in the 1890s, they began crossing the Brooklyn Bridge.

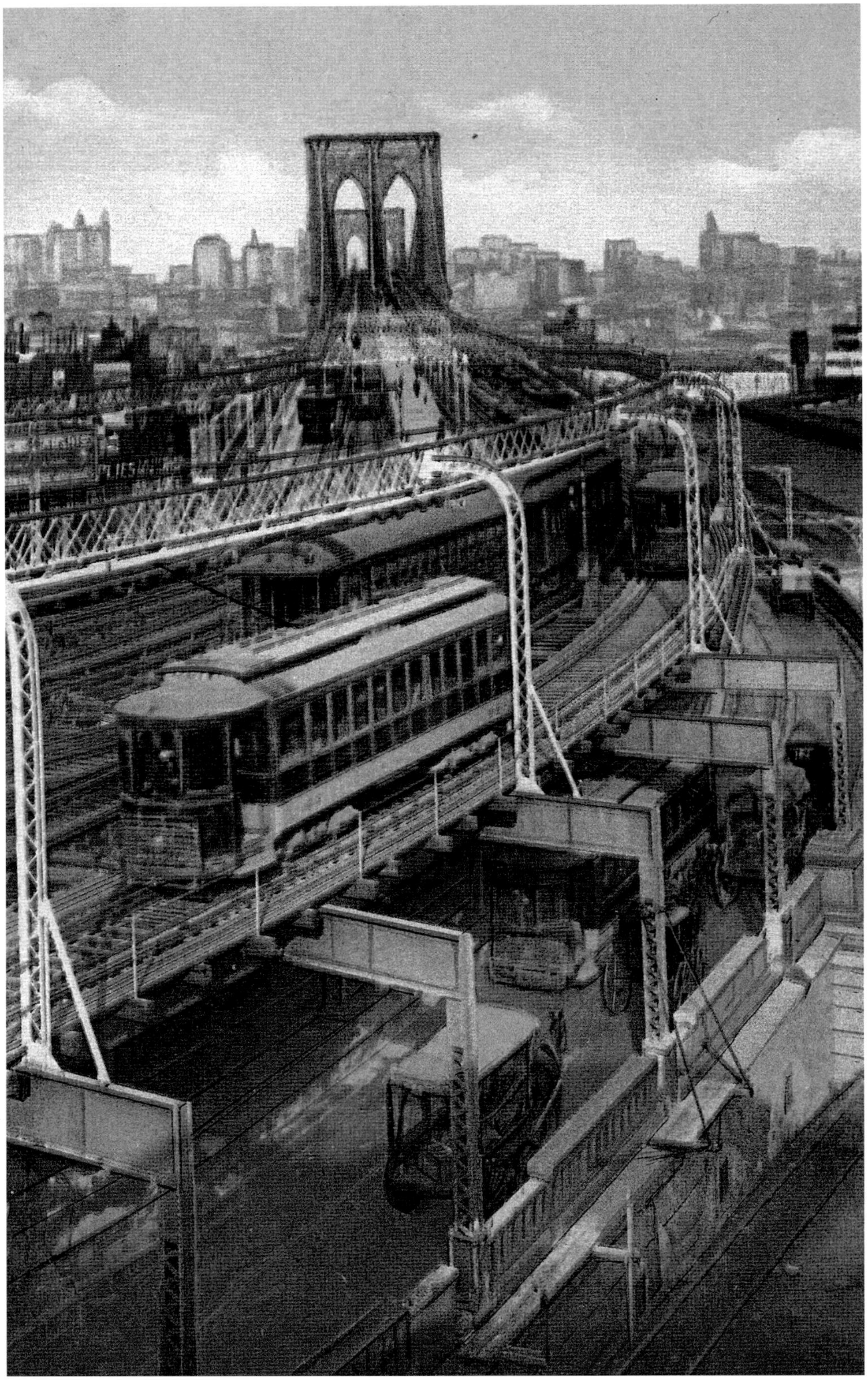

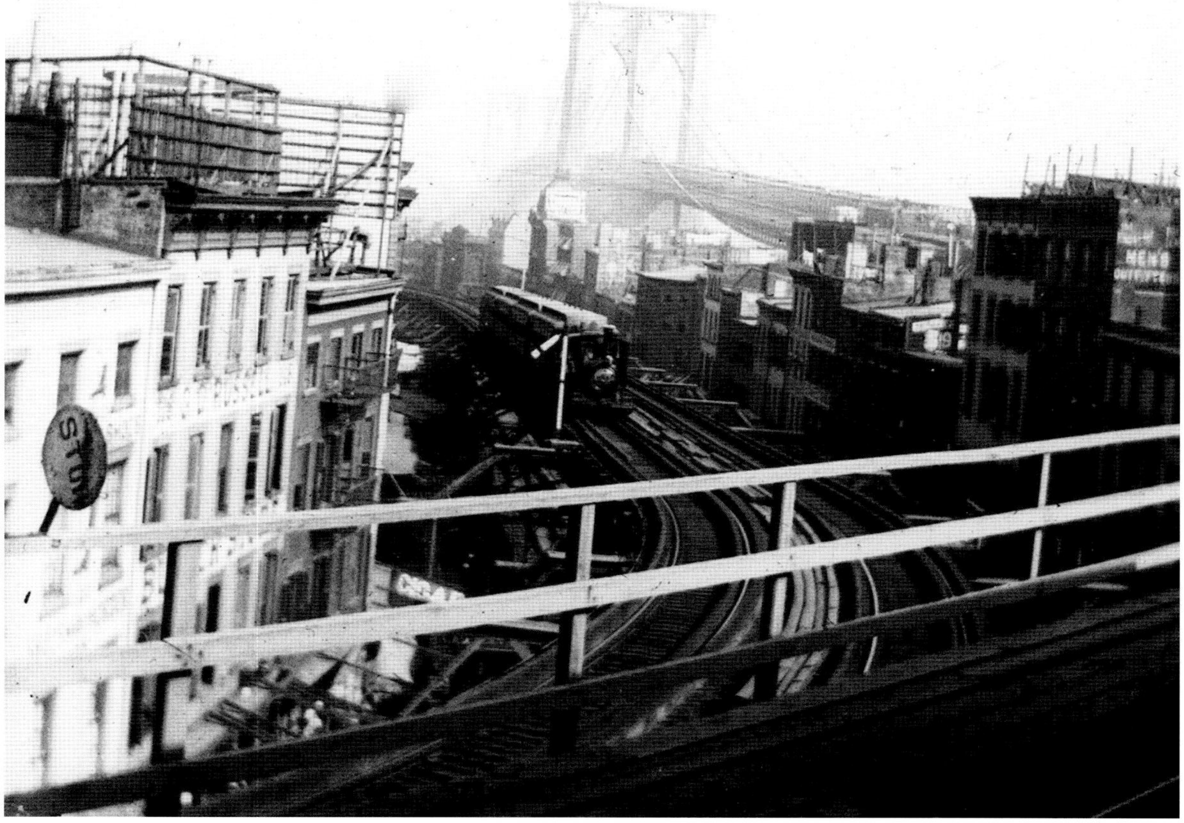

Road to Fulton Ferry: In the foreground, a Manhattan roadway to the Fulton Ferry crosses over an elevated train track, possibly the Third Avenue Elevated. On the other side of residential buildings stands the Brooklyn Bridge.

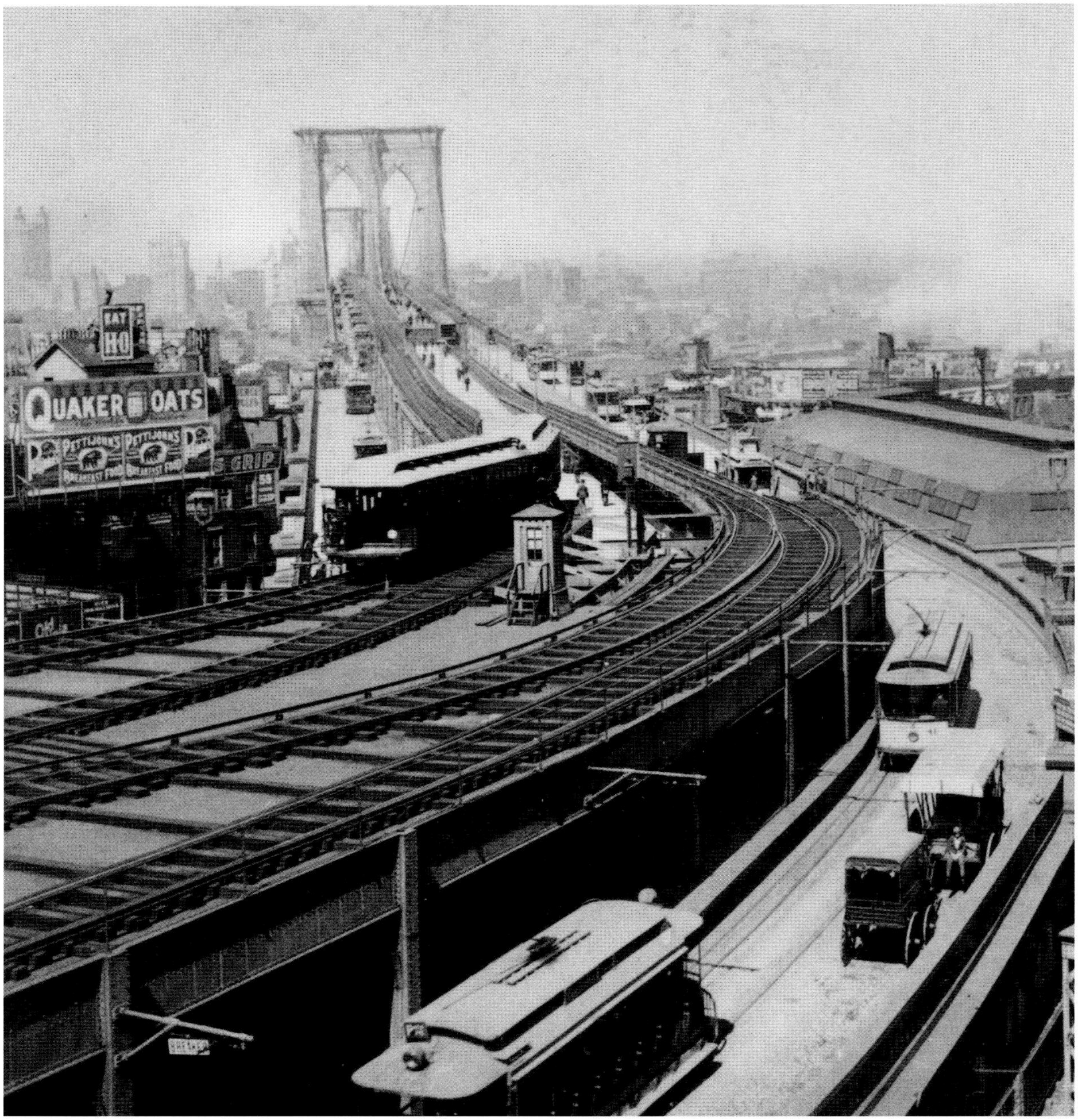

From New York to Brooklyn: In 1898, electricity moved both elevated trains and trolley cars over the bridge, yet horses and wagons can still be seen in the traffic lane. In the distance, pedestrians cross on the Promenade. That year, the City of Brooklyn consolidated with Staten Island, the Bronx, Manhattan, and Queens to become the city of New York as we know it today.

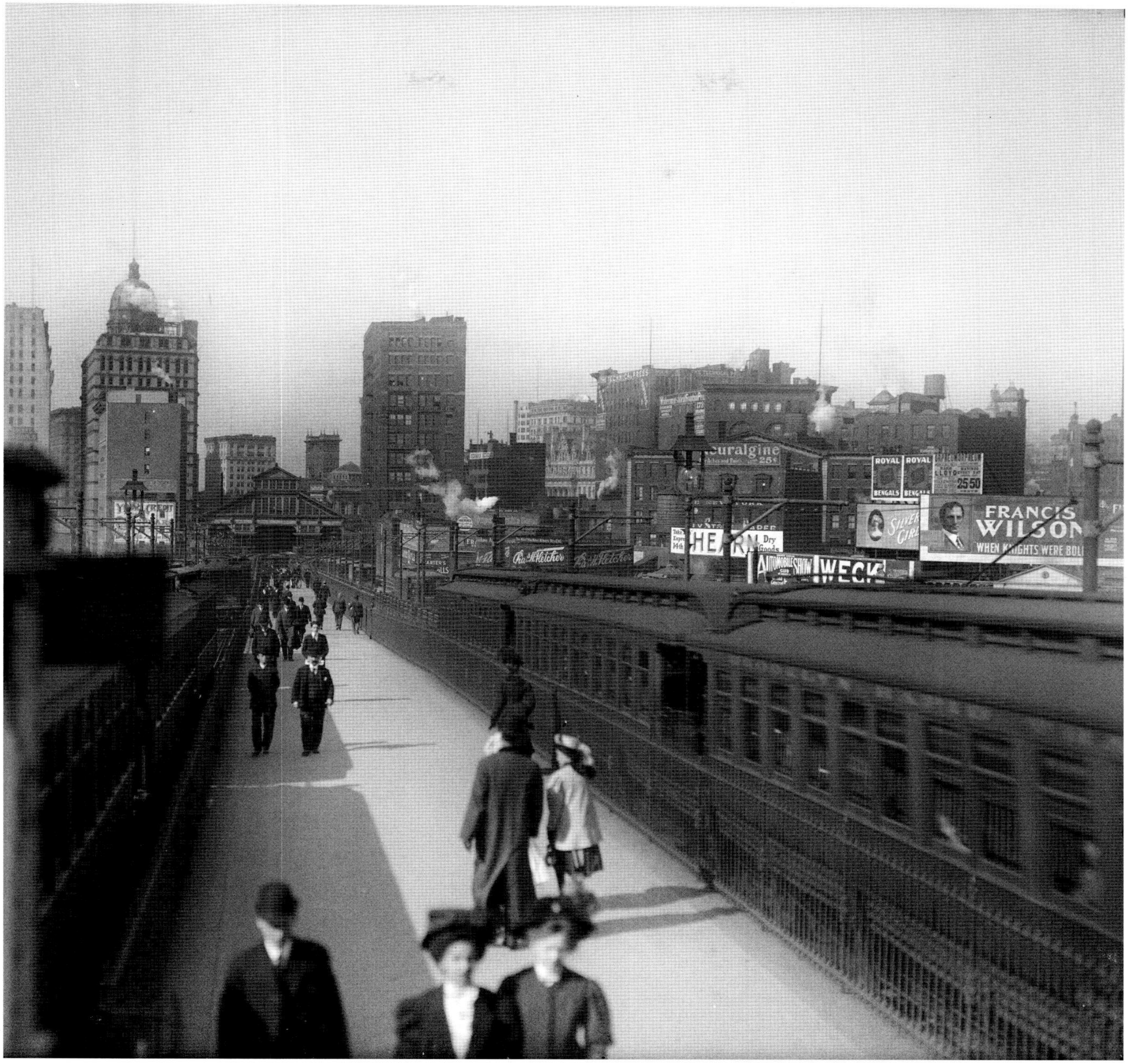

Walking to Brooklyn: By the turn of the century, the buildings in Manhattan rose higher and higher. Still, walking across the Brooklyn Bridge translated into a pleasant experience. In this view, elevated train tracks line both sides of the walkway, and new billboards advertise their products from nearby rooftops.

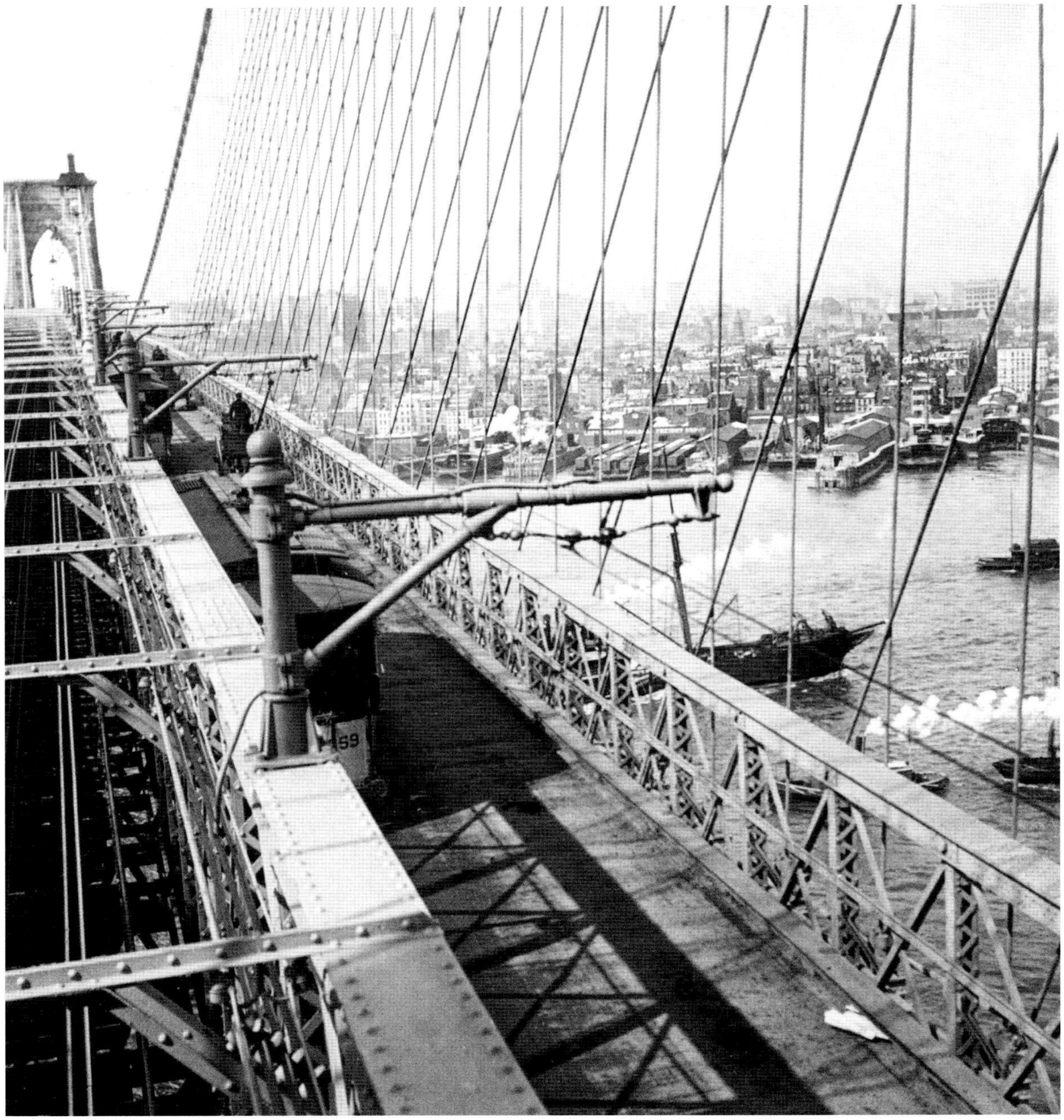

A glimpse of New York City: From the walking path, the burgeoning city spreads farther north while shipping traffic busily sails beneath the bridge. Sailing ships have given way to steamships. On the bridge, a trolley car rides the outer lane, and train tracks lie underneath the path.

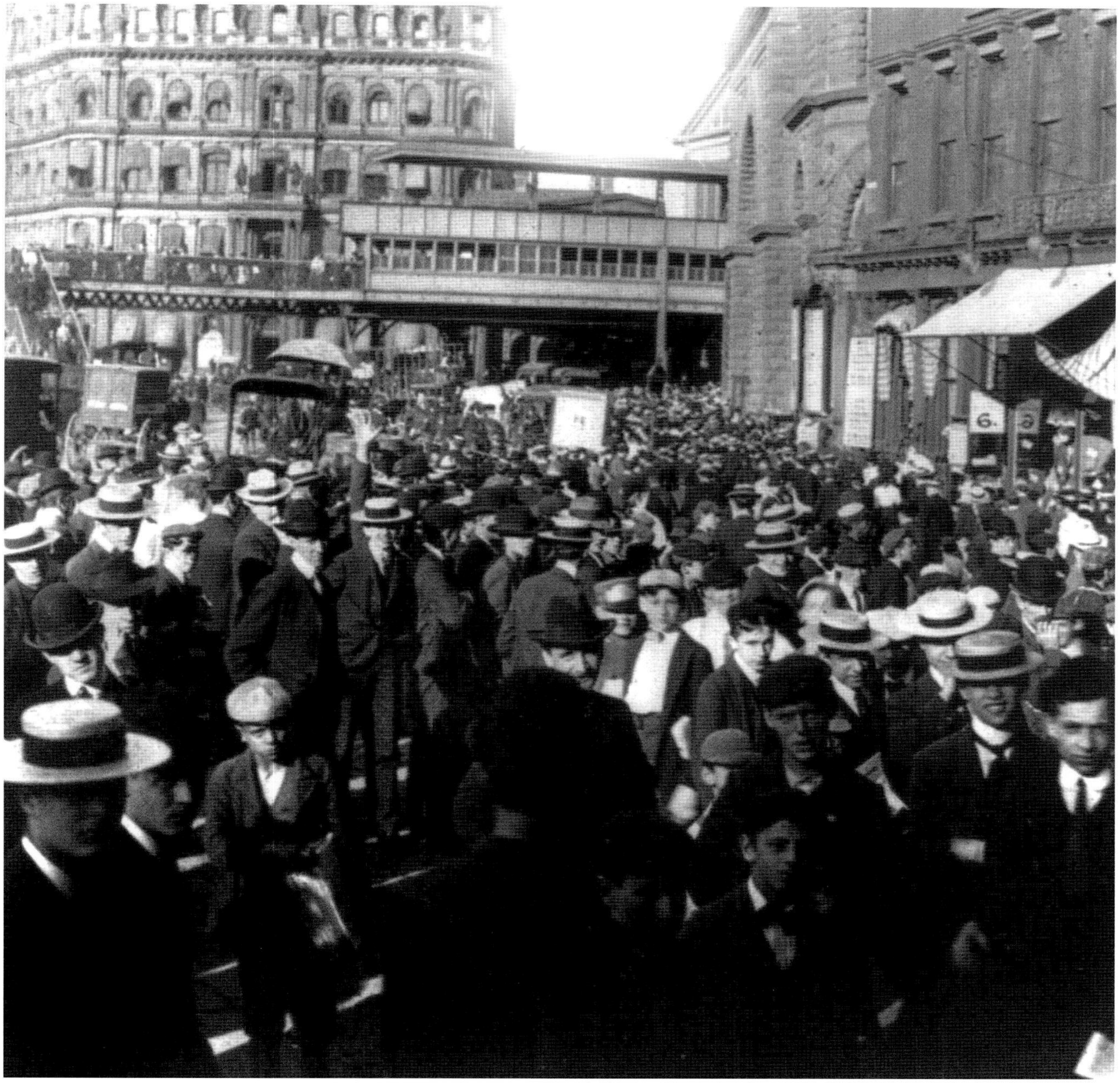

Rush hour crowds: Mostly men on their way home to Brooklyn mill around the Manhattan entrance to the Brooklyn Bridge terminal in 1904. In the left foreground, a newsie hawks his newspapers.

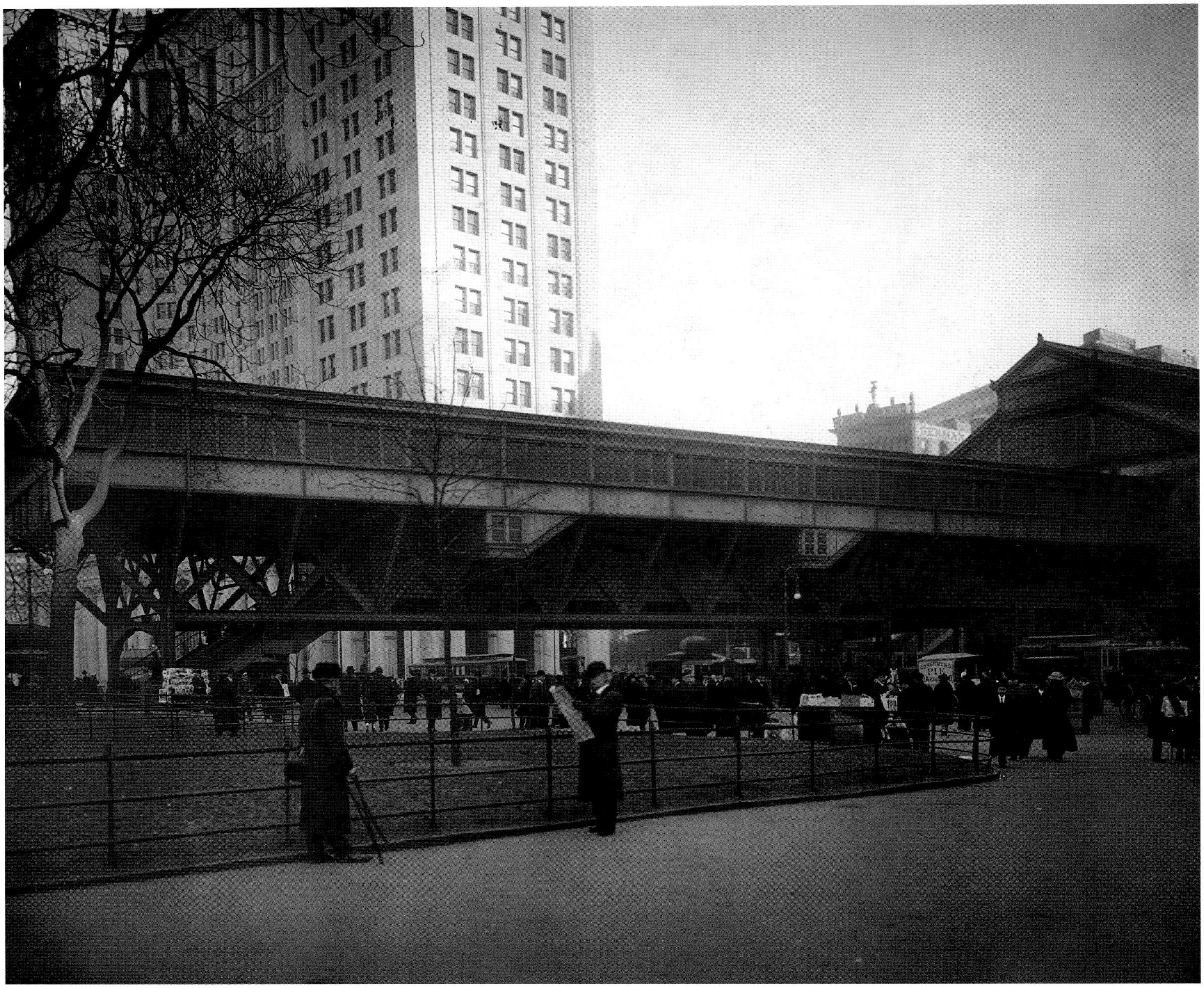

Exterior of Brooklyn terminal: This photo displays a rail terminal at the end of the Brooklyn approach near today's Cadman Plaza.

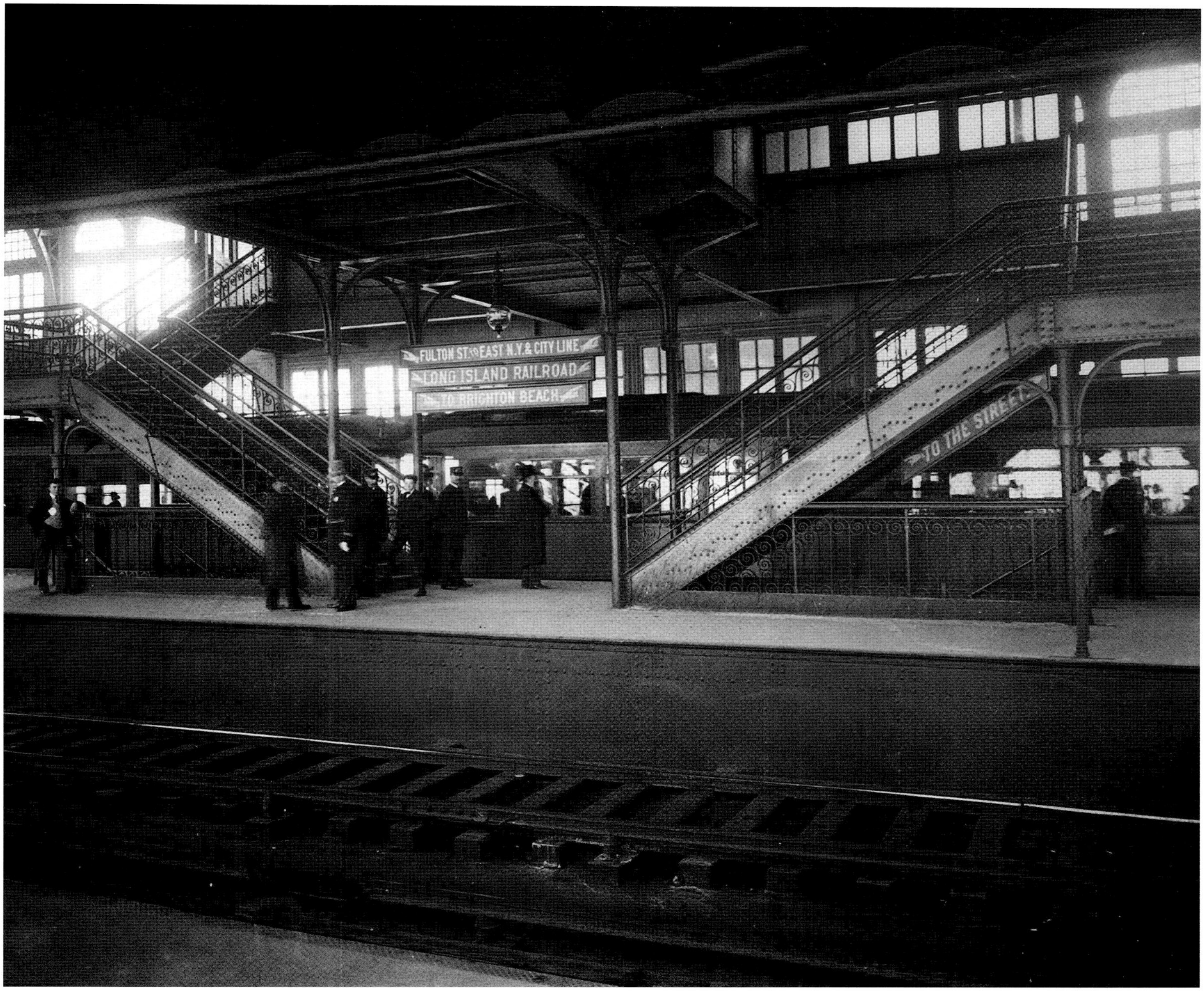

Terminal interior: The interior of the terminals turned out to be as busy as the outside. The terminals were constantly being expanded to cope with the huge crowds.

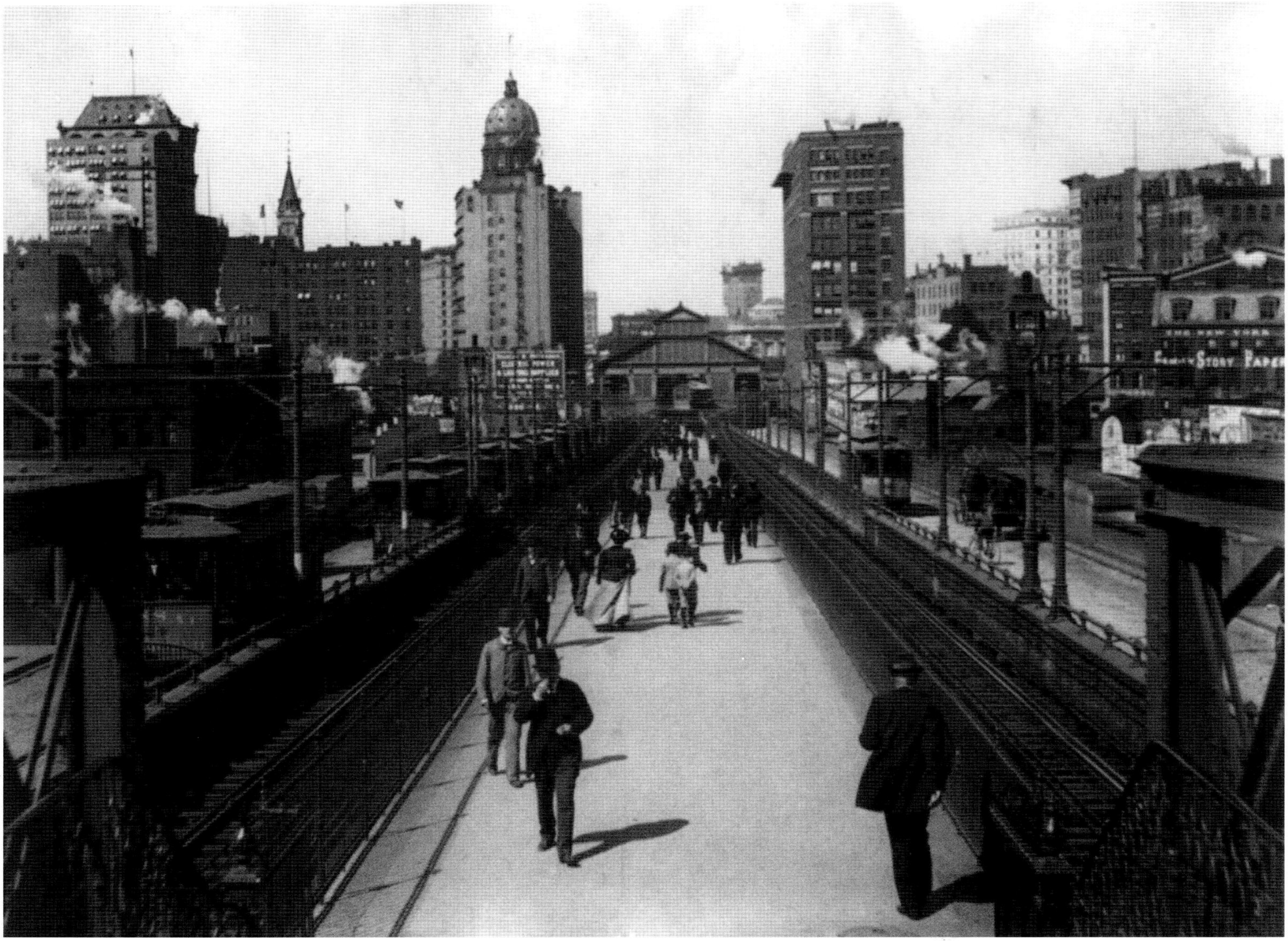

Walkers parade between the terminal and the bridge on the walkway, which is banked by trolleys on the left and a wagon on the far right. To the left of the terminal building in this 1898 picture are the round top of the New York World building, the spire of the New York Tribune building, and the mansard-type roof of the New York Times building.

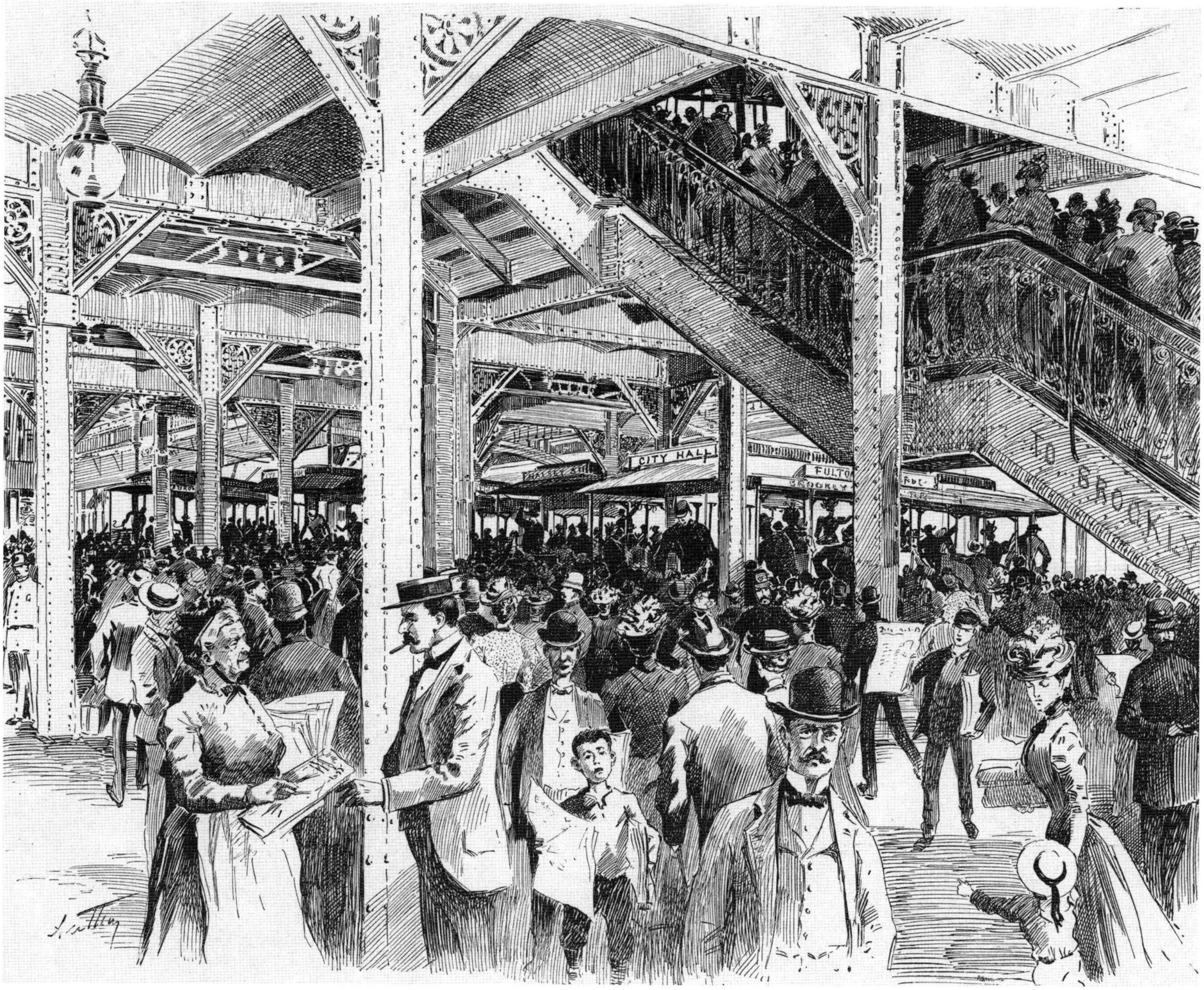

Daily scene at New York terminal: In this drawing of the inside of the Manhattan terminal, crowds climb the stairs to catch trains and to walk across the Promenade. The trolleys on this level are lined up according to their Brooklyn destinations.

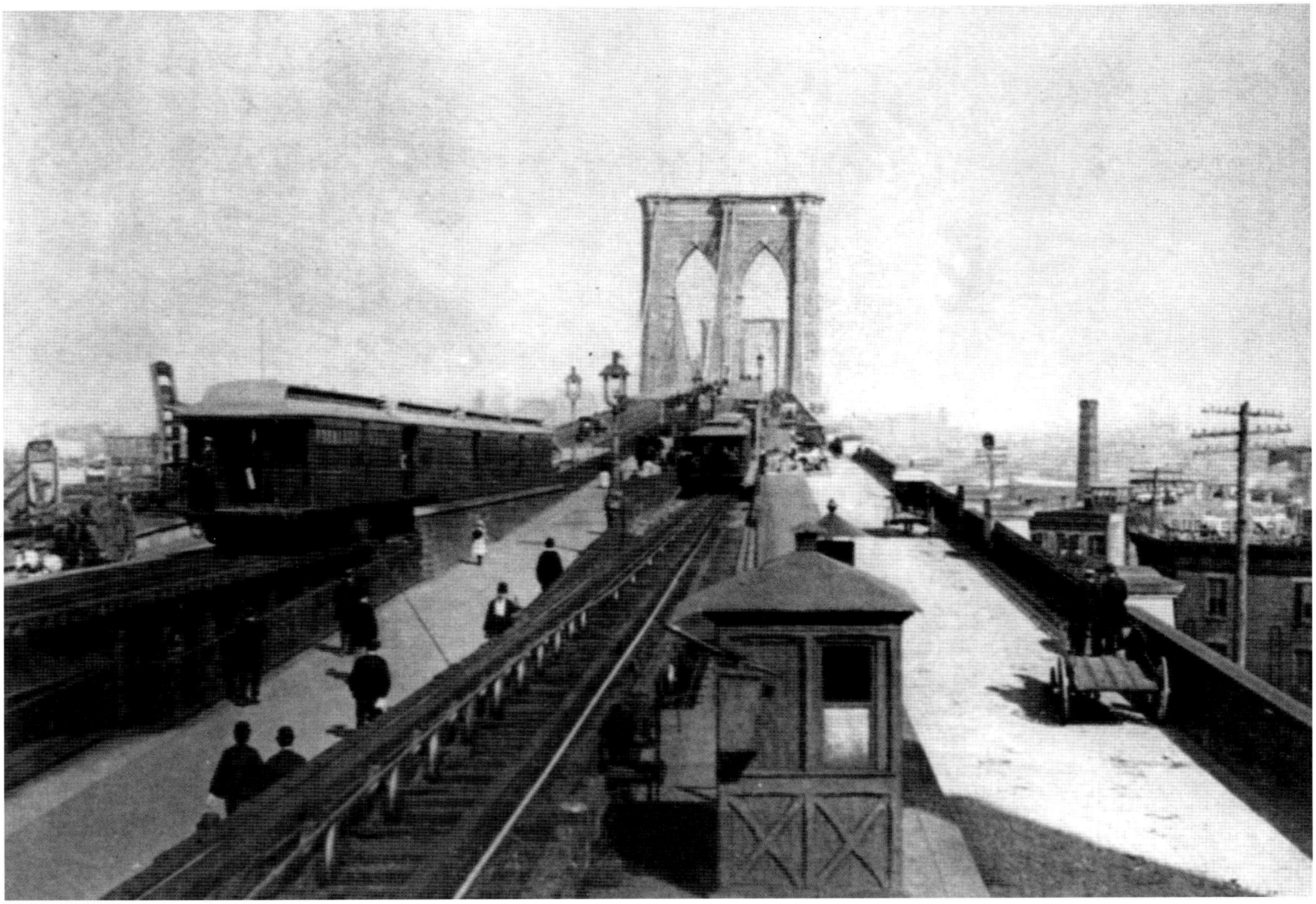

Fulton elevated trains: As transportation across the bridge improved, cable cars were replaced by the Fulton elevated trains, which continued into Manhattan so that passengers didn't need to disembark at the terminal.

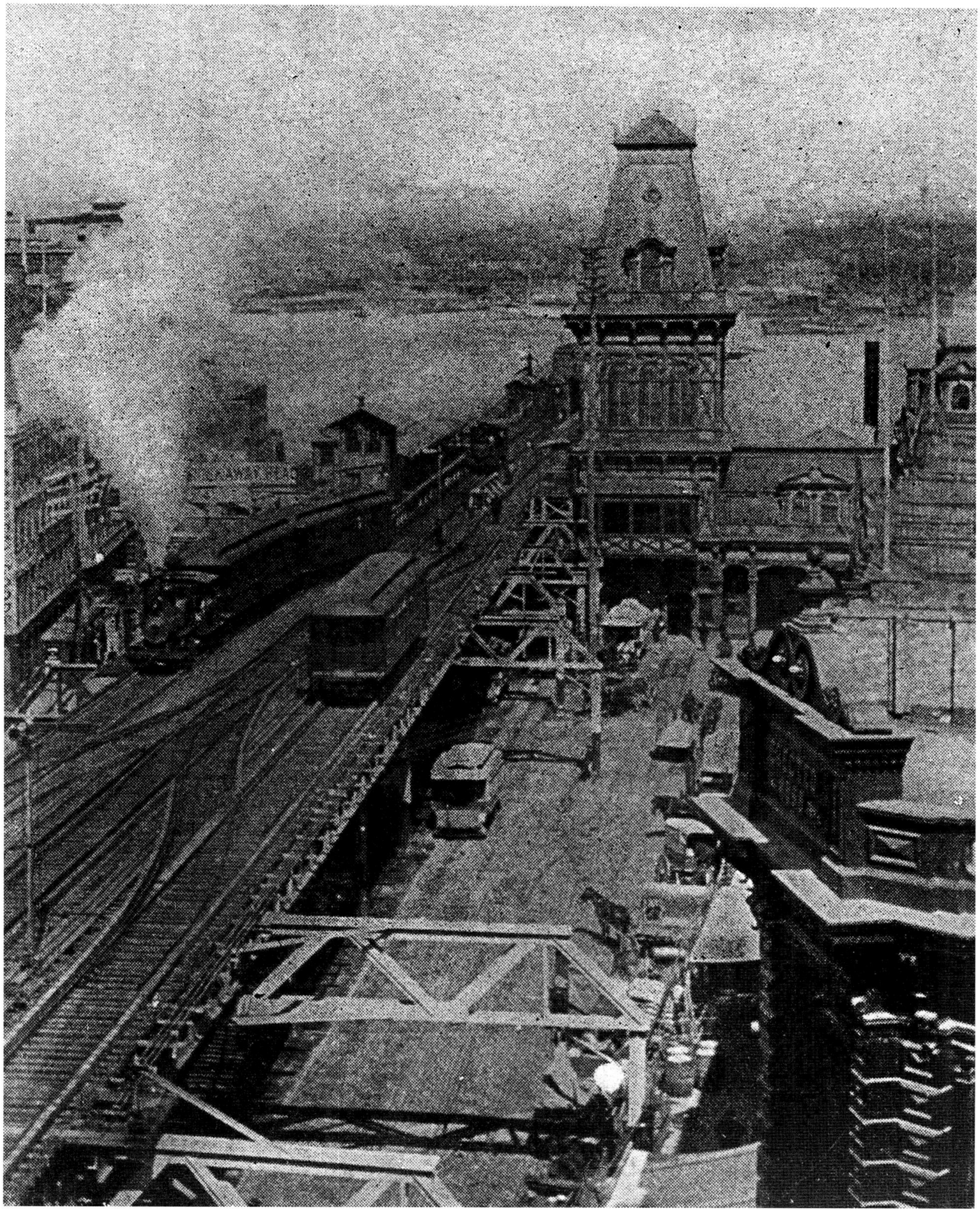

Elevated train: An elevated train at Brooklyn's Fulton Ferry.

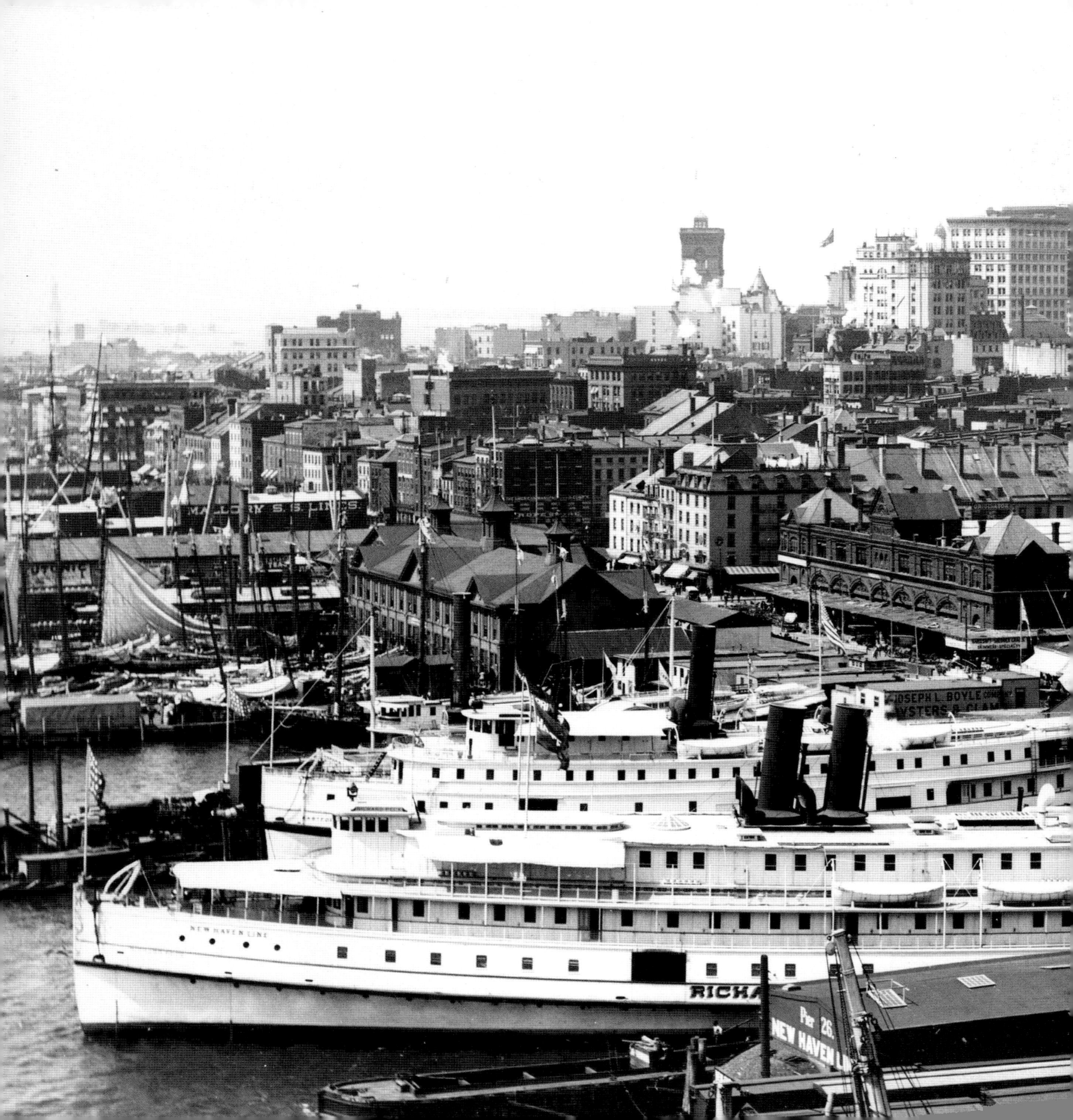
MALLORY S.S. LINES
NEW HAVEN LINE
Pier 26.

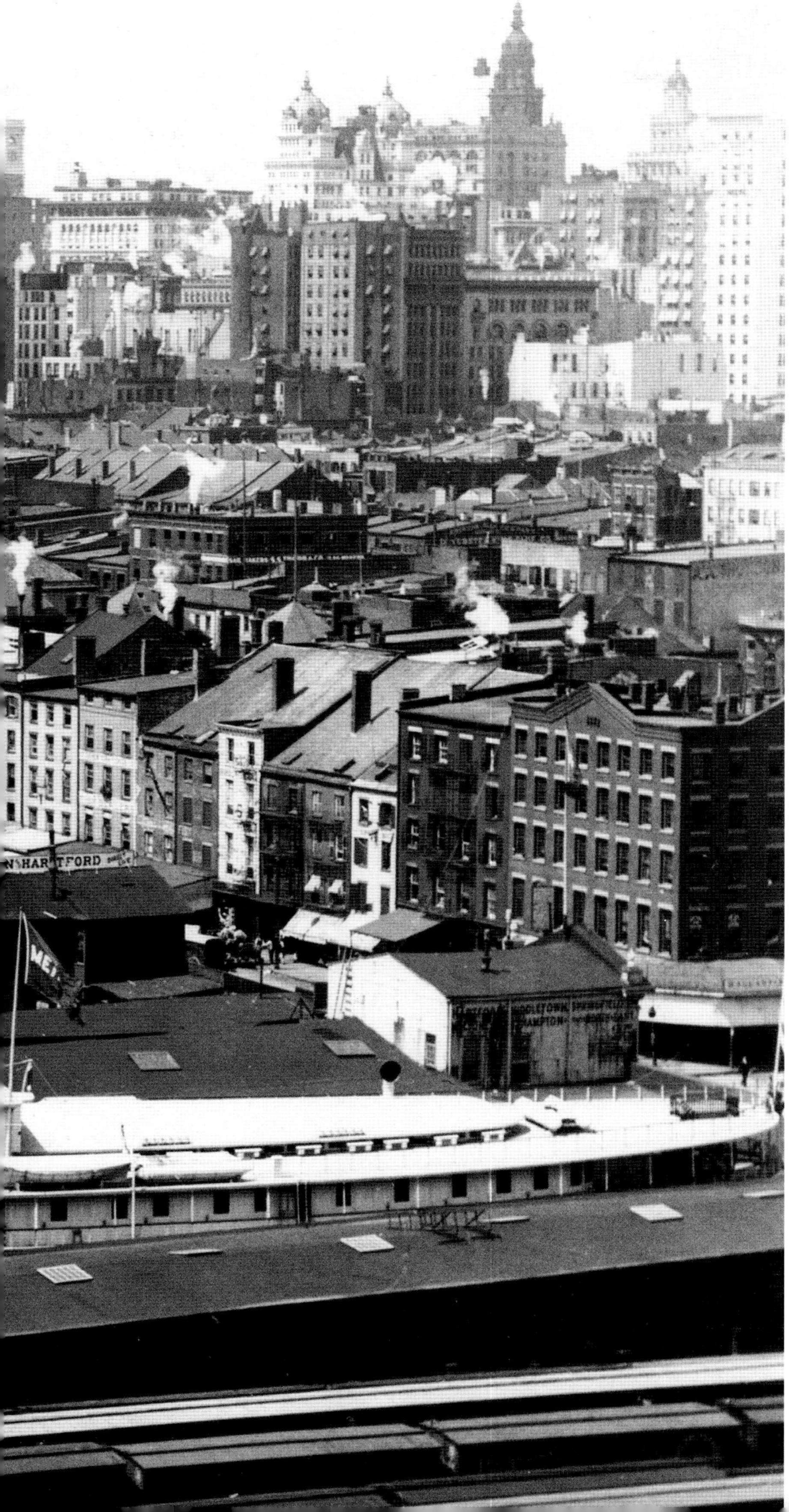

East River docks: Two passenger liner steamships are docked approximately where the Fulton Fish Market later stood. The long, dark building on South Street in the center is the current South Street Seaport. The financial district in the distance was beginning to grow by the time this 1897 photo was taken.

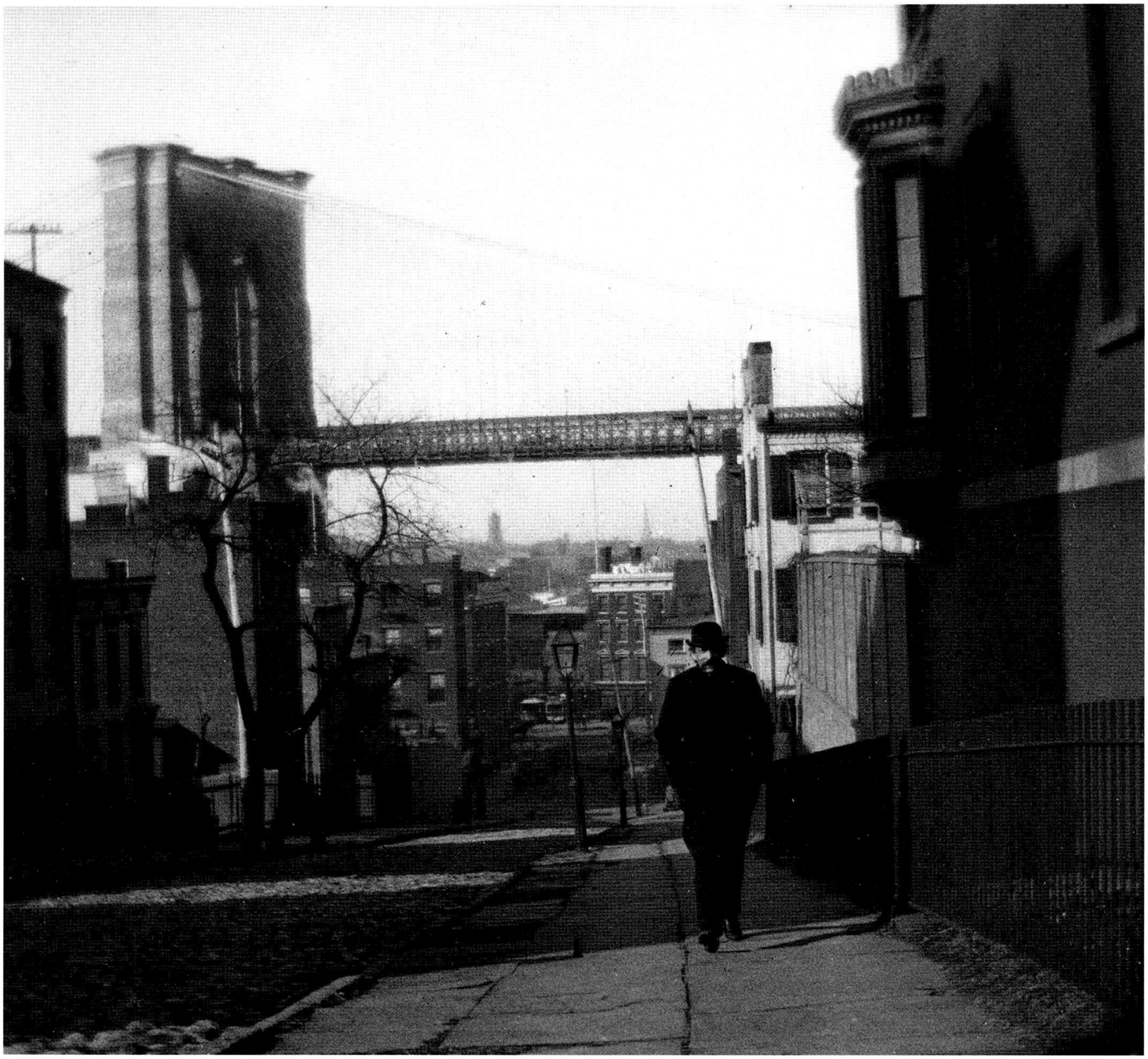

Walking from the Fulton Ferry: A lone man trudges up Columbia Heights toward Middagh Street from the Fulton Ferry in the 1890s. The Brooklyn Bridge tower looms over the neighborhood. The building at left in the foreground and the one at Fulton Street on the down the hill still exist, but the others have been demolished.

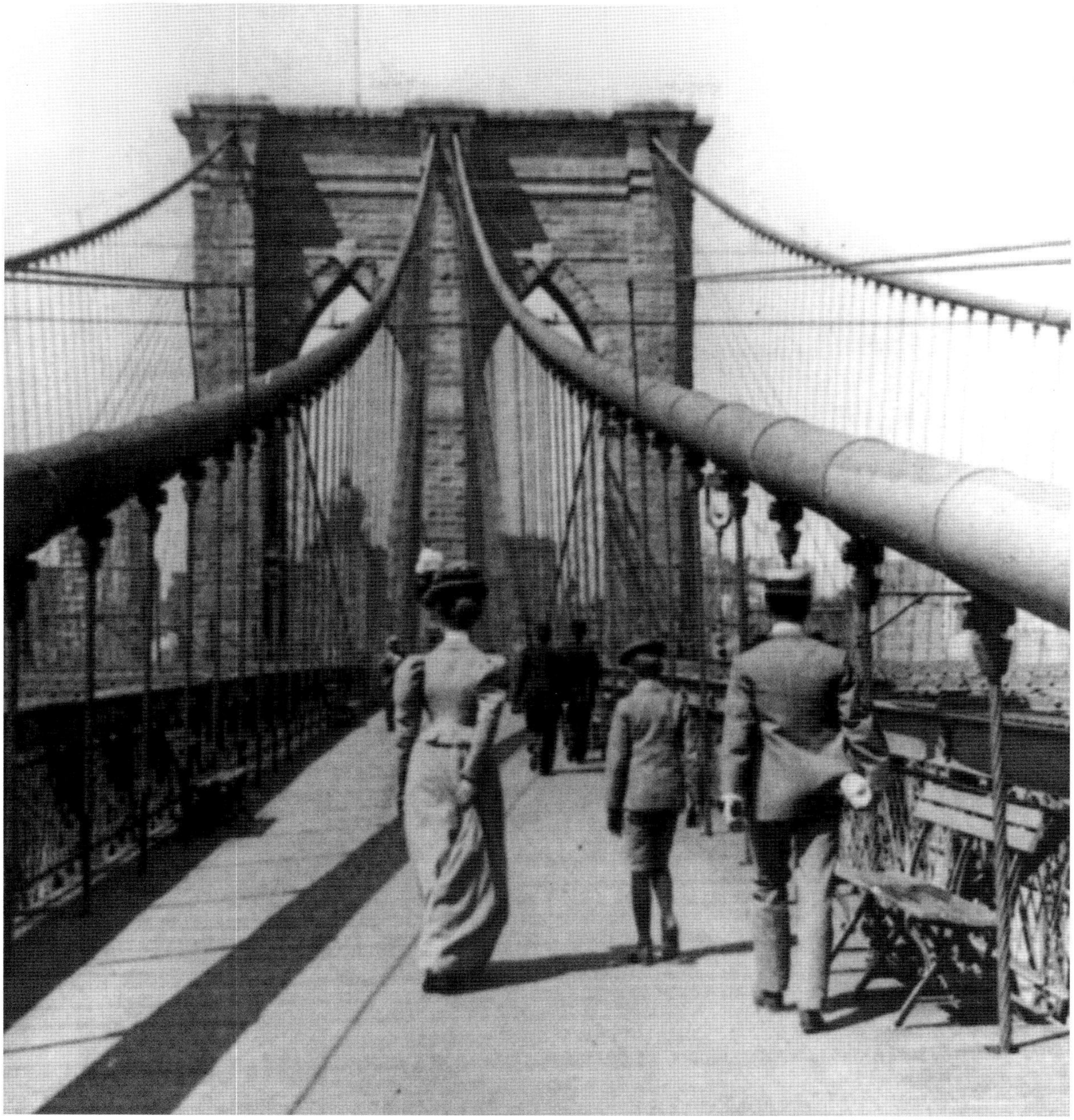

Strolling on the Promenade: The trio in this 1899 photo includes a woman holding up her long skirt so it wouldn't drag; a boy dressed in hat, suit, and knickers; and a man in a straw boater.

Bridge pedestrians: Two young boys in knickers stop to talk while a man waits in this 1901 photo. Other pedestrians in the background cross to Brooklyn. Notice the dip in the walkway.

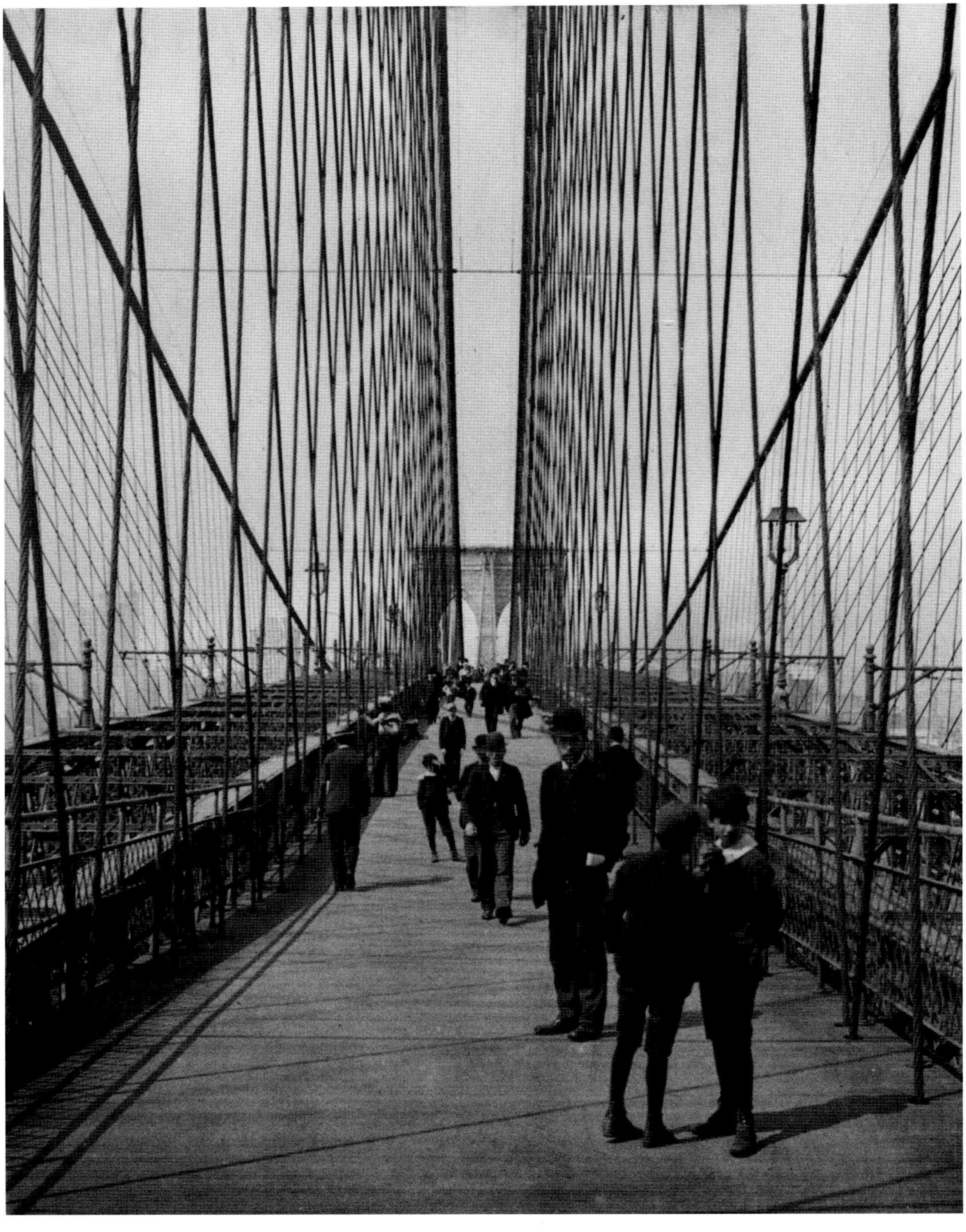

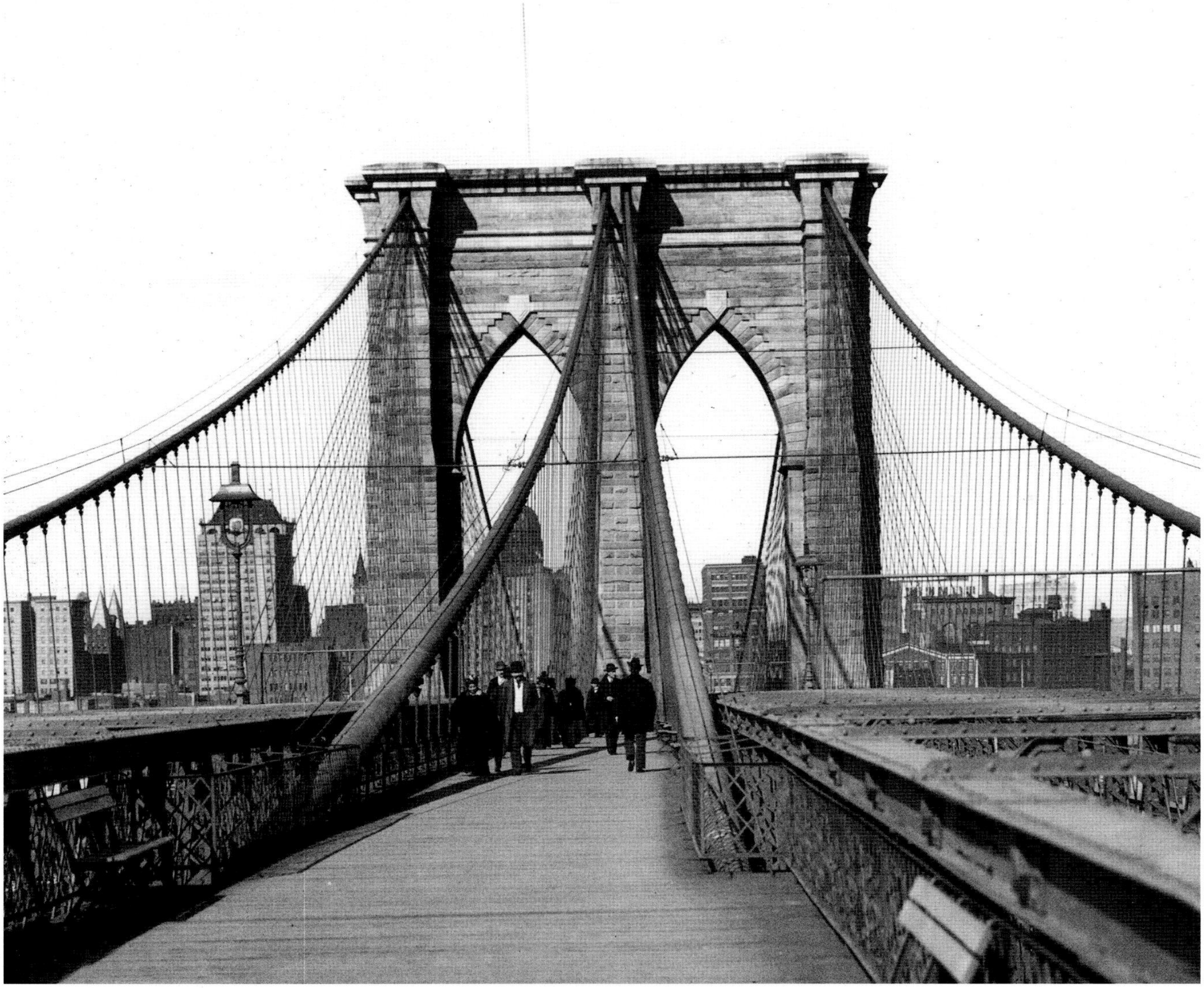

To the other side: This group crosses the Brooklyn Bridge near one of the towers. The towers were constructed of pink granite from a quarry in Branford, Connecticut.

Crossing over in 1904: People from all walks of life cross the bridge in this view, including a father and two children followed by a man dressed in a formal top hat.

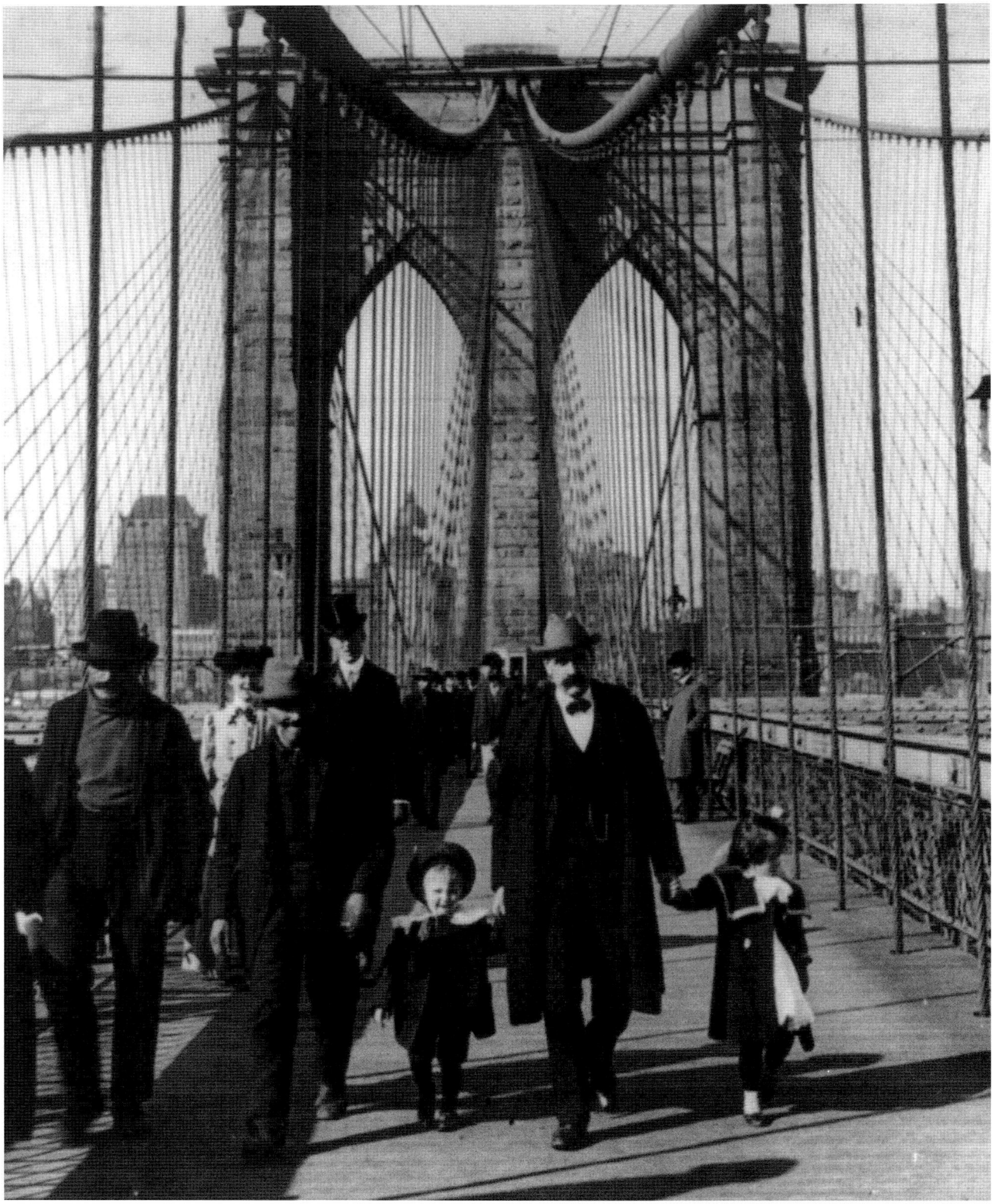

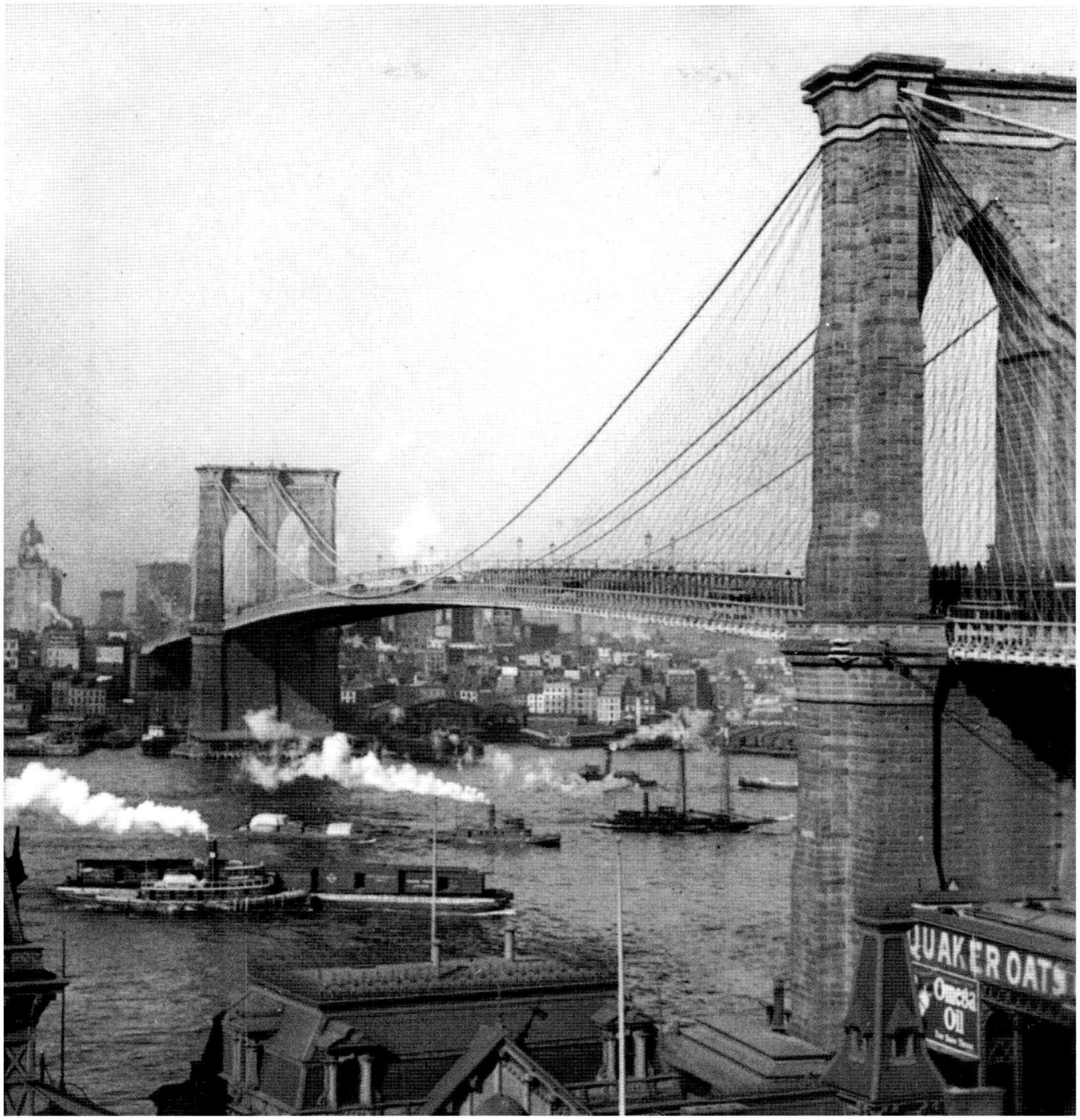

This famous close-up shows the Brooklyn Bridge against a Manhattan background (ca. 1901). Tugs ply the East River, and an ad for Quaker Oats is painted on a building next to the old Fulton Ferry House.

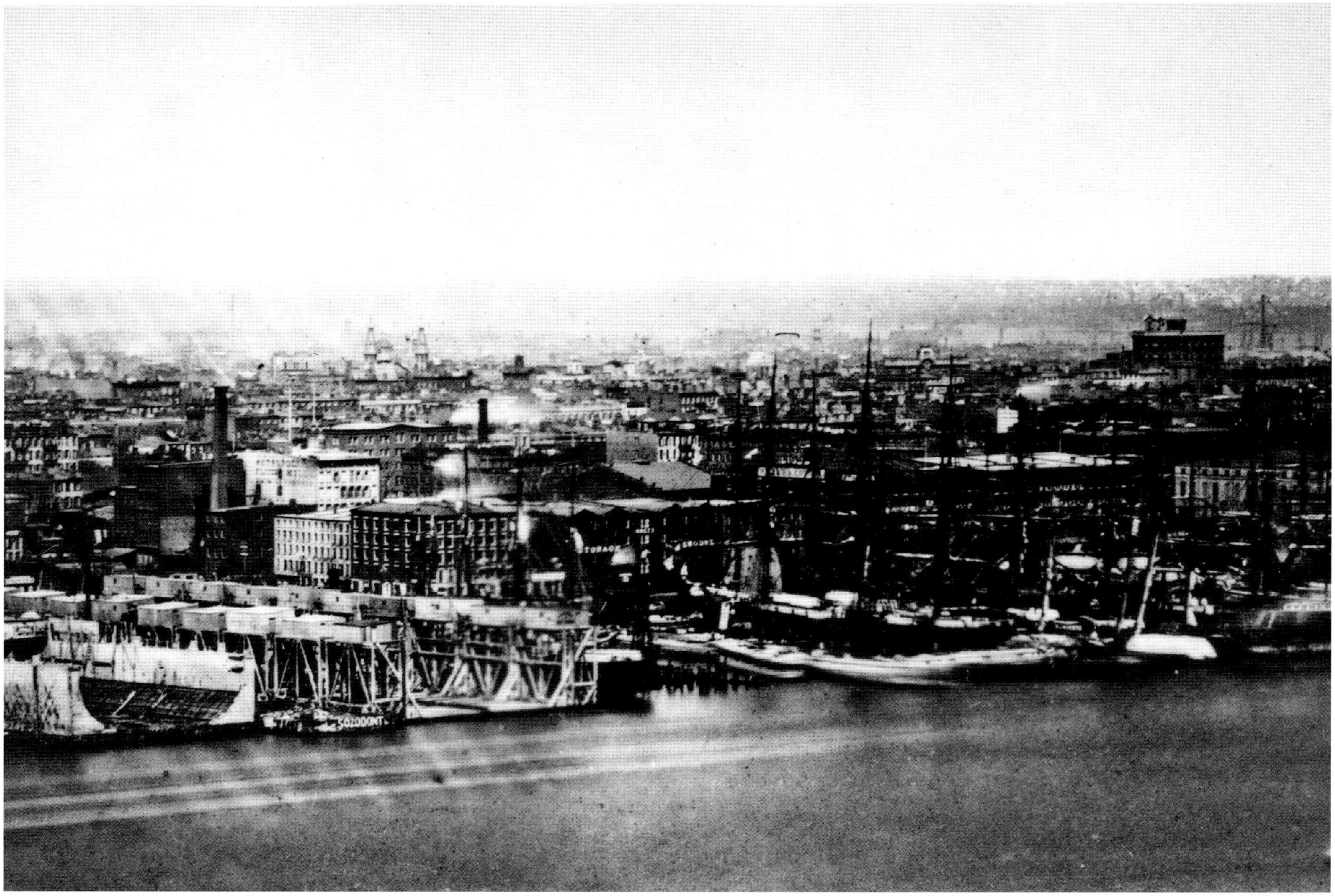

Brooklyn Navy Yard: Founded in 1801, Brooklyn Navy Yard, also known as New York Naval Shipyard, lay to the east of the Brooklyn Bridge but antedated it by almost three-quarters of a century. Edison used the bridge in its early days as a subject for his short films. He also filmed the shipyard and made movies of warships passing under the bridge.

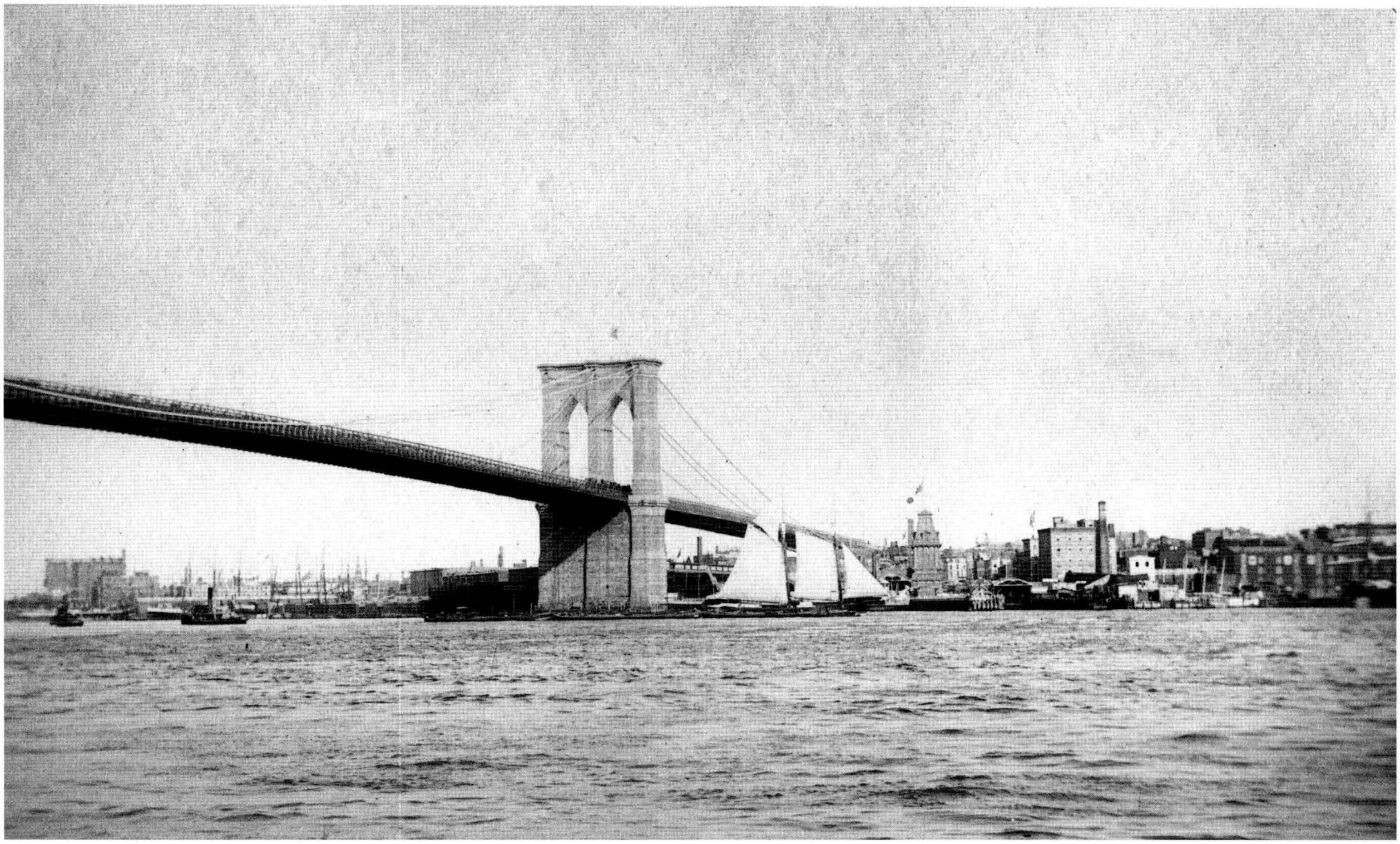

Bridge and boat: A two-masted schooner sails under the Brooklyn Bridge in this early photograph.

Brooklyn Bridge: In this 1896 view of the Manhattan tower, barges and a dock are visible beneath the bridge while an excursion boat passes under the span.

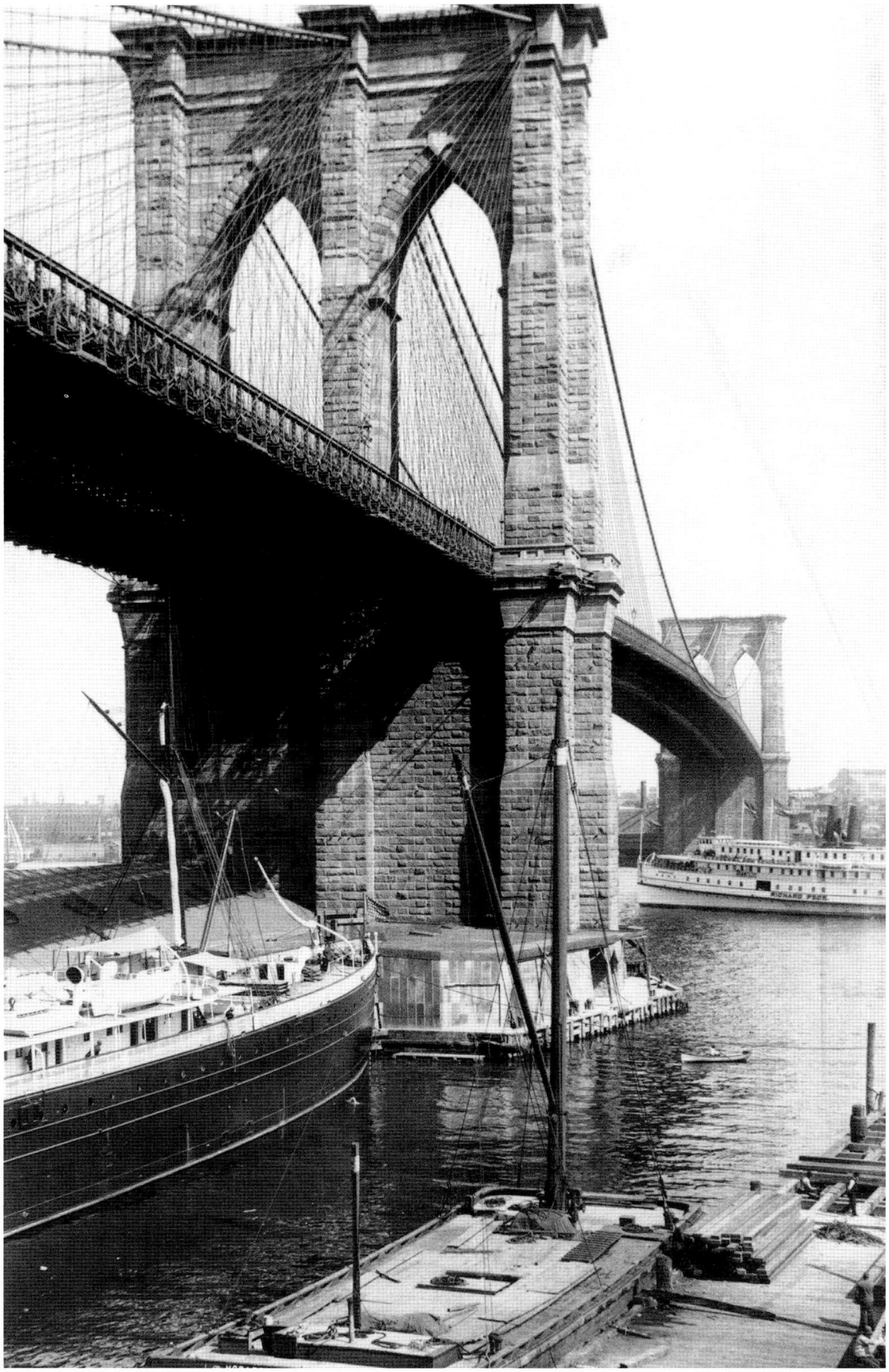

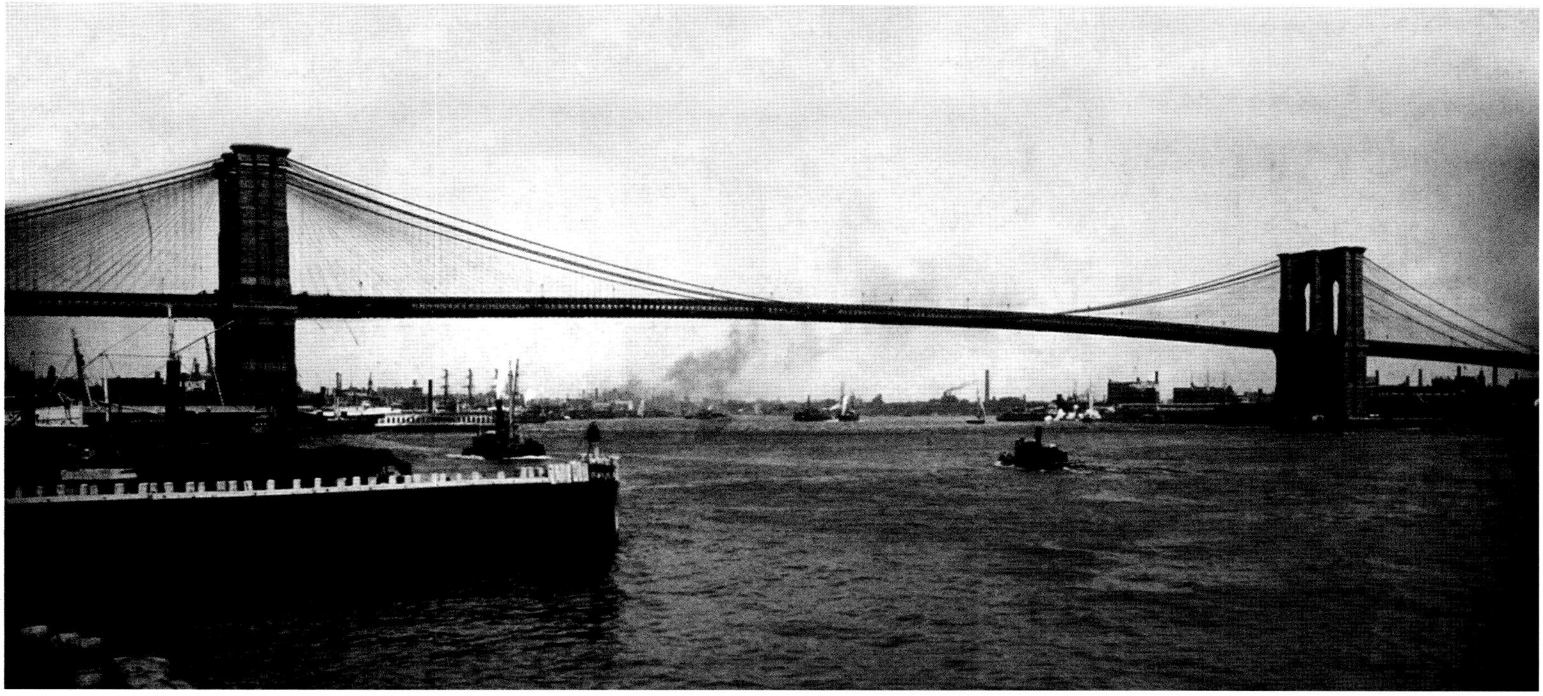

Brooklyn Bridge from Manhattan: This 1896 photo shows the bridge, Brooklyn's shoreline, and, on the left, a ferry preparing to sail to the other shore.

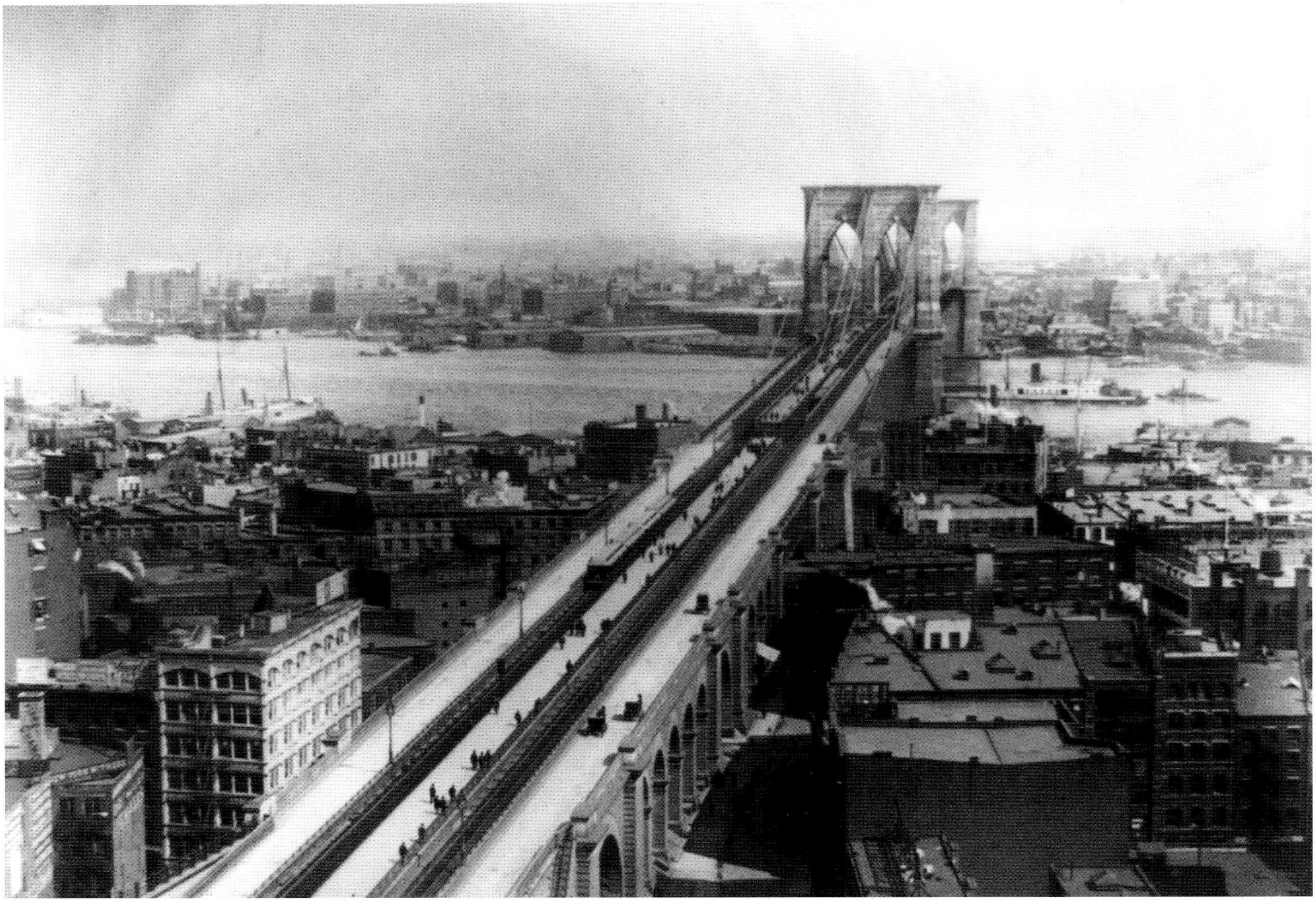

Bird's-eye view: Taken from a tall Manhattan building in 1897, this aerial view depicts the Brooklyn Bridge and its approach with carriages and carts in the outer lanes, trains in the next lanes, and pedestrians in the center.

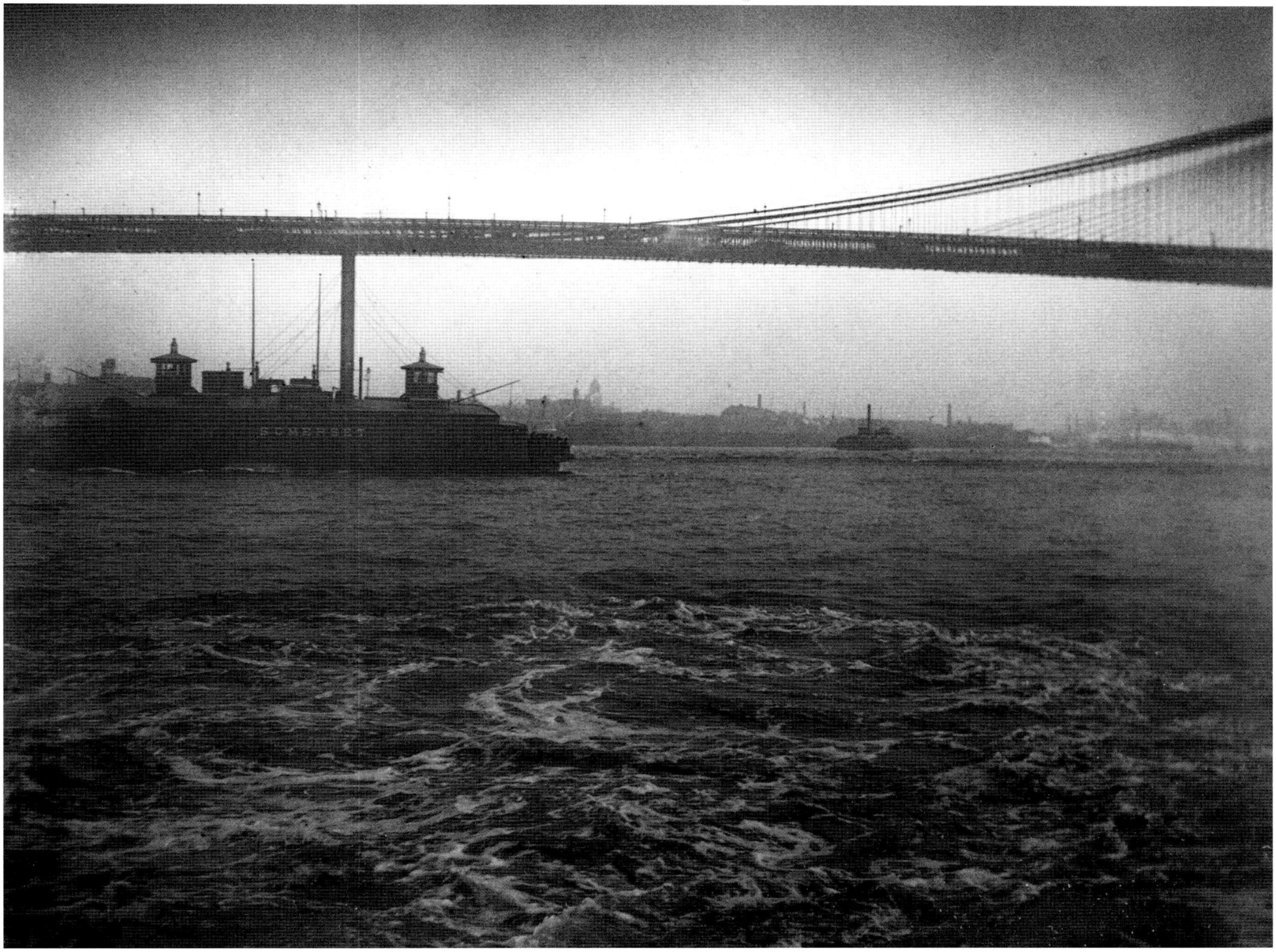

Ferry docks in Brooklyn: In 1900, the ferry *Somerset* docks at the Fulton Ferry dock.

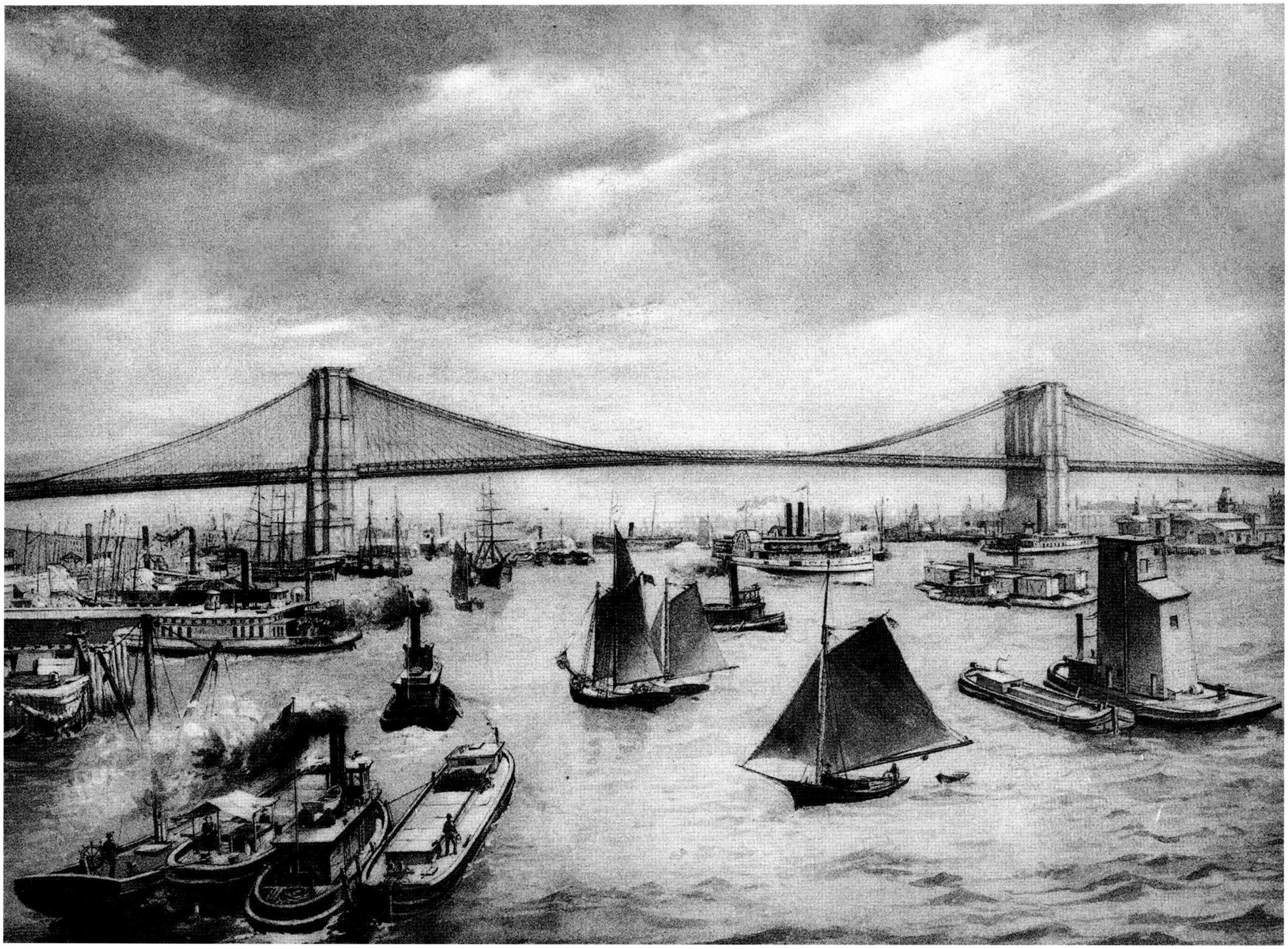

Heavy shipping: Commercial shipping crowds the waters beneath the Brooklyn Bridge in this early view.

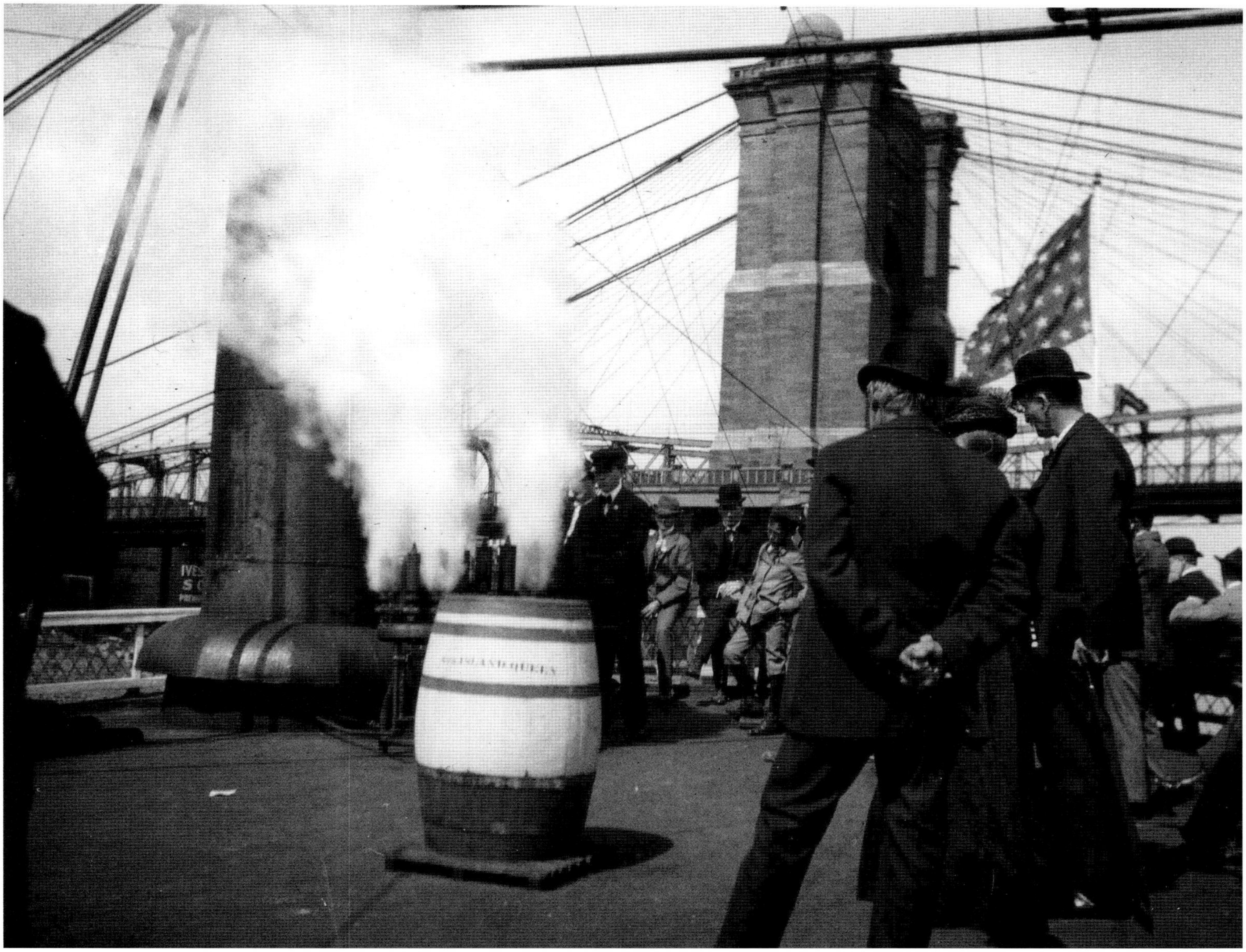

Excursion boat passing under Brooklyn Bridge: Men on the upper deck of an excursion boat are watching and listening to a steam calliope (on the other side of the barrel). The flag is flying from the stern of the boat as it passes one of the Brooklyn Bridge towers.

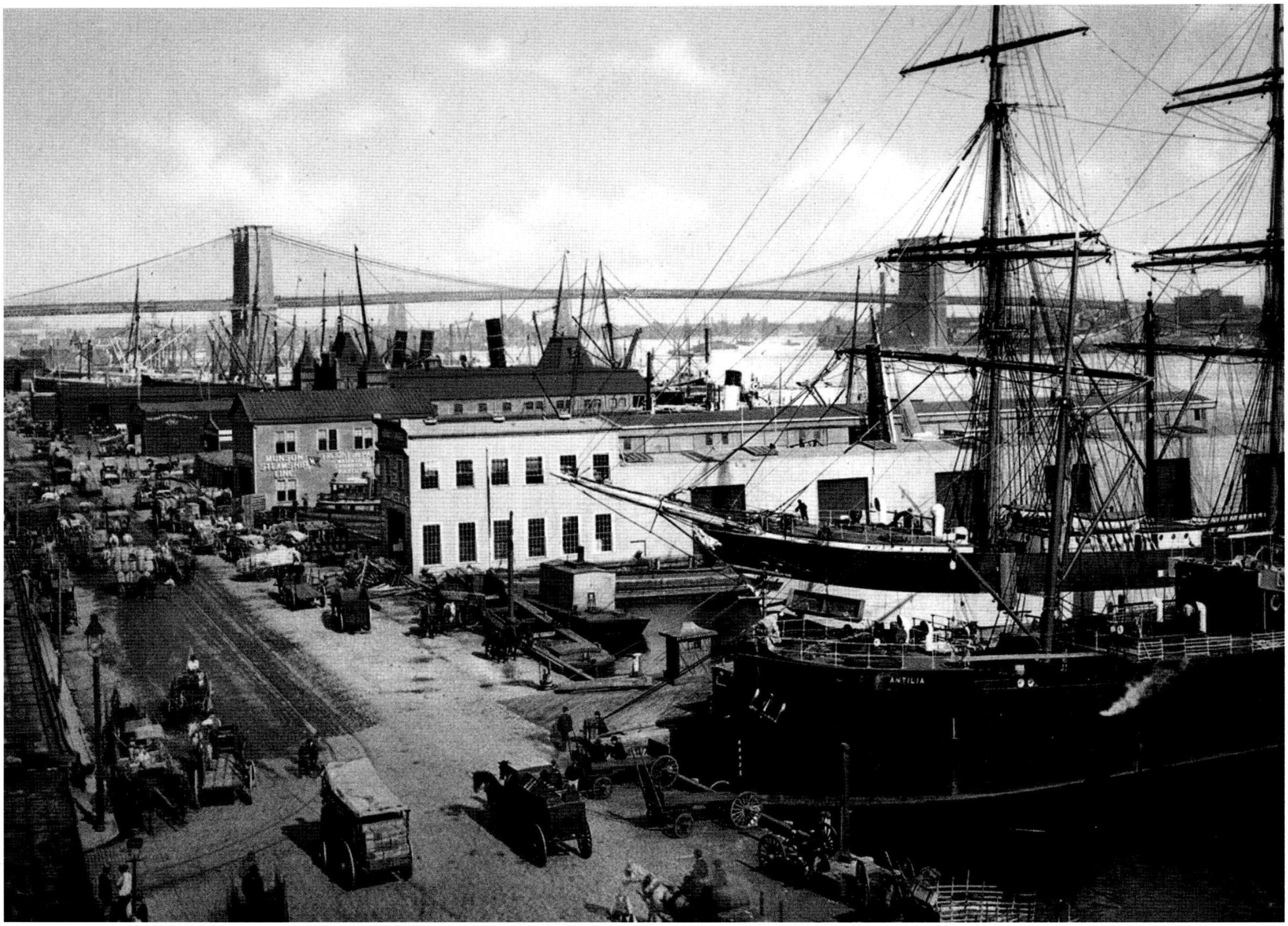

South Street activity: A busy seaport in 1901 with sailing ships, wagons, workers and sailors, and, in the background, the Brooklyn Bridge. In the foreground lies the schooner *Antilia* out of Barbados.

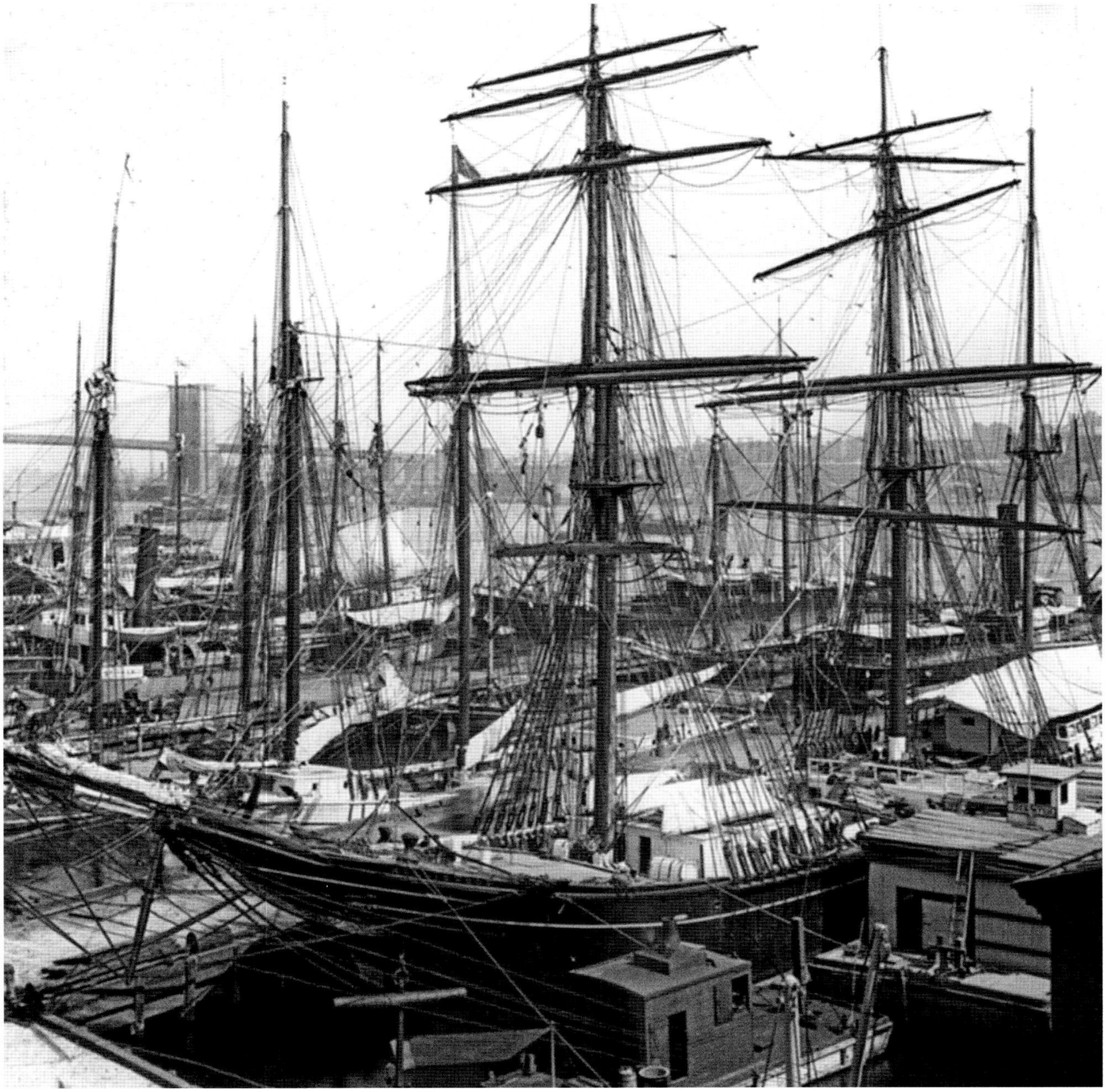

A network of masts: Sailing schooners are berthed along the Manhattan shore as the Brooklyn Bridge peeks through their ropes.

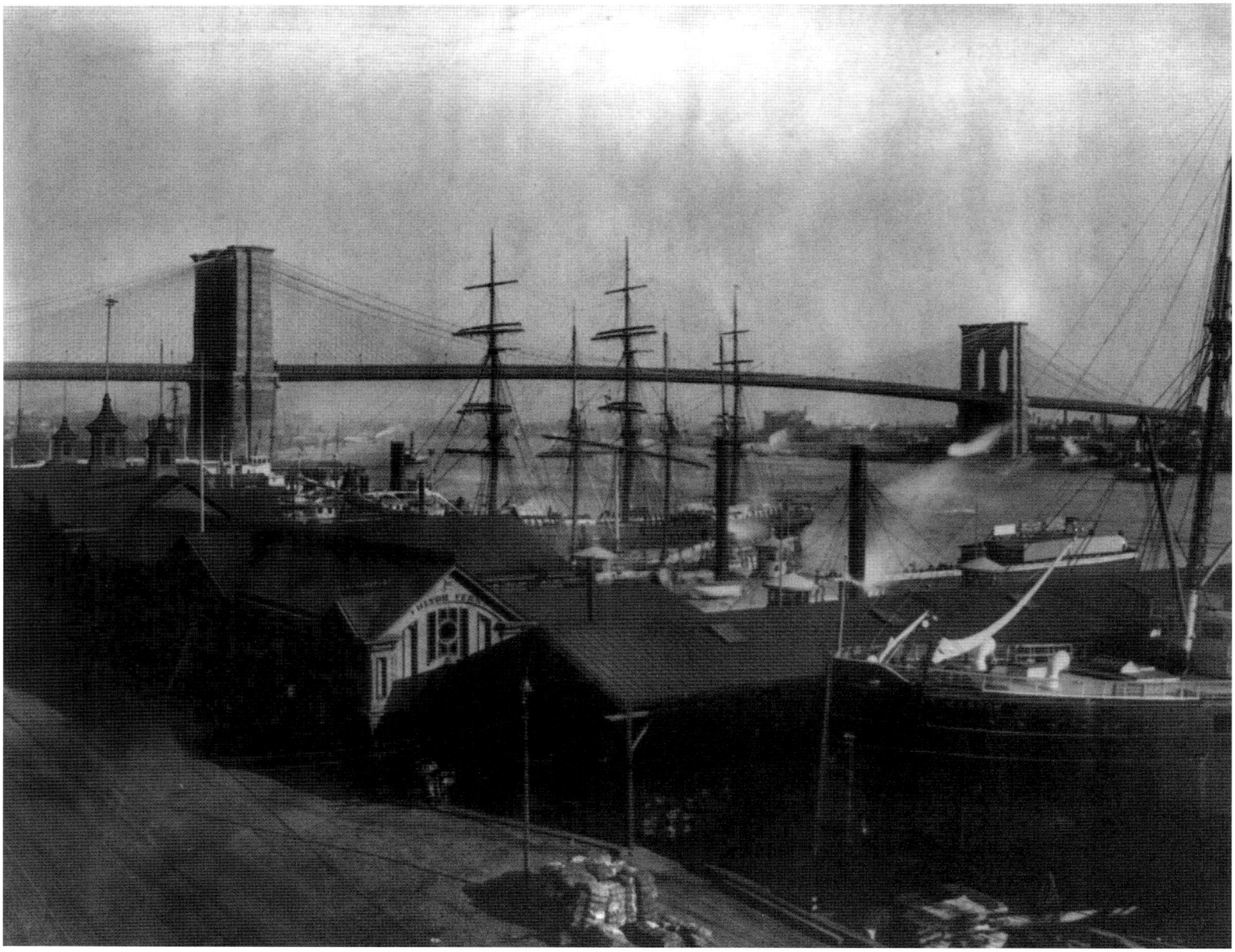

Brooklyn docks: This picture provides a view of the bridge from the Brooklyn docks where sailing ships are moored. Prows of ships hug the shore abutting the warehouses, while cargo lies on the wharves.

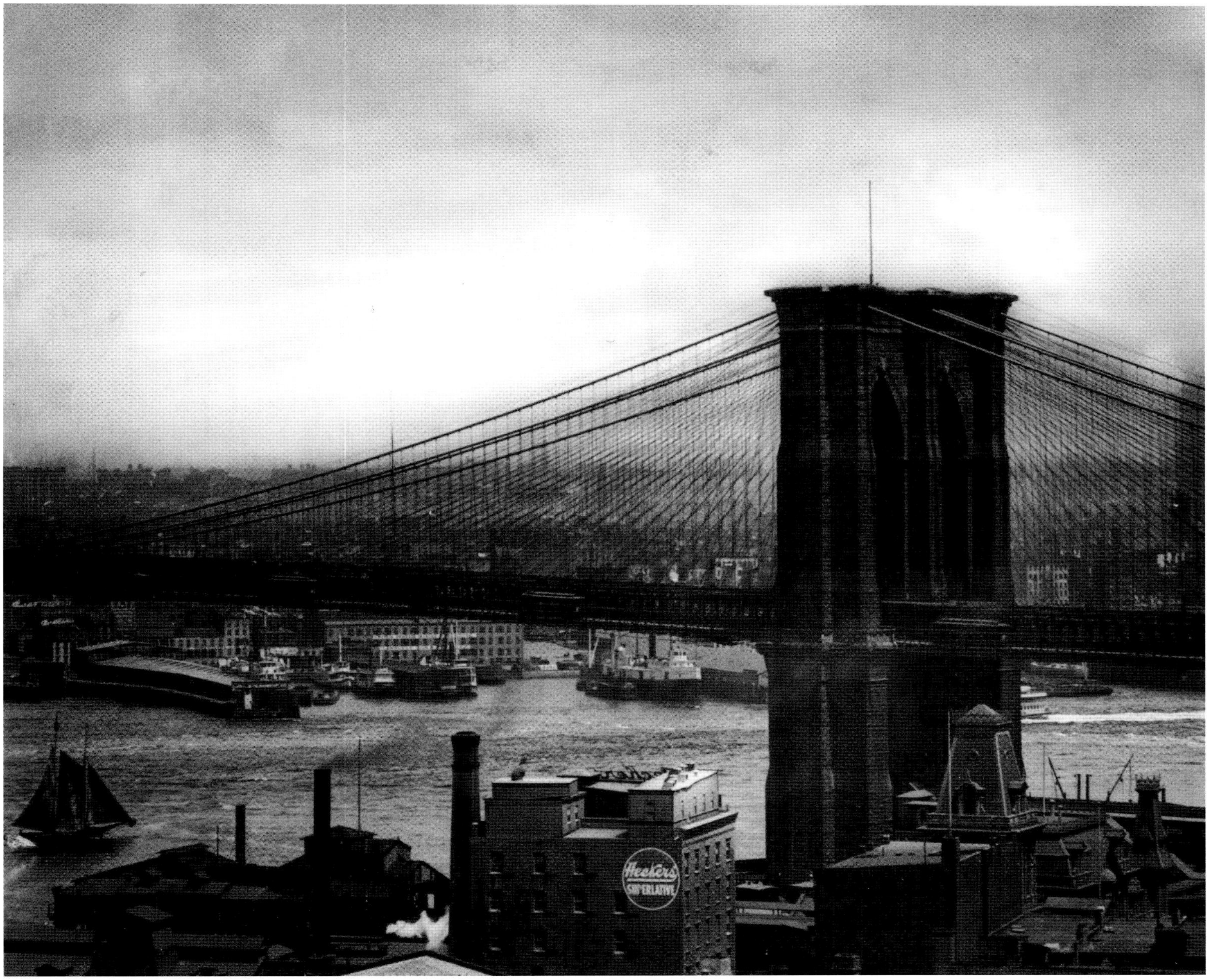

From a Brooklyn point of view: The Brooklyn tower, Brooklyn waterfront, and commercial buildings in 1903. Manhattan buildings are visible through the webbing of the bridge as a sailboat glides beneath it.

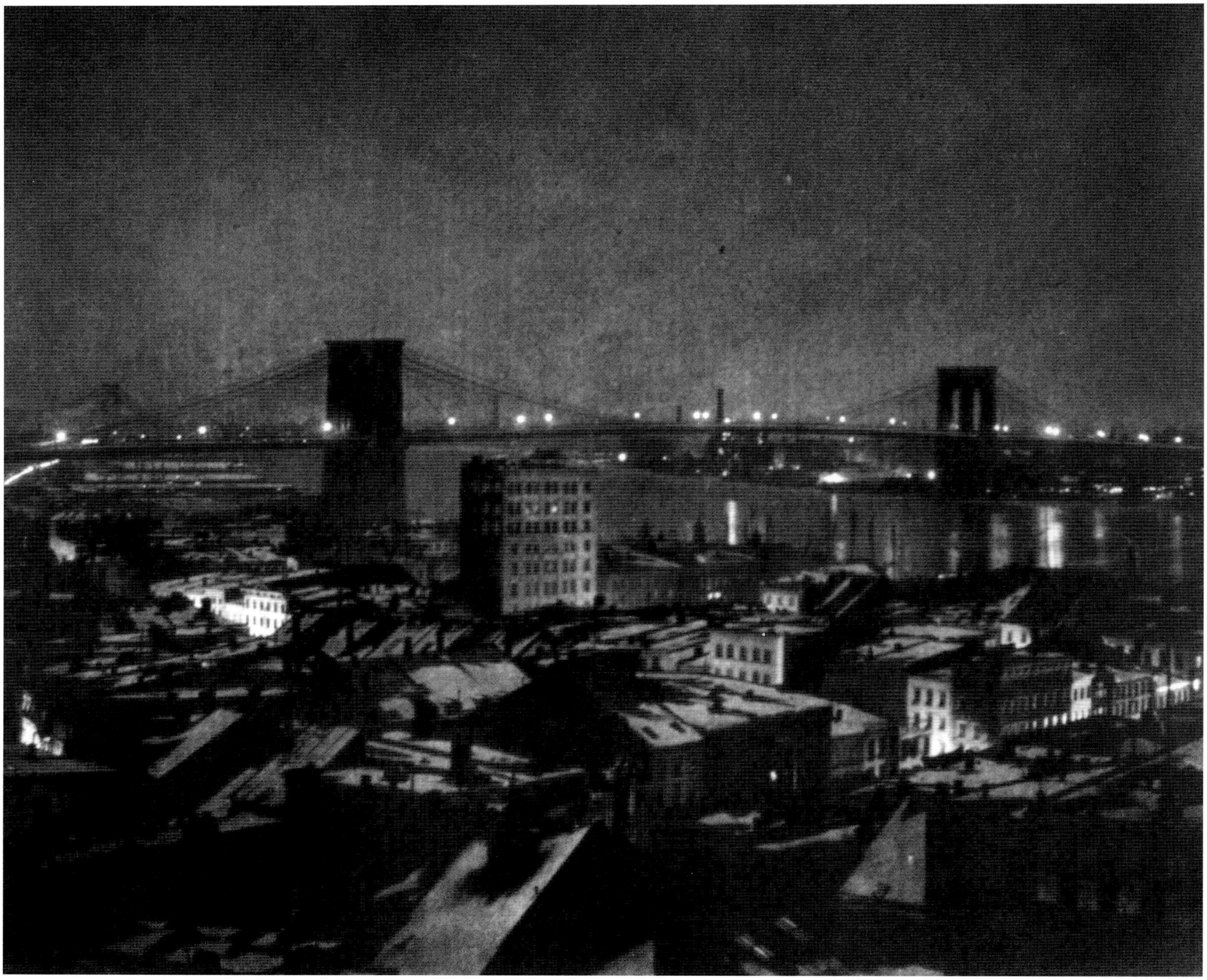

Bridge at night: The Brooklyn Bridge captured at night before the "necklace of lights" illuminated the cables. The only lights on the bridge line its roadway. In the foreground, buildings are illuminated by street lighting. This 1903 photogravure came from the studio of famed photographer Alfred Steiglitz but was taken by one of his staff.

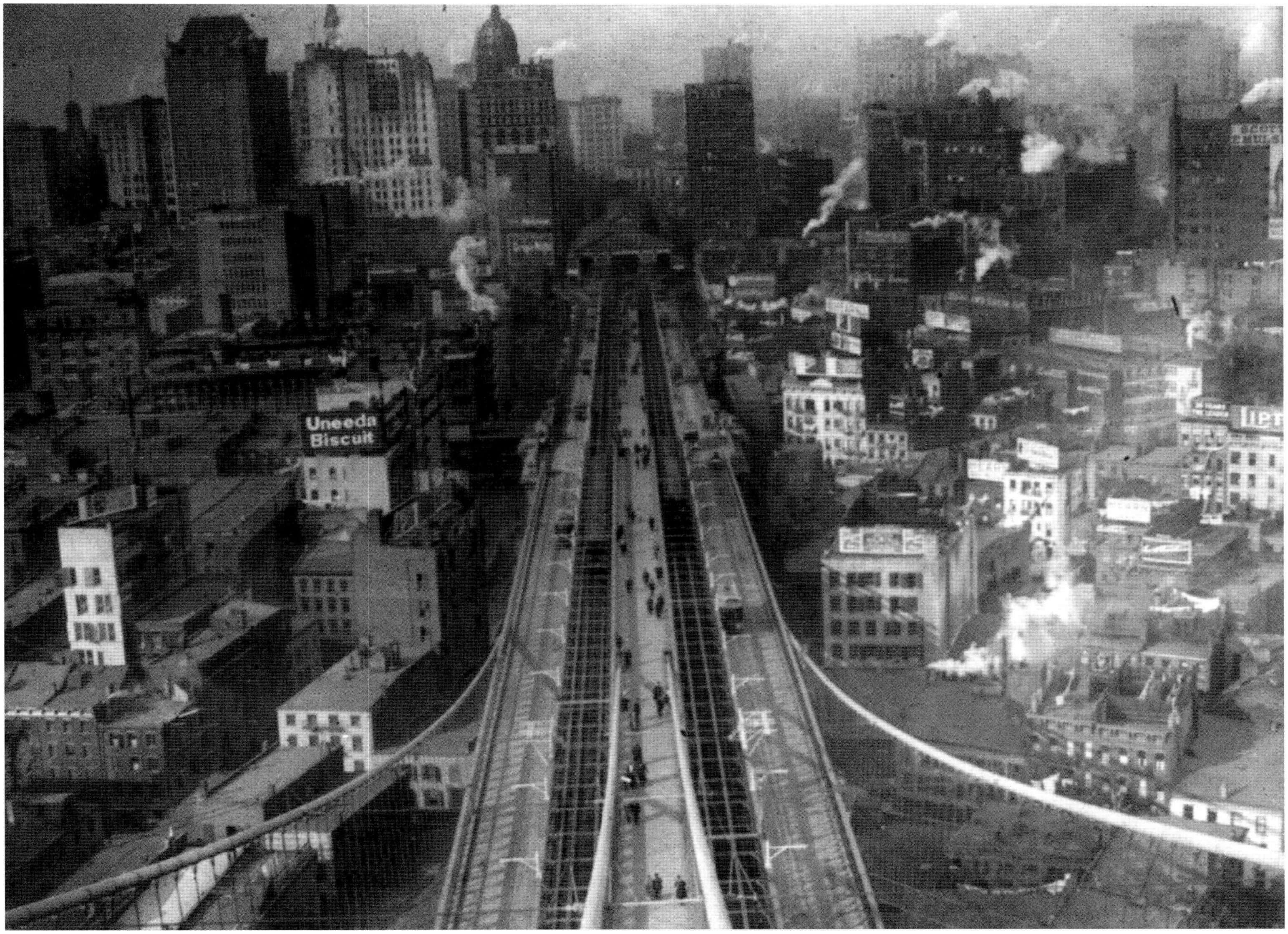

Lower Manhattan from the top of the bridge: A view from the New York tower shows the Manhattan approach and early twentieth century buildings in Lower Manhattan. The dome in the distance belongs to the New York World building, site of today's Pace University.

River traffic: A transport steamboat carrying both passengers and cargo passes under the Brooklyn Bridge. This is the *Maryland,* owned by the New York, New Haven & Hartford Railroad Company, which also carried railroad cars.

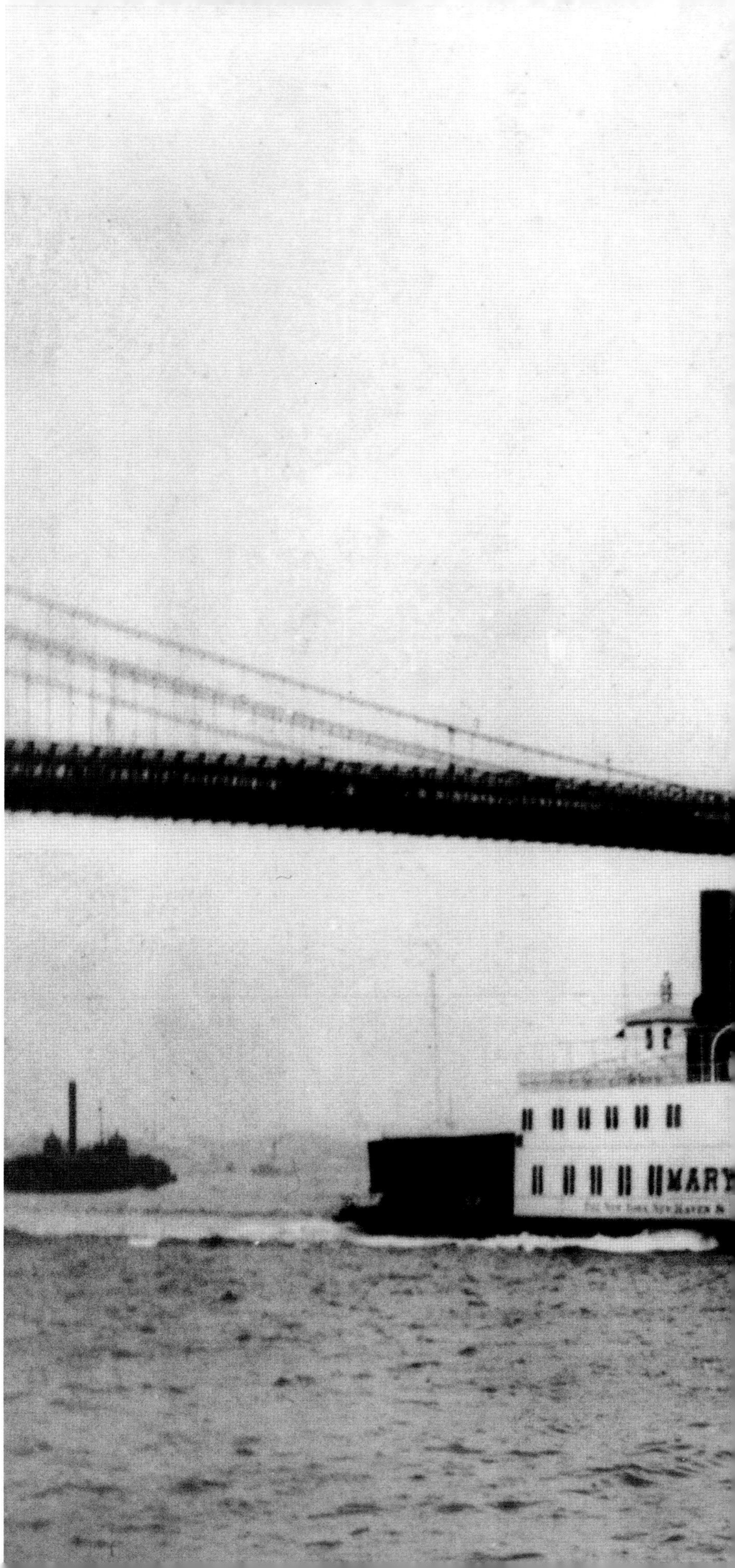

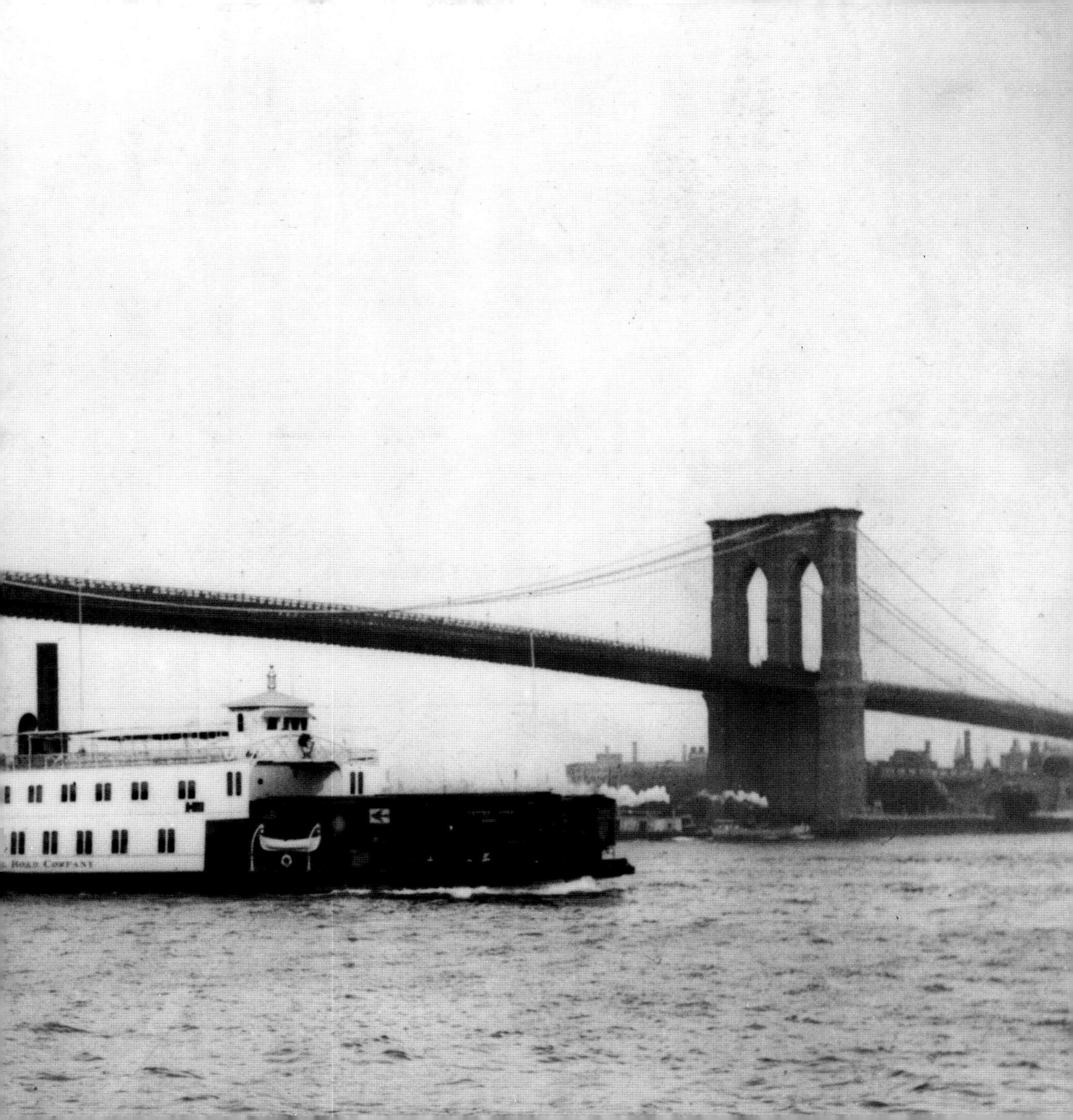
ROAD COMPANY

Skyscrapers as seen from Brooklyn: With the tower of Brooklyn's Fulton Ferry House in the foreground, Manhattan's 1904 spreading skyline and East River piers are seen beyond the Brooklyn Bridge. Next to the ferry house are tracks for the Brooklyn Union Elevated Railroad trains, and an excursion boat sails in the middle of the East River.

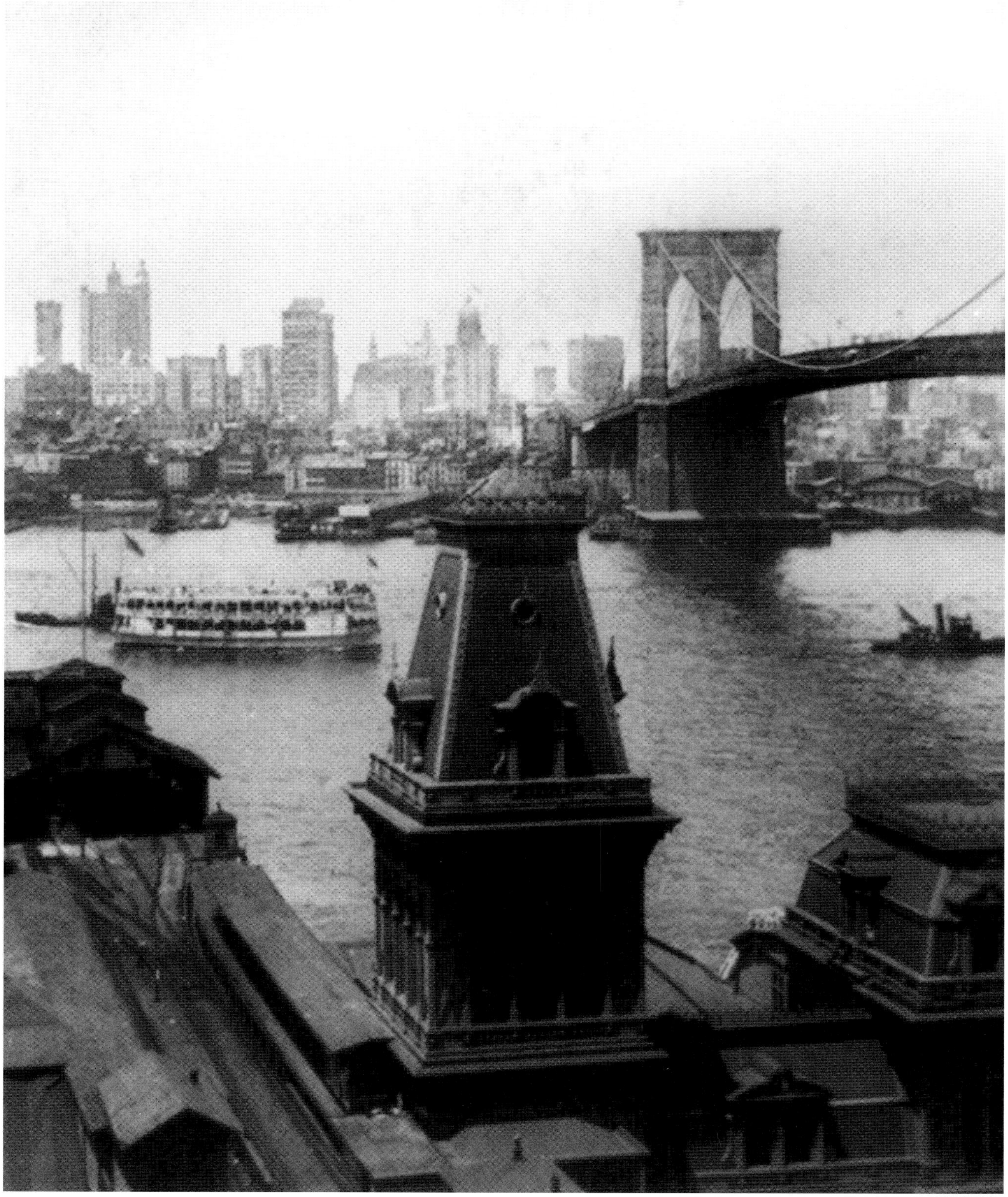

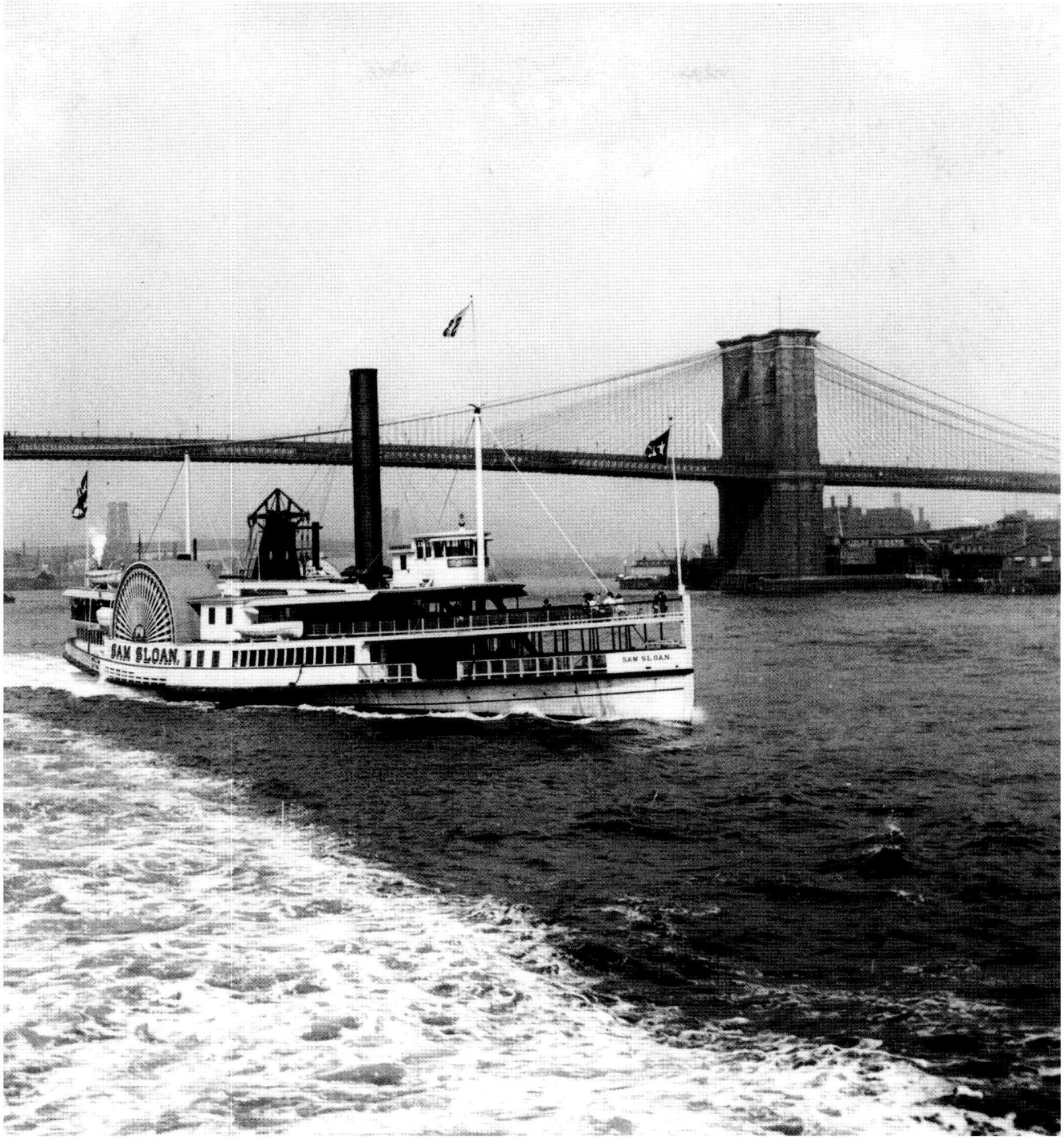

The Brooklyn Bridge upriver: The Brooklyn Bridge, upriver from the photographer, stands stately as the side-wheeler excursion steamboat *Sam Sloan* leaves the dock for a cruise in 1904. The wake at lower left is from the boat that the photographer is on.

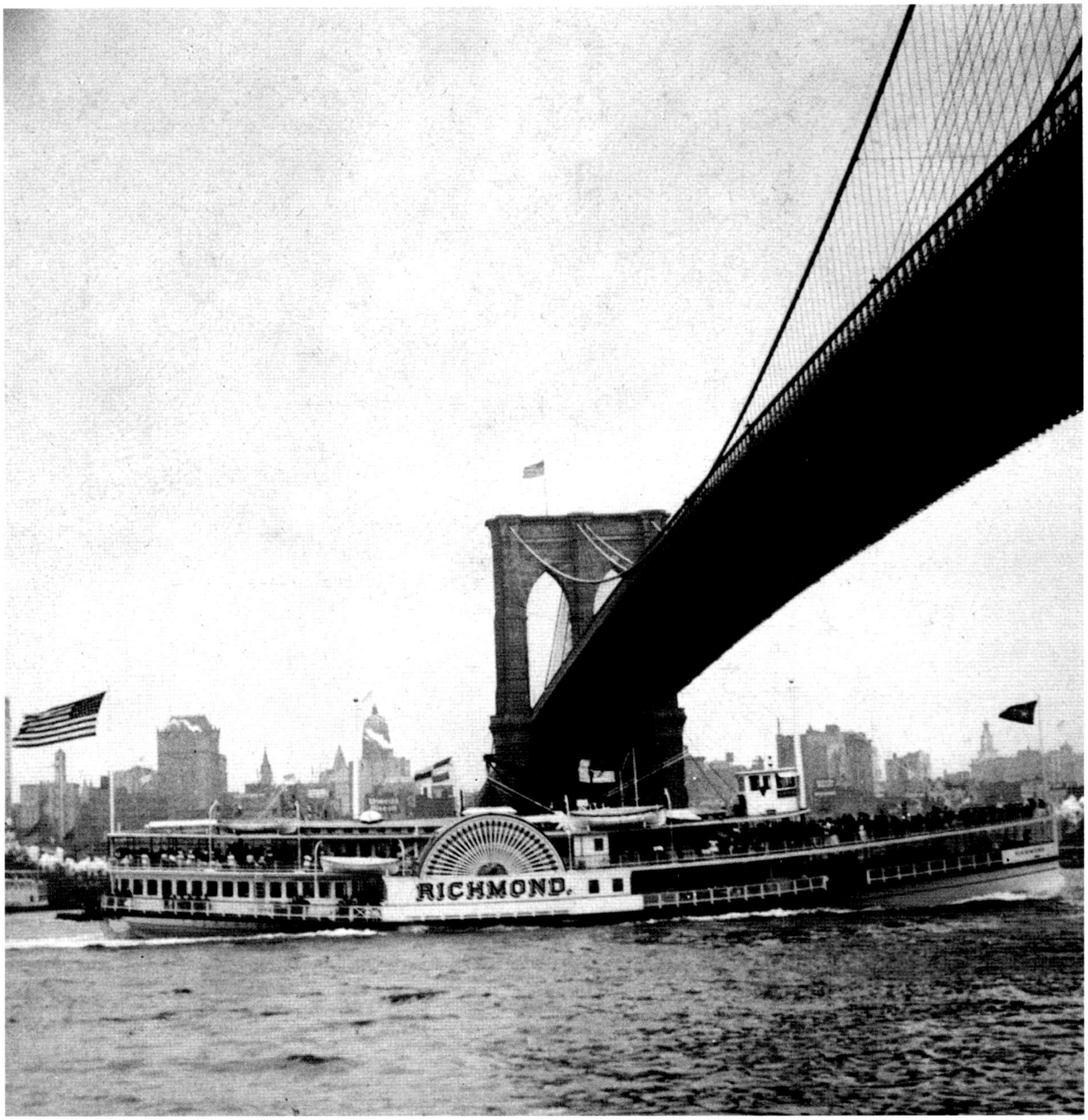

Down under the bridge: Heading north, the steamboat side-wheeler *Richmond* passes under the Brooklyn Bridge in 1905, carrying passengers to Harlem.

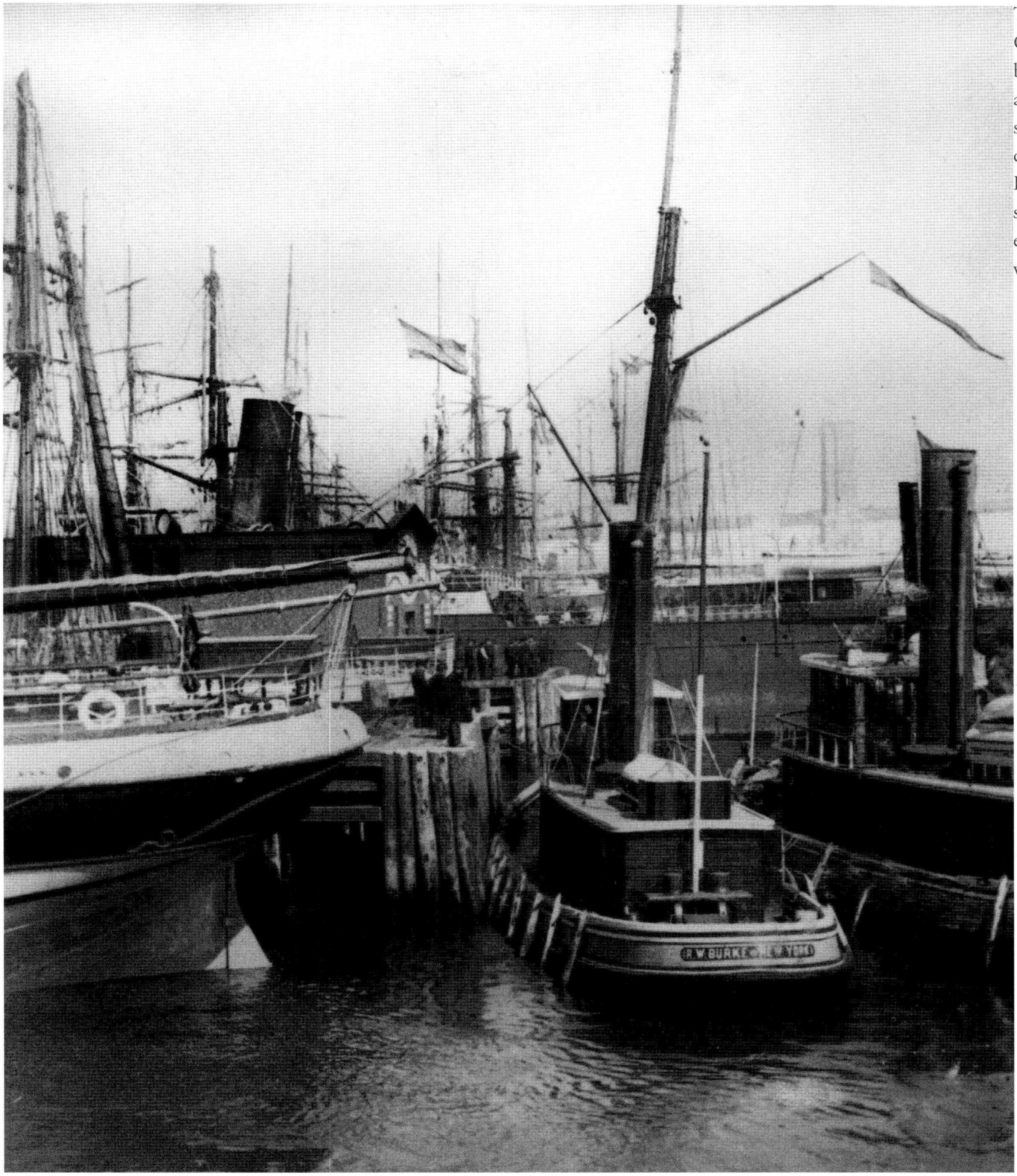

Traffic jam: Commercial boats, tugs, and steamer smokestacks crowd into the East River docks, symbolizing the era's reliance on waterborne travel.

Following Spread: Looking north: A faint image of the new Williamsburg Bridge (1903) can be seen in this photo of the Brooklyn Bridge, taken from the middle of the East River in 1905. Tugs and other ships crowd the waters.

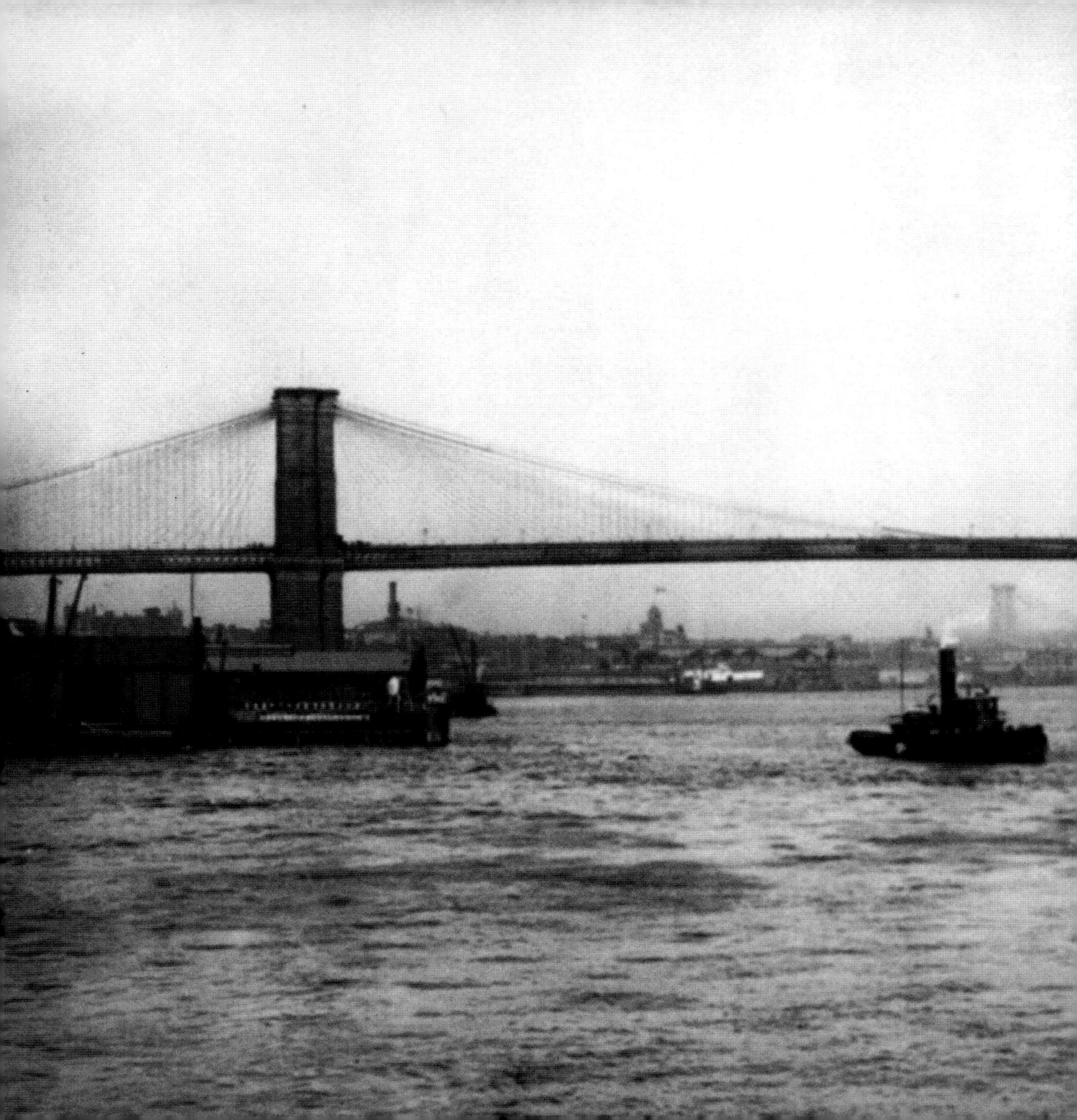

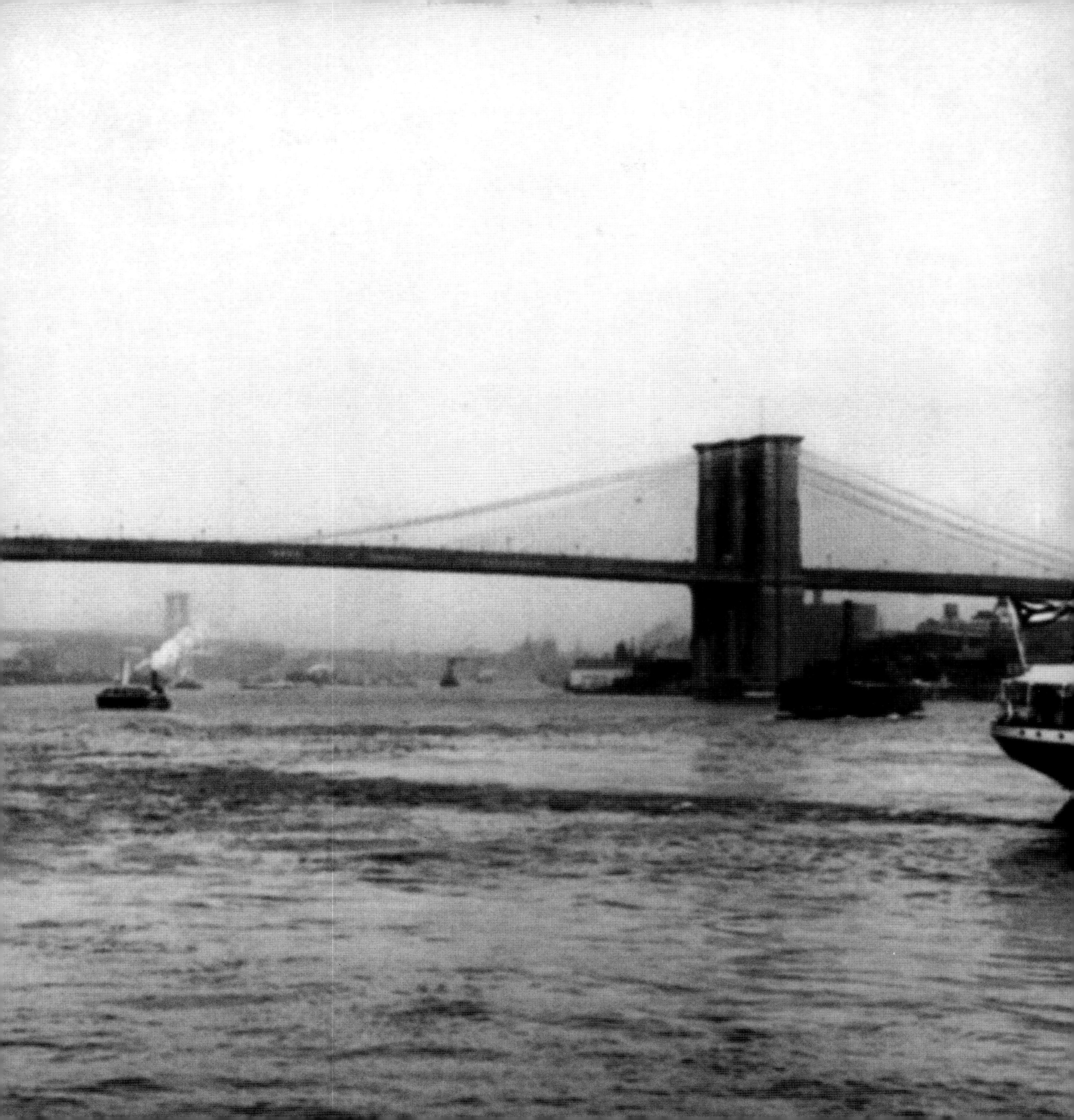

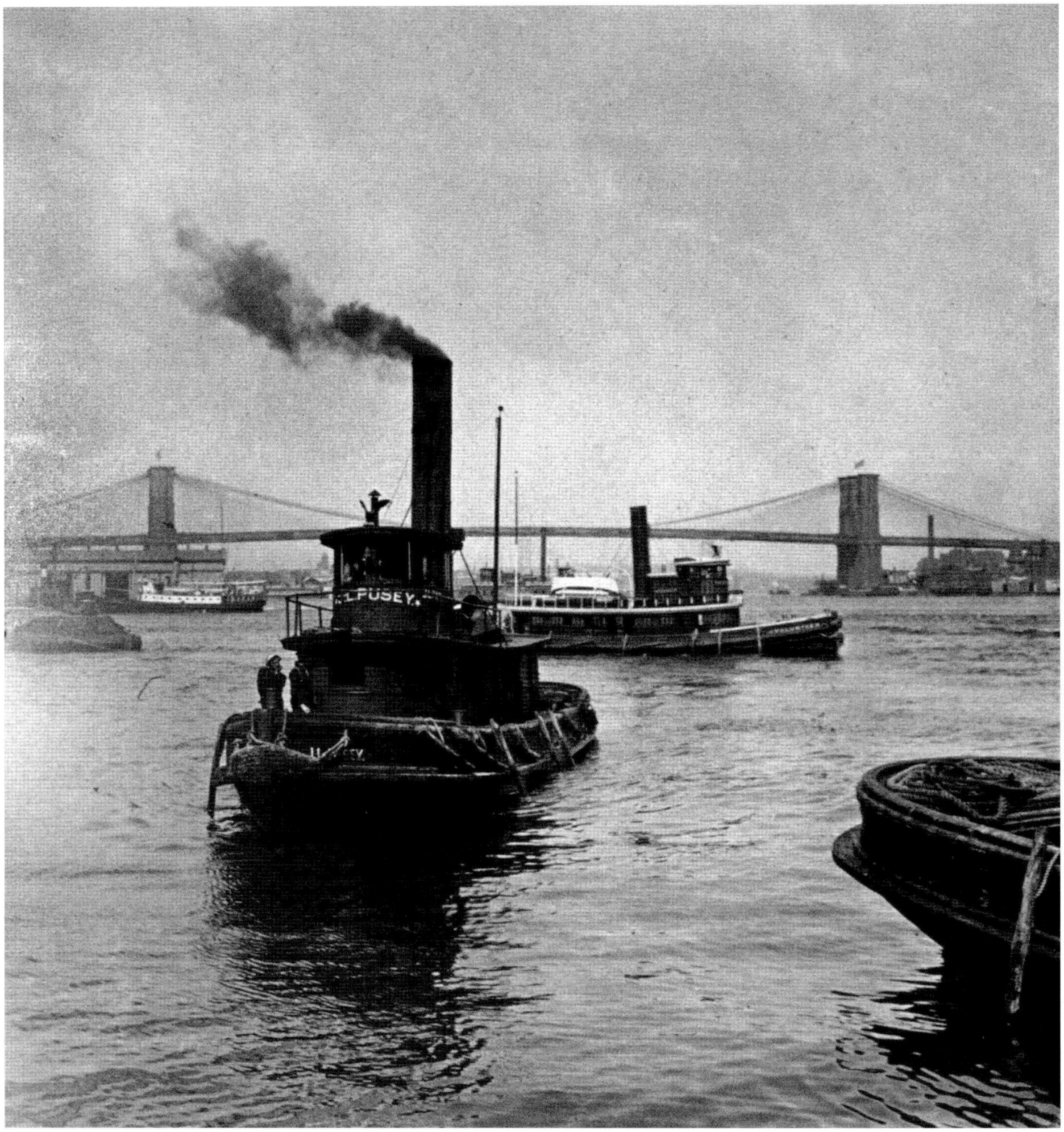

Busy harbor: Traffic on the East River accelerated after the Brooklyn Bridge was created; the bridge, along with the later consolidation of the boroughs, attracted new businesses to the city. Here tugs scurry across the river to their assigned jobs.

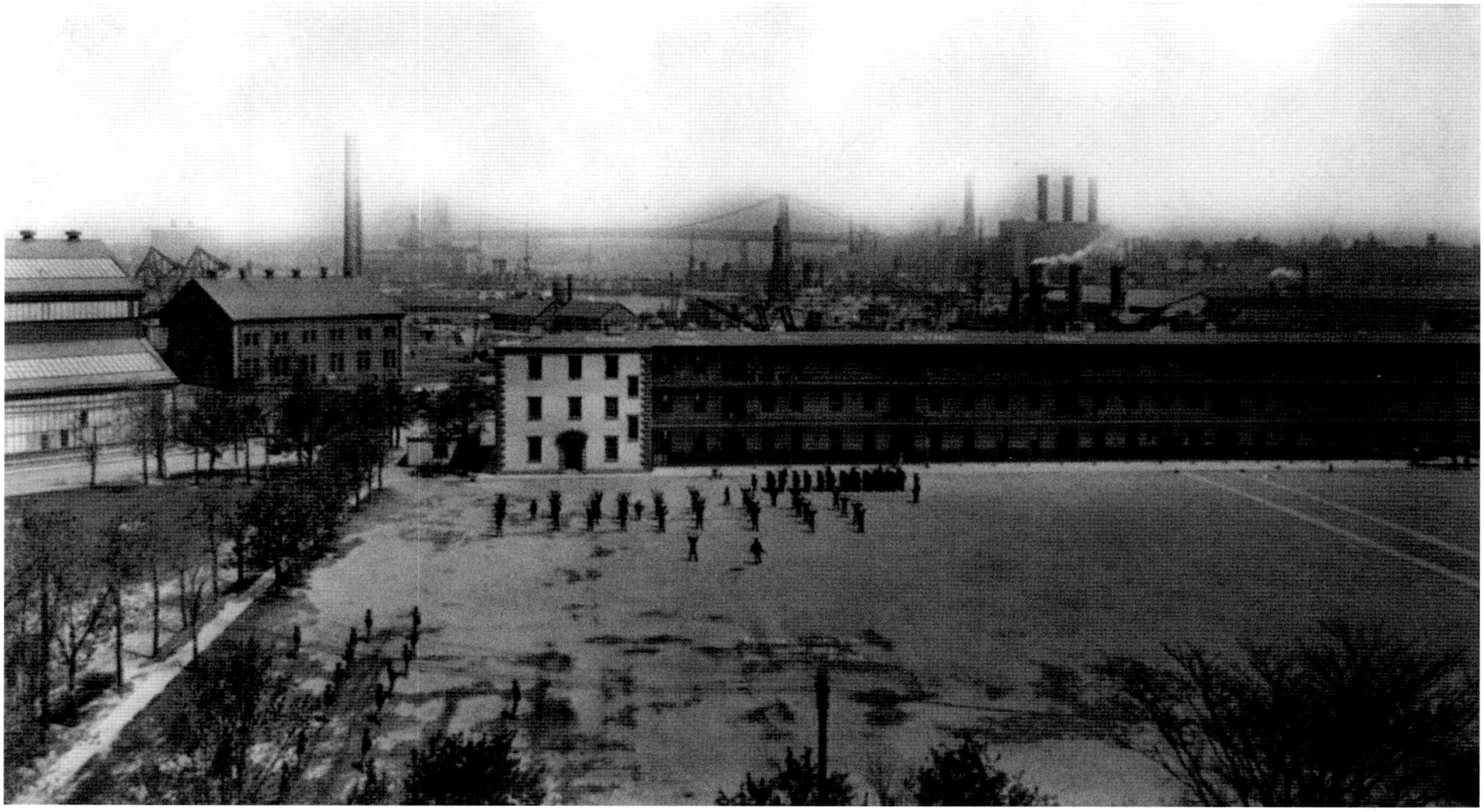

Navy Yard exercise yard: By 1909, sailors lined up in formation for exercises outside their barracks. Behind other naval buildings lies the bridge, visible on the horizon.

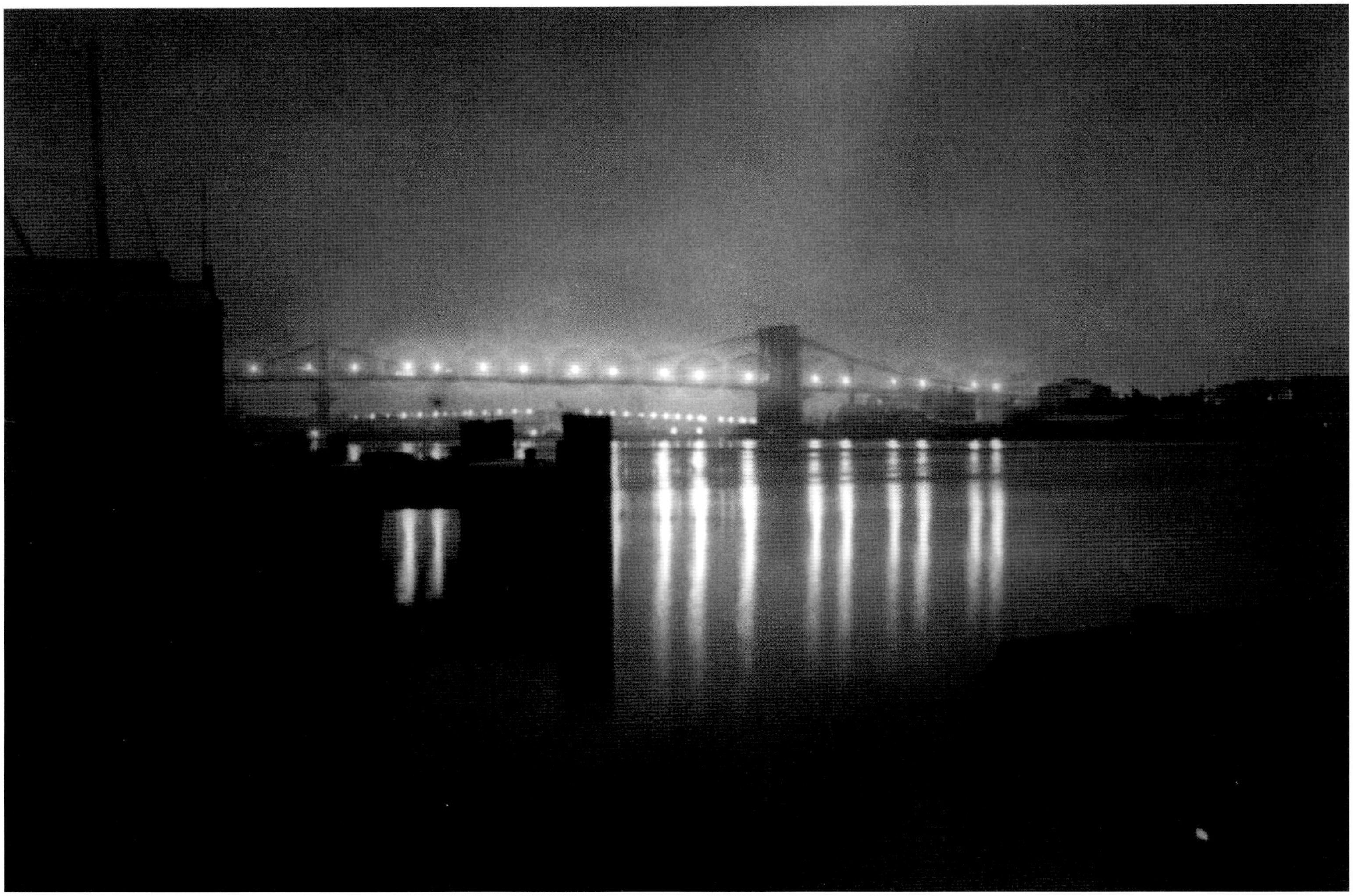

The Brooklyn Bridge electric lights reflect into the East River as seen in this 1910 photograph.

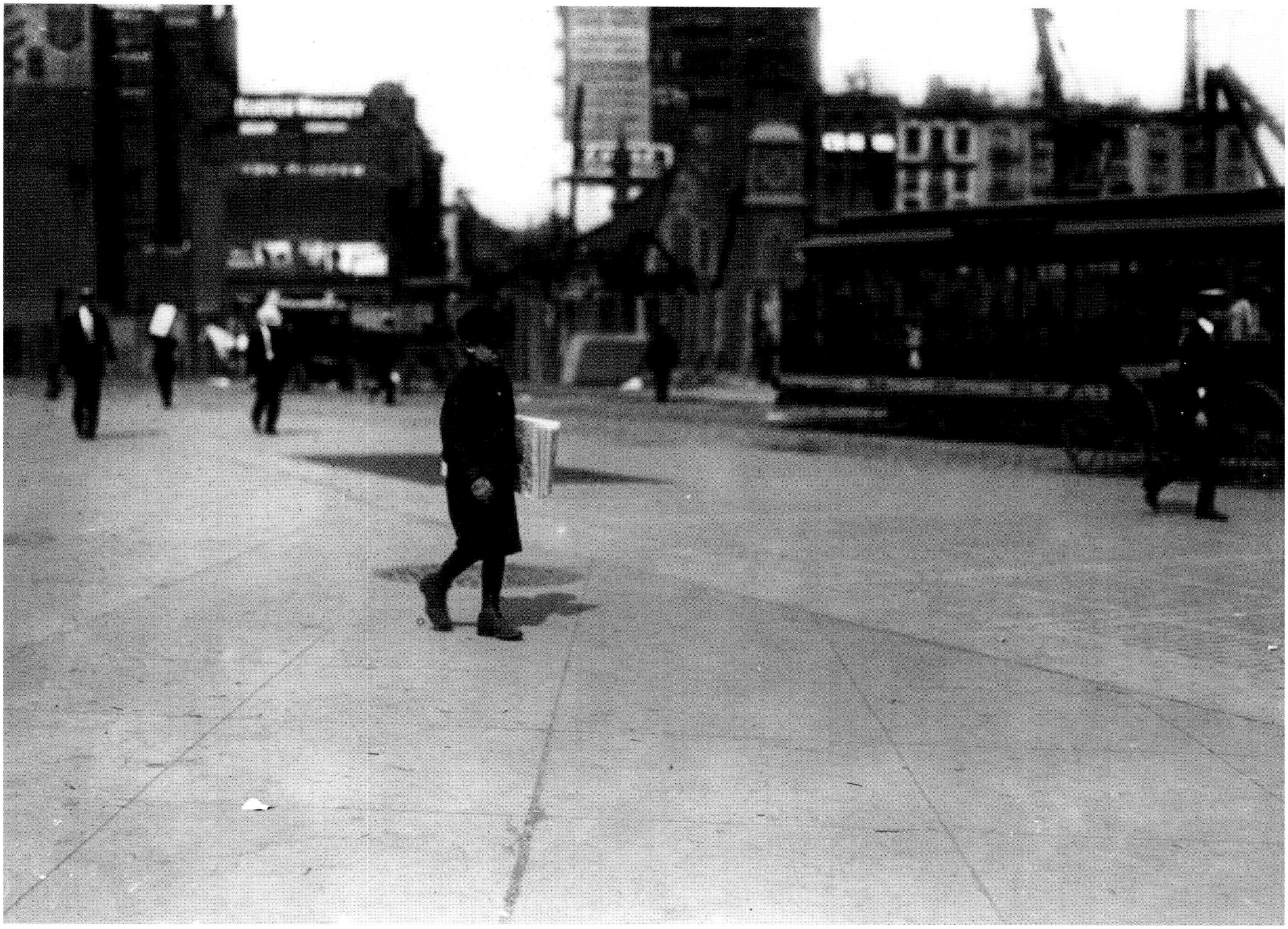

Brooklyn Bridge newsie: Newsboys hawked newspapers to crowds on the bridge. Generally, these newspaper-distributing jobs were offered to orphanages whose residents, as young as five years old, competed for sales. Three major newspapers were published on Park Row (or "Newspaper Row") just at the Manhattan end of the bridge. This 1910 photo by Lewis Hine includes a trolley on the right.

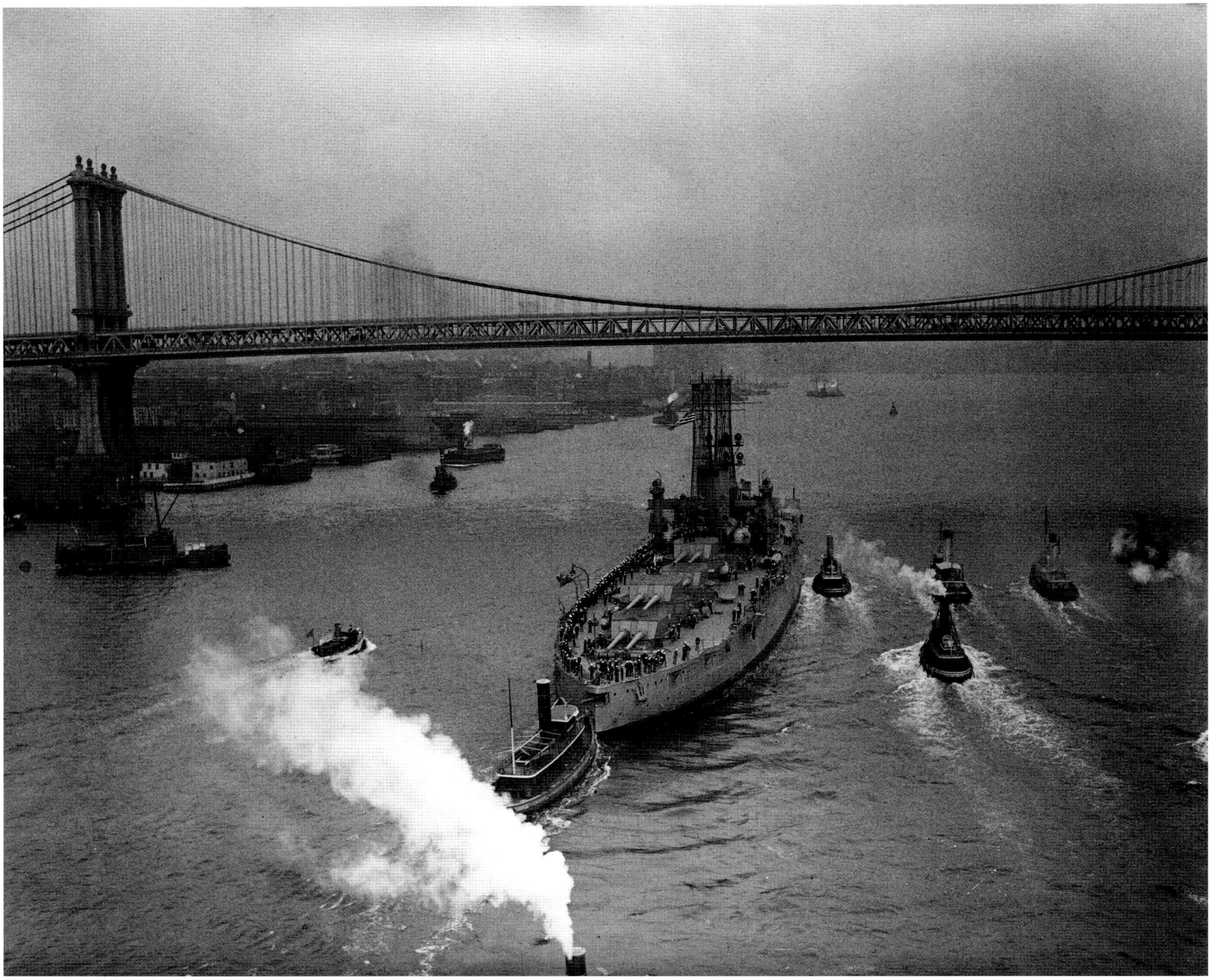

Big boats, too: The battleship *Wyoming* sails under the Manhattan Bridge, which opened just north of the Brooklyn Bridge on the last day of December, 1909. Launched in 1912, the *Wyoming* served in both world wars and is shown here on its way to the Brooklyn Navy Yard. A flotilla of tugboats accompanies her.

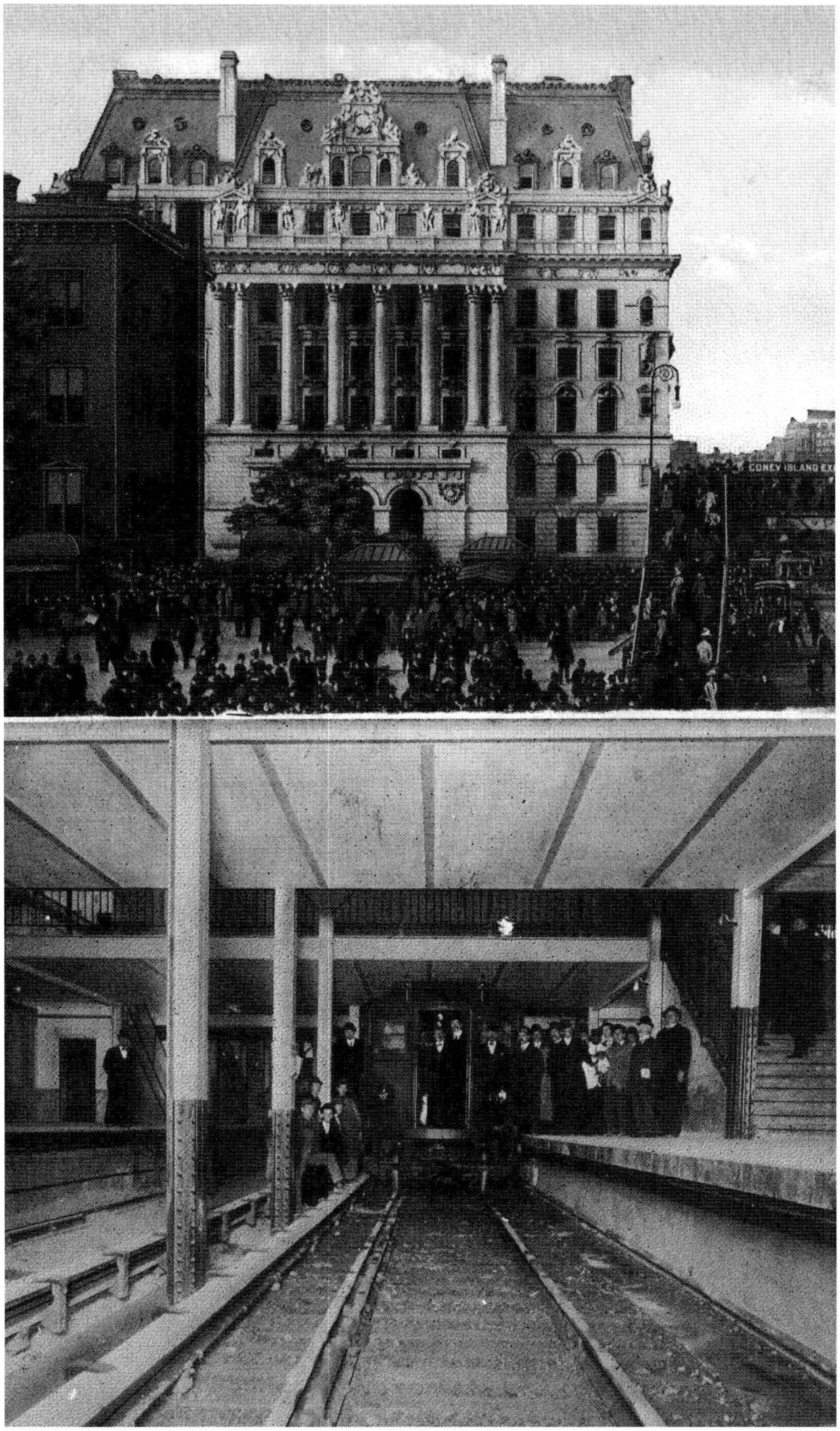

Brooklyn Bridge subway station: When New York City's 40-story Municipal Building was constructed by McKim, Mead & White in 1915, a subway station was built into its basement. Replacing trains that crossed the bridge above, the station still carried the name "Brooklyn Bridge." This composite cutaway shows both above and below ground.

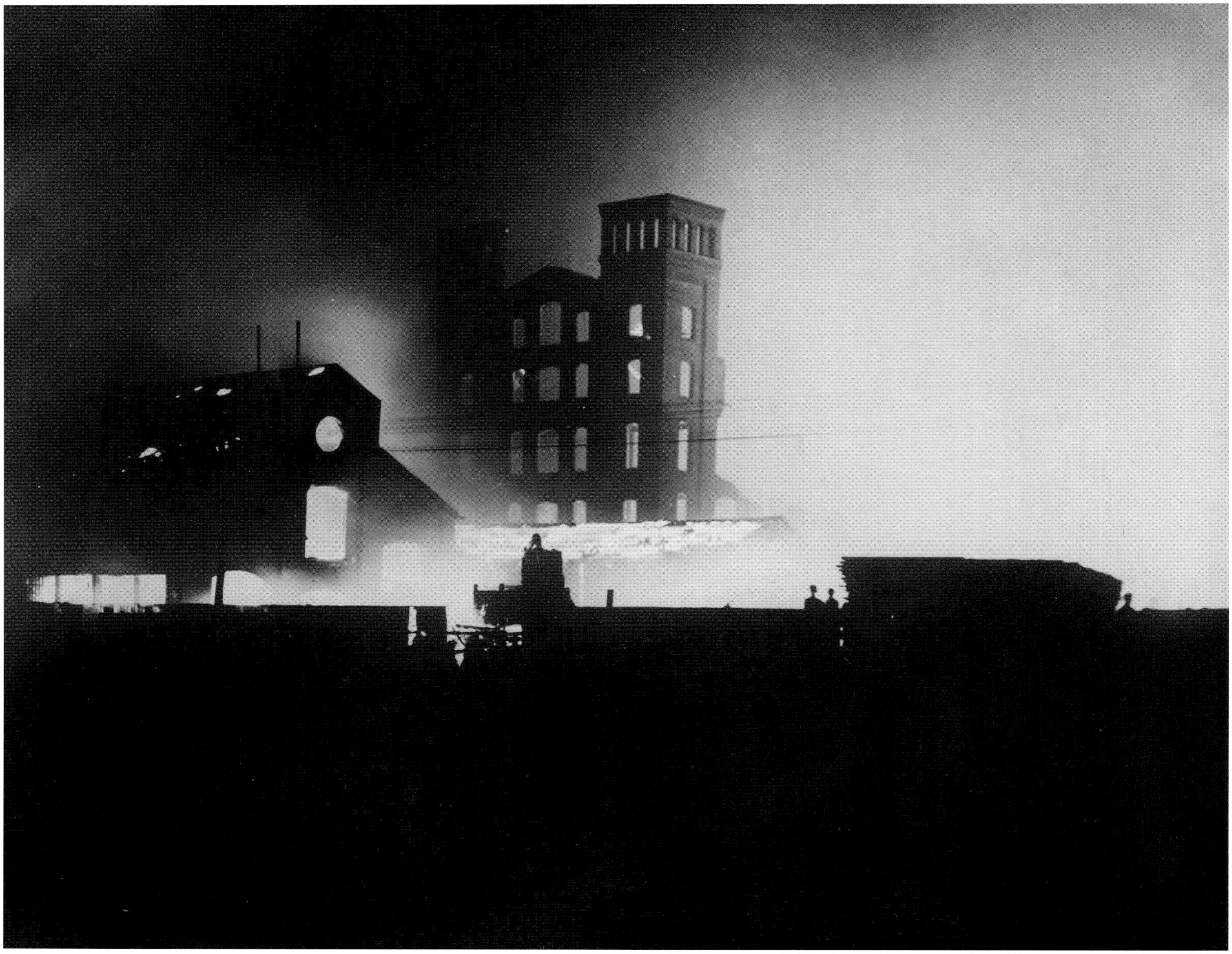

Roebling's Wire Factory on fire: By 1915, when a devastating fire broke out in the Trenton factory, the John A. Roebling Company had produced cables for the Williamsburg and Manhattan bridges as well as the Brooklyn Bridge. By this time the factory had expanded to include electrical and electric-related business. According to Washington Roebling's account, the fire that destroyed the Trenton plant was set by an arsonist.

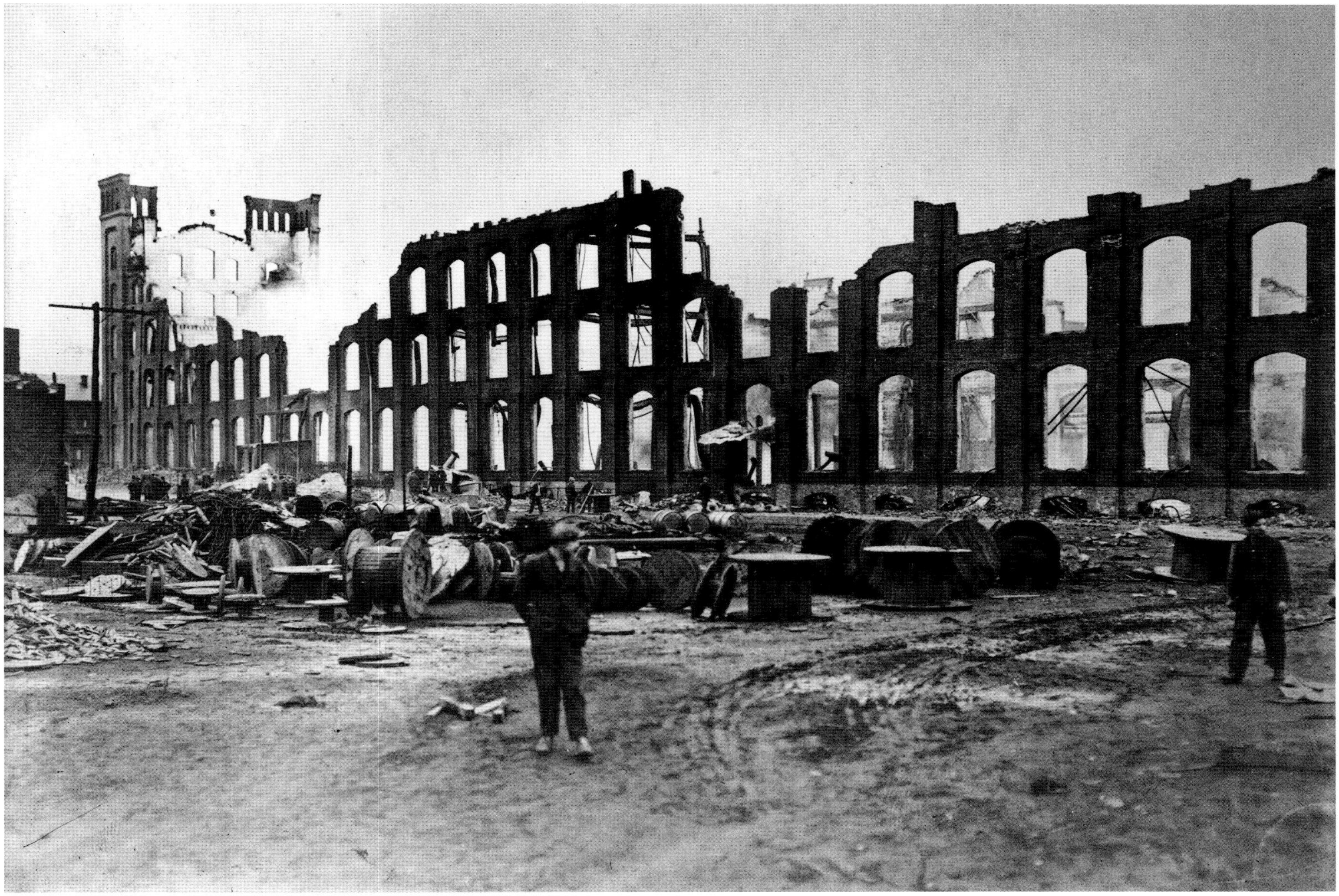

Remains of Roebling's factory: Requiring more than a year to rebuild, Roebling's factory went on to manufacture wire for World Wars I and II, used in developing submarine nets and parts for airplanes and mechanized vehicles. The company also made cables for both the George Washington and Golden Gate bridges.

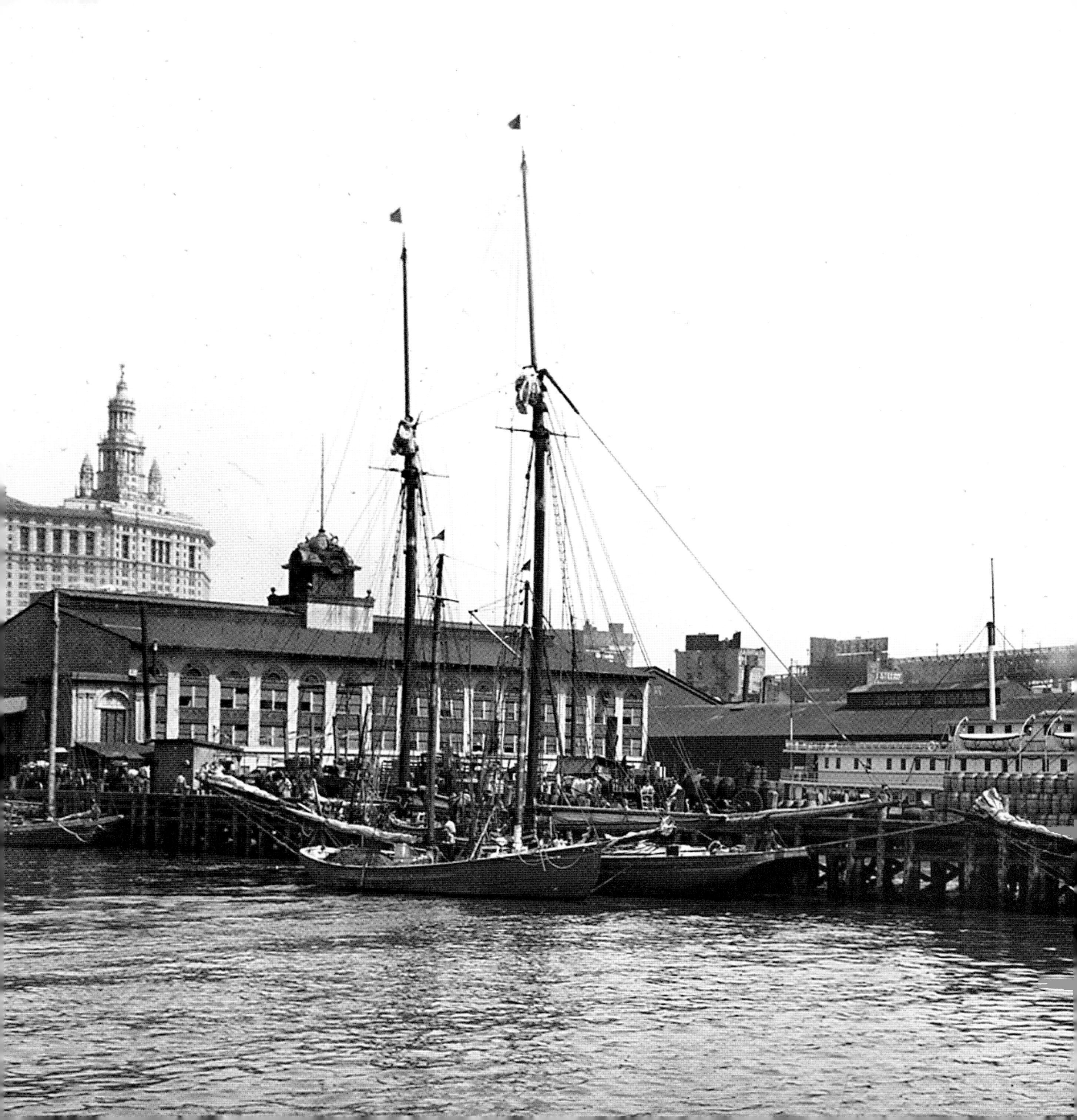

Fulton Fish Market, 1916: A fish market has stood on Manhattan's eastern shore since 1835, back when fishermen brought their catch right to the fishmongers' stalls. In this 1916 photo, fishing boats line up at the dock, the Brooklyn Bridge rises to the right, and toward the left lies the 1915 Municipal Building.

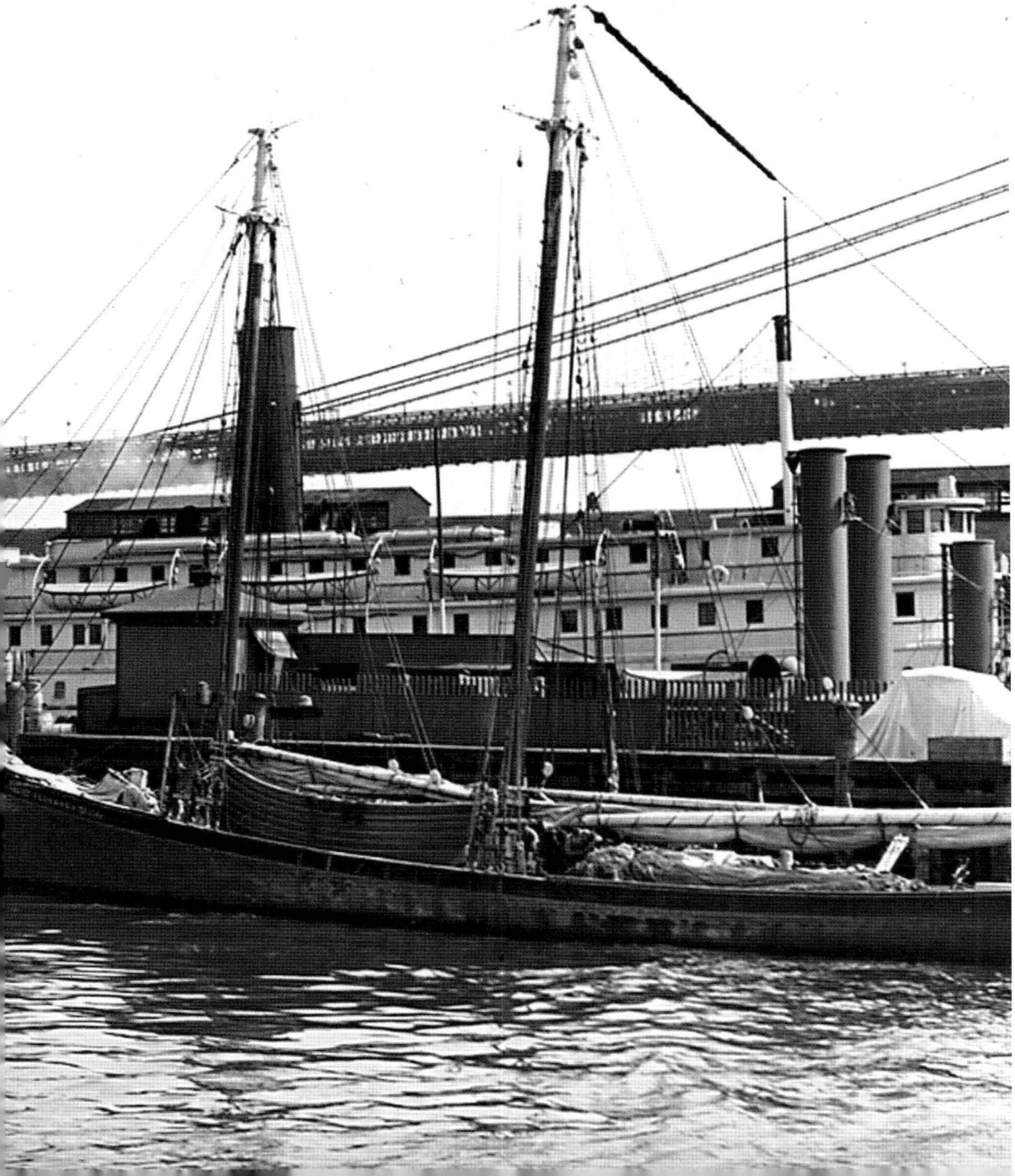

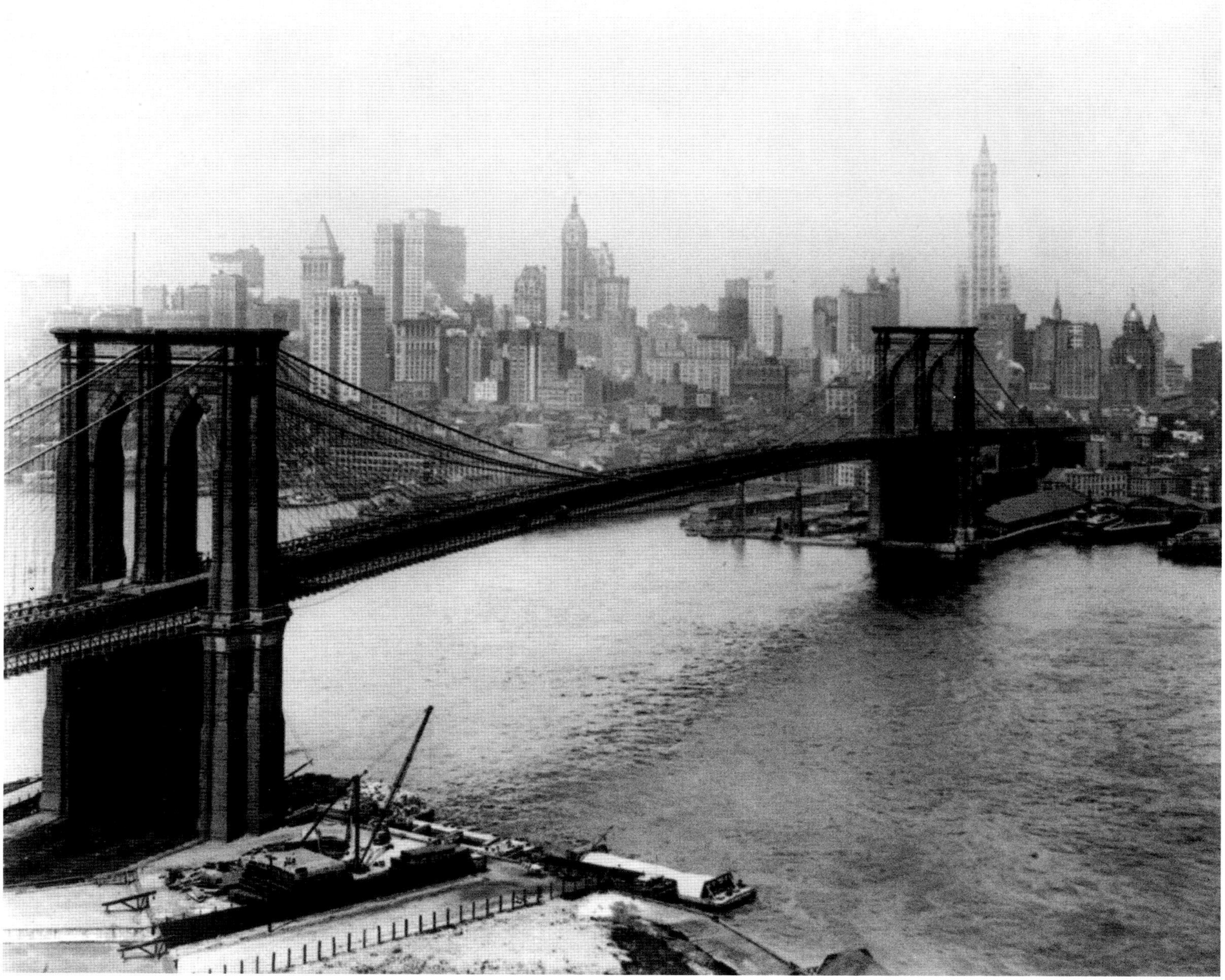

Classic pose: As the city grew, the Brooklyn Bridge acquired a sheen of seniority. This 1915 photograph shows the sweeping curve of its roadway and the magnificence of its towers.

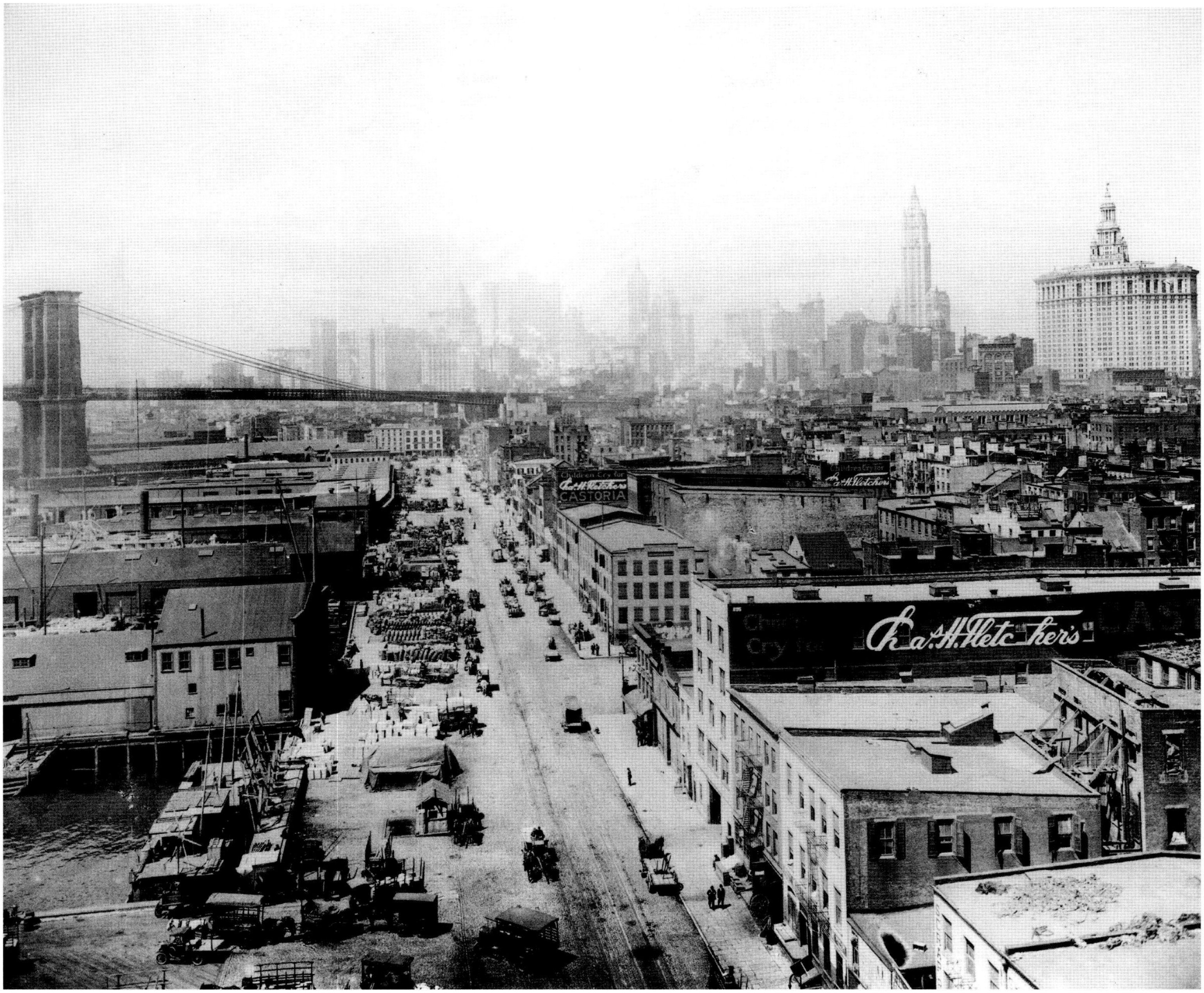

South Street looking south: This 1917 view shows wharves and sheds but few ships. The Brooklyn Bridge dominates the left of the frame in the background. To the right is the Manhattan Municipal Building and the faintly visible Woolworth Building.

Following Spread: Brooklyn Bridge overlooking Brooklyn: This 1919 photo features a broad view of Brooklyn and its bridge. The Manhattan Bridge is clearly visible to the north.

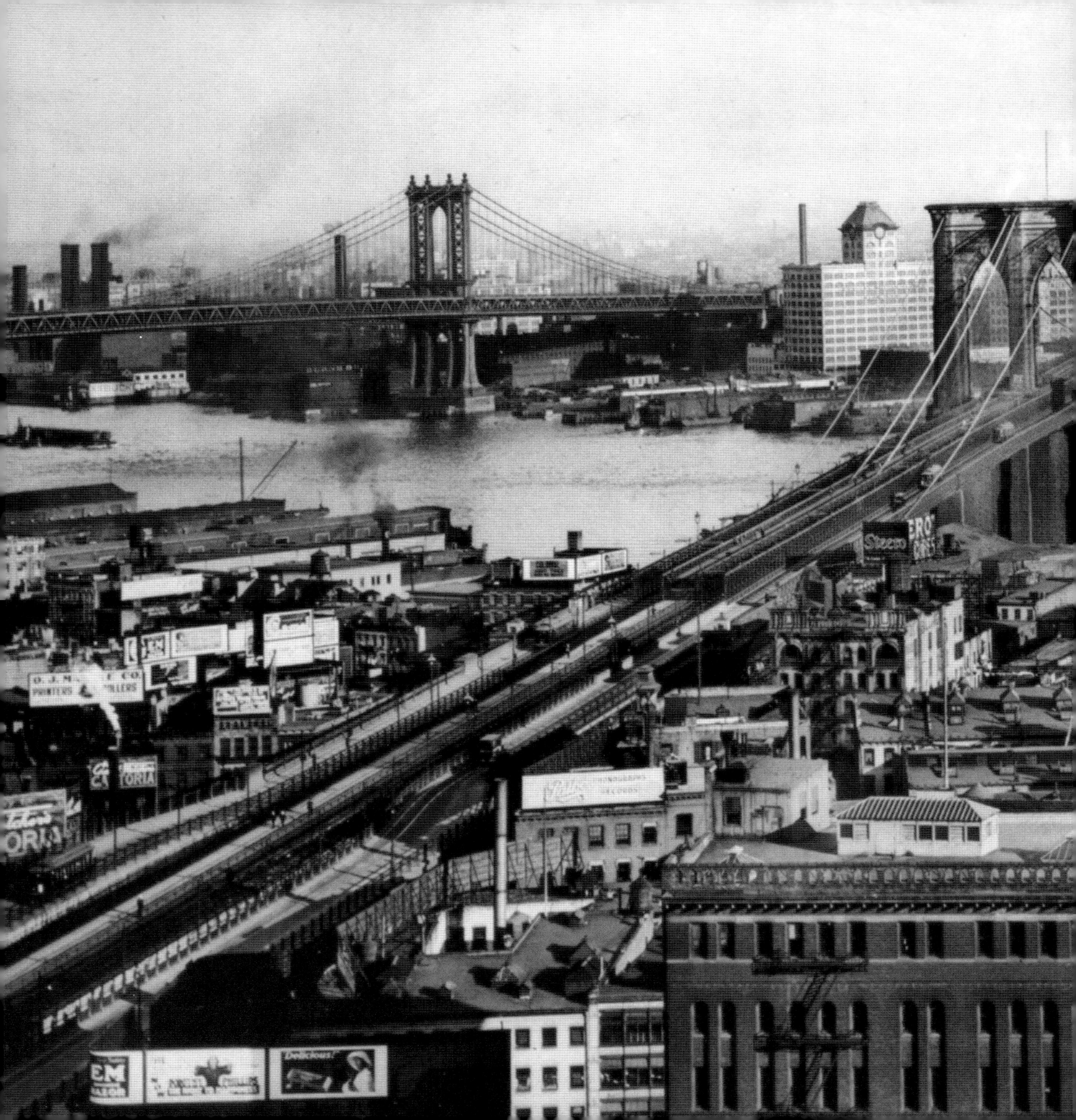
Steero
PRINTERS
Delicious!

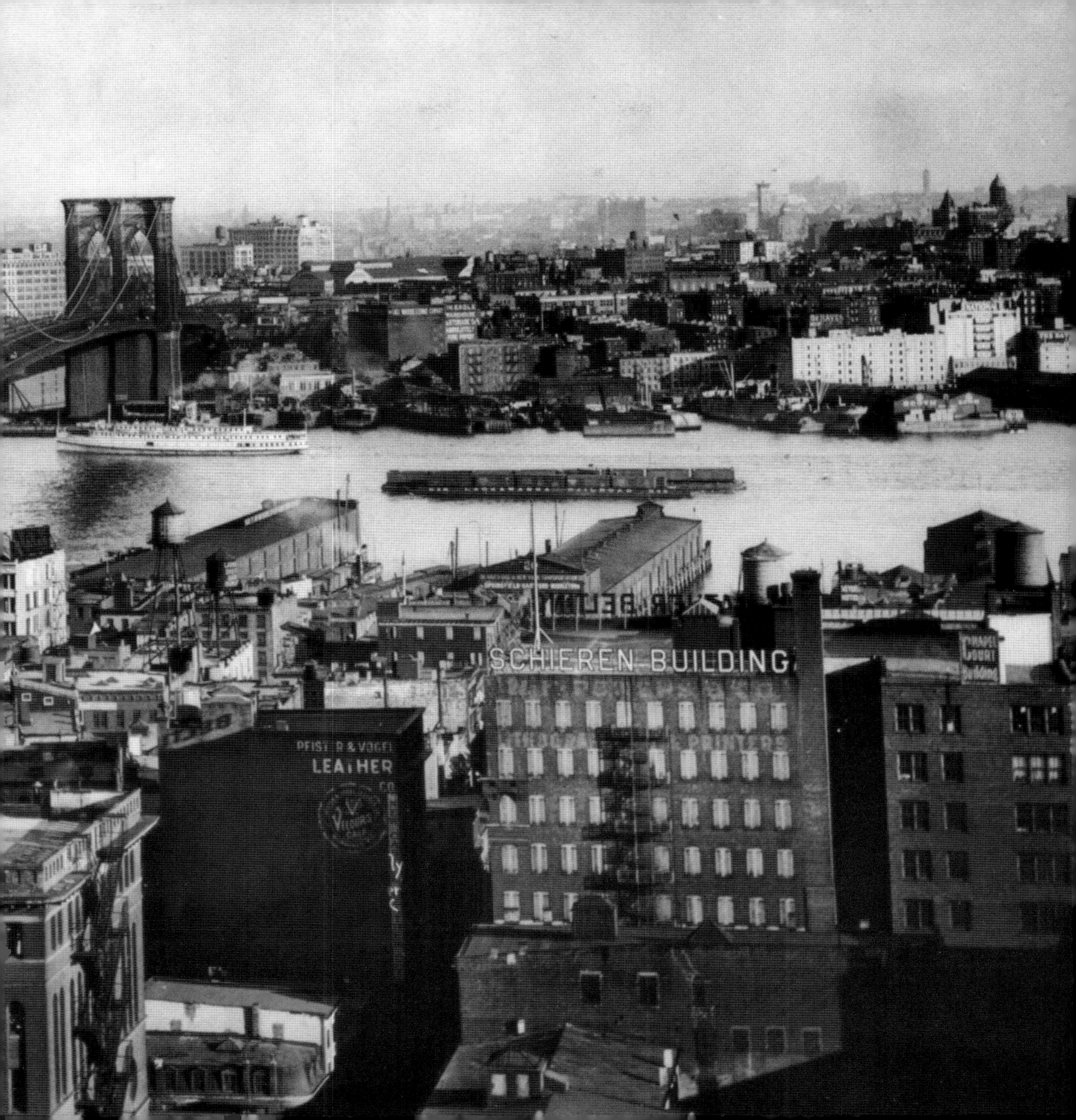
SCHIEREN BUILDING
PRINTERS
PFISTER & VOGEL
LEATHER

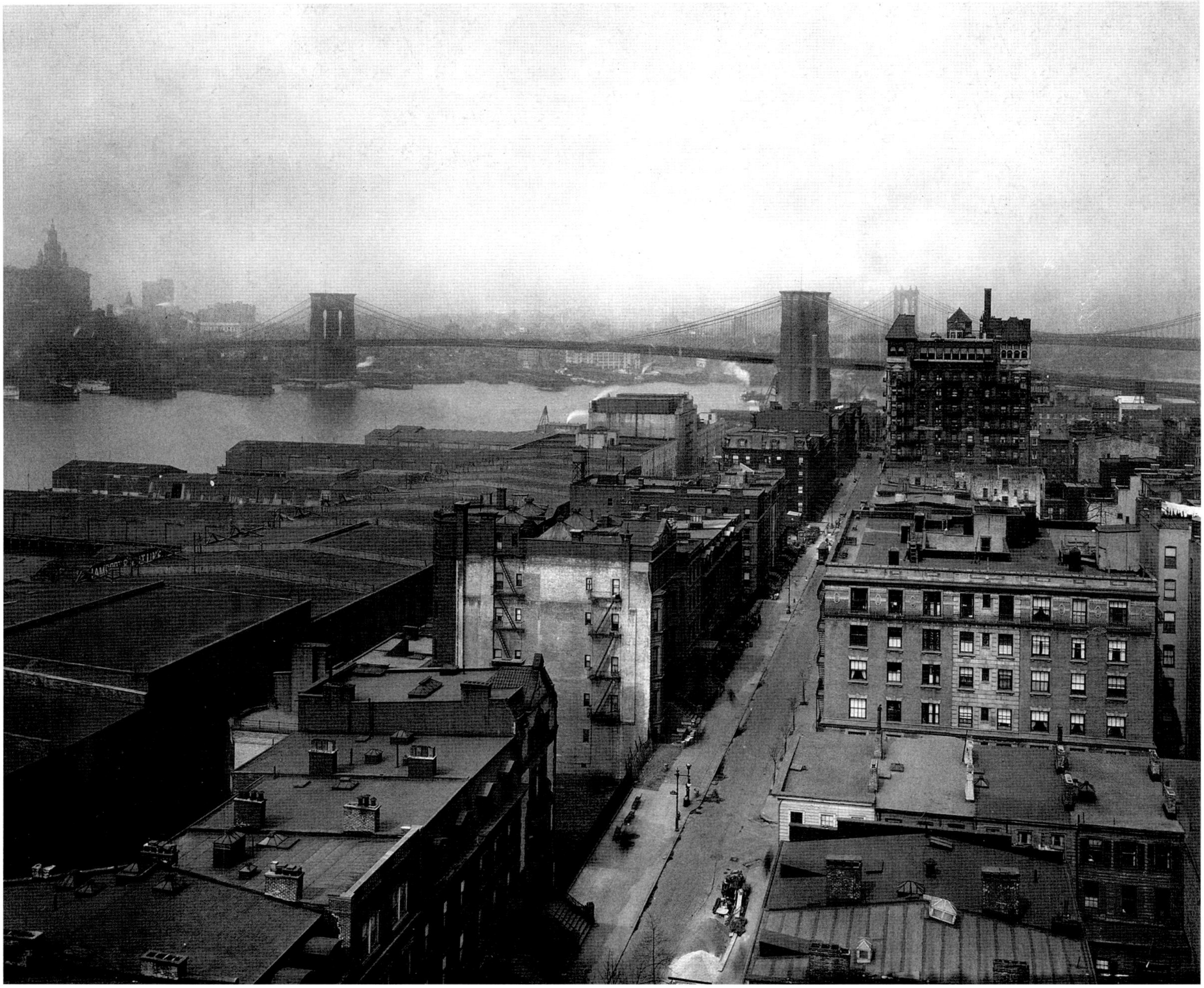

Commerce in Brooklyn Heights: An elevated view of Columbia Heights facing north toward the Brooklyn Bridge. In this 1920 photo, warehouses and piers descend to the river's edge, and rooftops of brownstone houses line the street. In the distance to the right, the tallest building is the Hotel Margaret, a magnificent 16-story residence built in 1889- but destroyed by a fire in the 1980s.

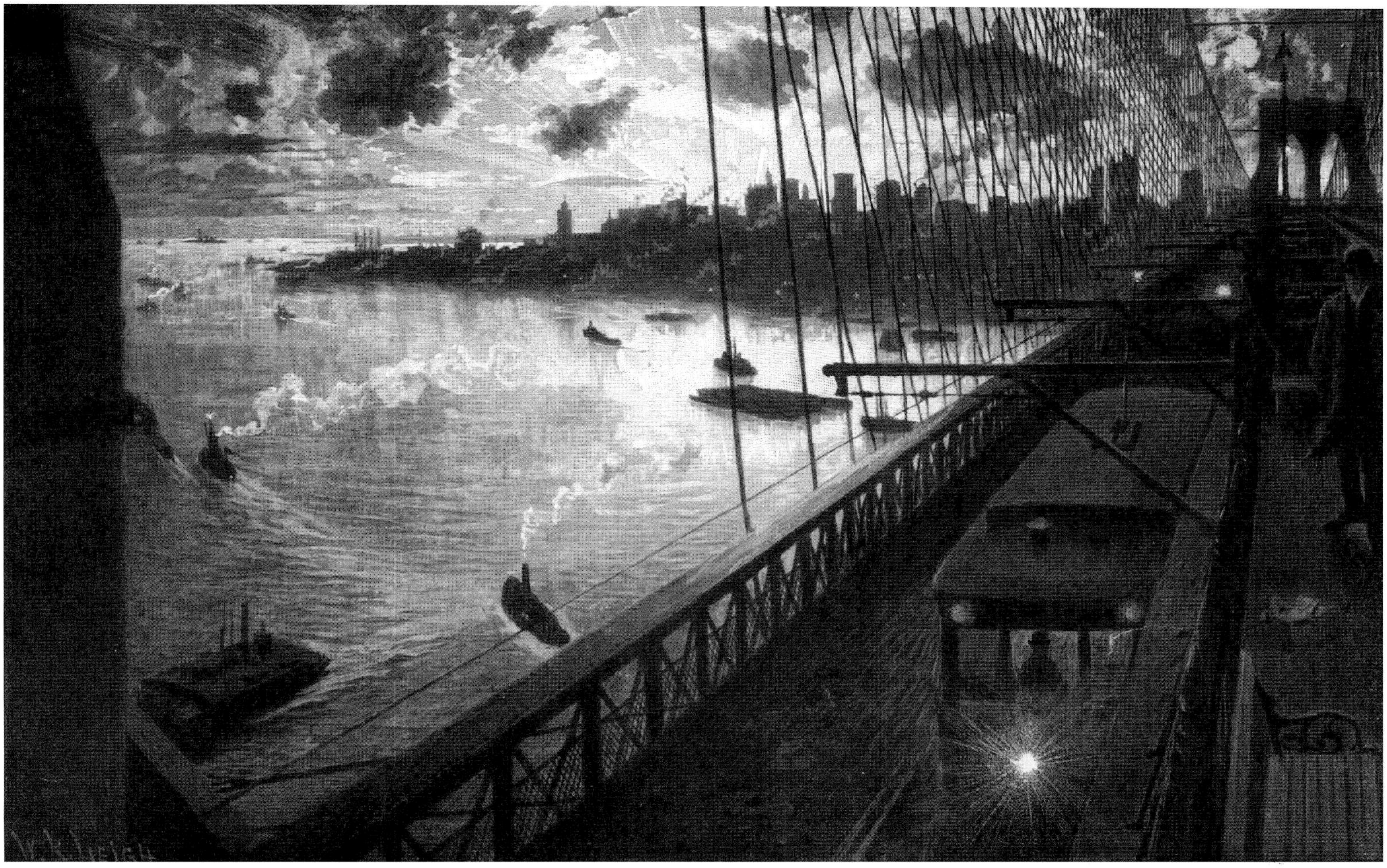

Trolley crossing bridge at night: At dusk, the silhouettes from the Brooklyn Bridge and a trolley car set off the shadowy Manhattan coast.

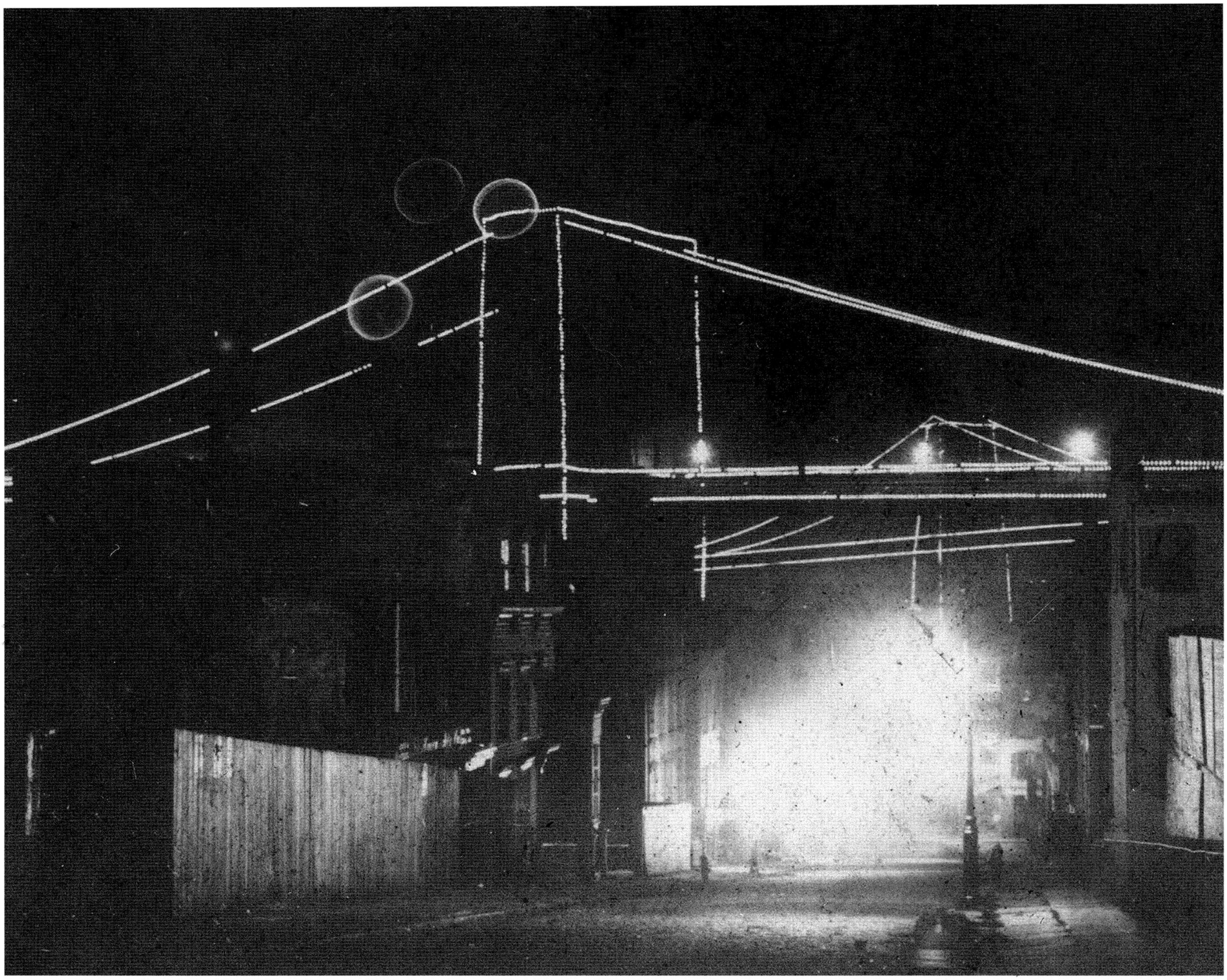

Classic night shot: The Brooklyn Bridge shines in its glory. The outline of the bridge is seen through electric lighting, and the newly installed lights shine brightly from their posts. The illuminated Manhattan Bridge can be seen behind the lights of the Brooklyn Bridge.

From Shore to Shore: A Great City

(1921–1982)

As Brooklyn adapted to a modernizing world, so did its transportation in and around the bridge. Ferry terminals were demolished by the 1920s, leaving stray pilings in their places. With the arrival of modern New York City, roadways replaced railways and the familiar became nostalgic. Beneath the Brooklyn Bridge, South Street Seaport lured sailing ships as memories. For a while, caverns beneath the bridge became art galleries and great vaulted spaces for public performances.

During this fifty-year time period, modern media capitalized on the bridge's beauty and symbolism. Early films used her as an object to capture and as a base from which to photograph action in the waters beneath. Famous names commented on the bridge's significance to the world. F. Scott Fitzgerald, in his Hollywood days, even wrote a script for the movie *Brooklyn Bridge;* several television series and numerous films reproduced the world-famous name and picture. Songs have been written and sung in honor of the great bridge.

The year 1933 marked the 50th anniversary of the Brooklyn Bridge. As with the public opening of the bridge in 1883, city officials, citizens, and visitors celebrated the magnificent structure, although this time in a more modern scene as they were surrounded by ships, trolleys, automobiles, and skyscrapers. Soon, the Brooklyn Bridge Plaza was built, and the last train on the bridge was replaced by the subway in the 1940s. Pedestrians enjoyed the new bridge walkway as well as the new Brooklyn-Queens Expressway, which permitted automobiles to merge smoothly from the expressway to the bridge. Although the city was changing from shore to shore, its bridge still remained the dominating feature spanning the great East River.

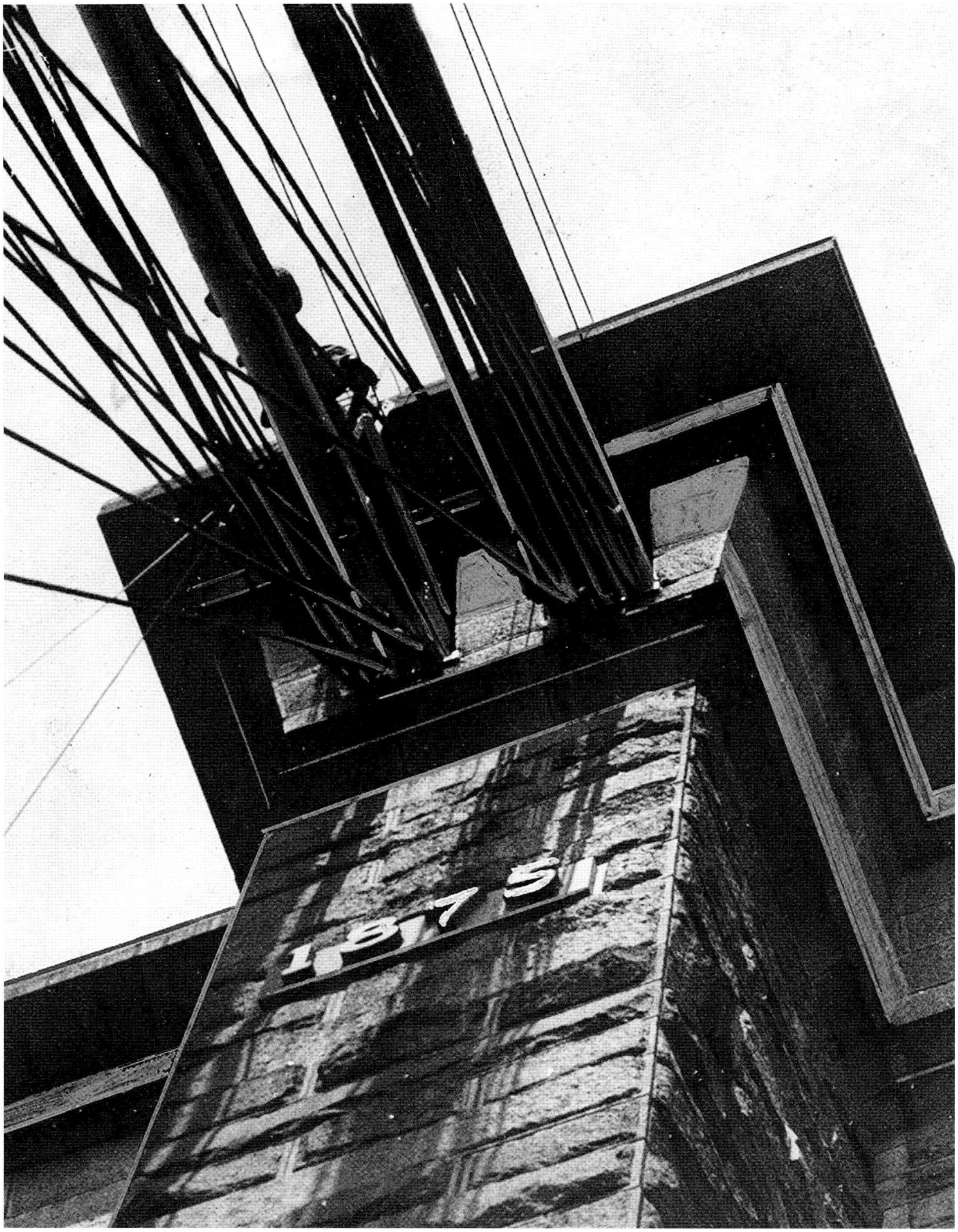

Bridge tower: This 1922 photo is focused in on the Brooklyn tower of the Brooklyn Bridge. The tower prominently displays its year of completion.

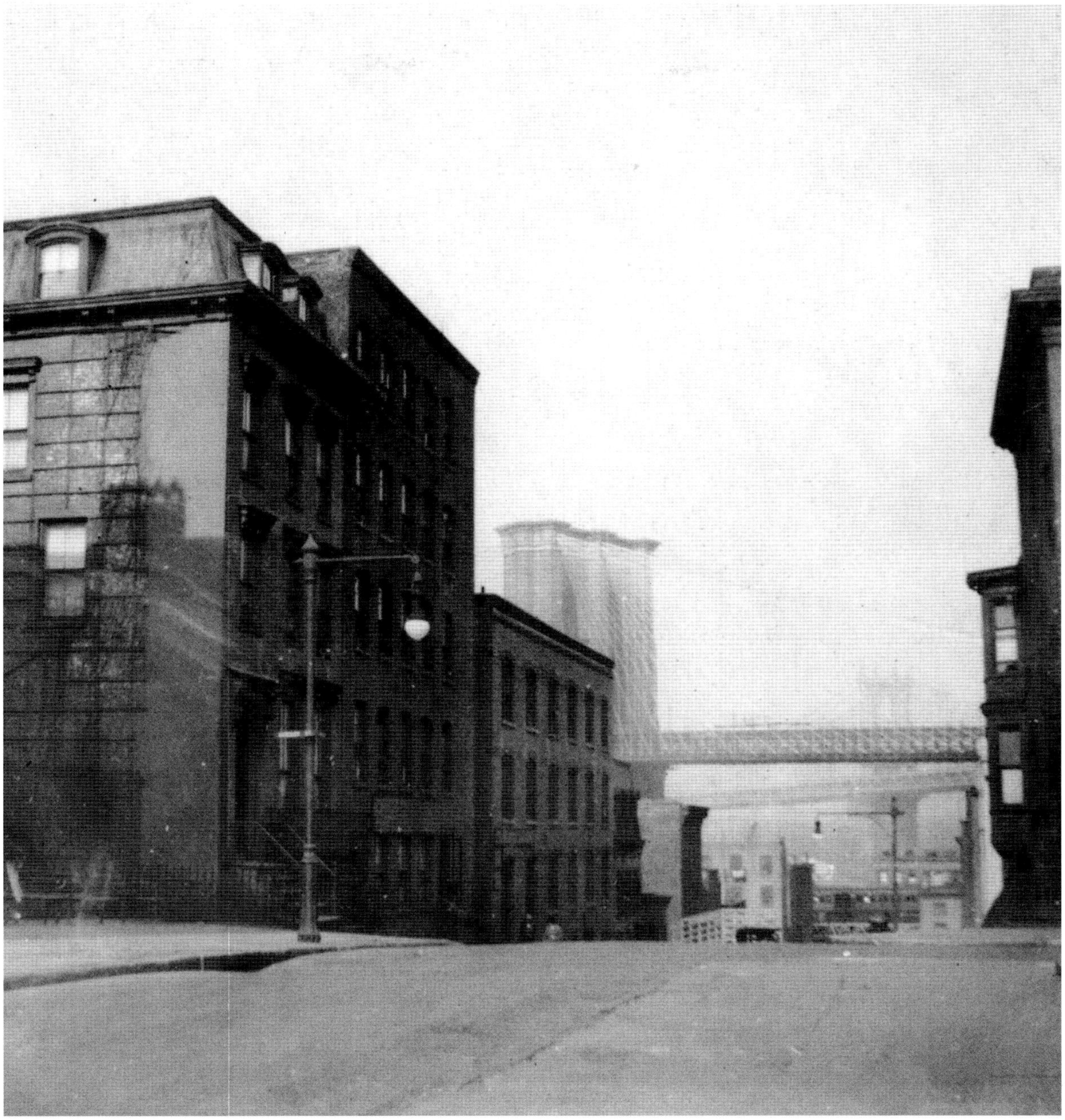

Columbia Heights, 1923: Leading to both the Brooklyn and Manhattan bridges as well as the ferry, this street features No. 62, the mansard-roofed house on the left that was built in 1850. Today the land is a parking lot.

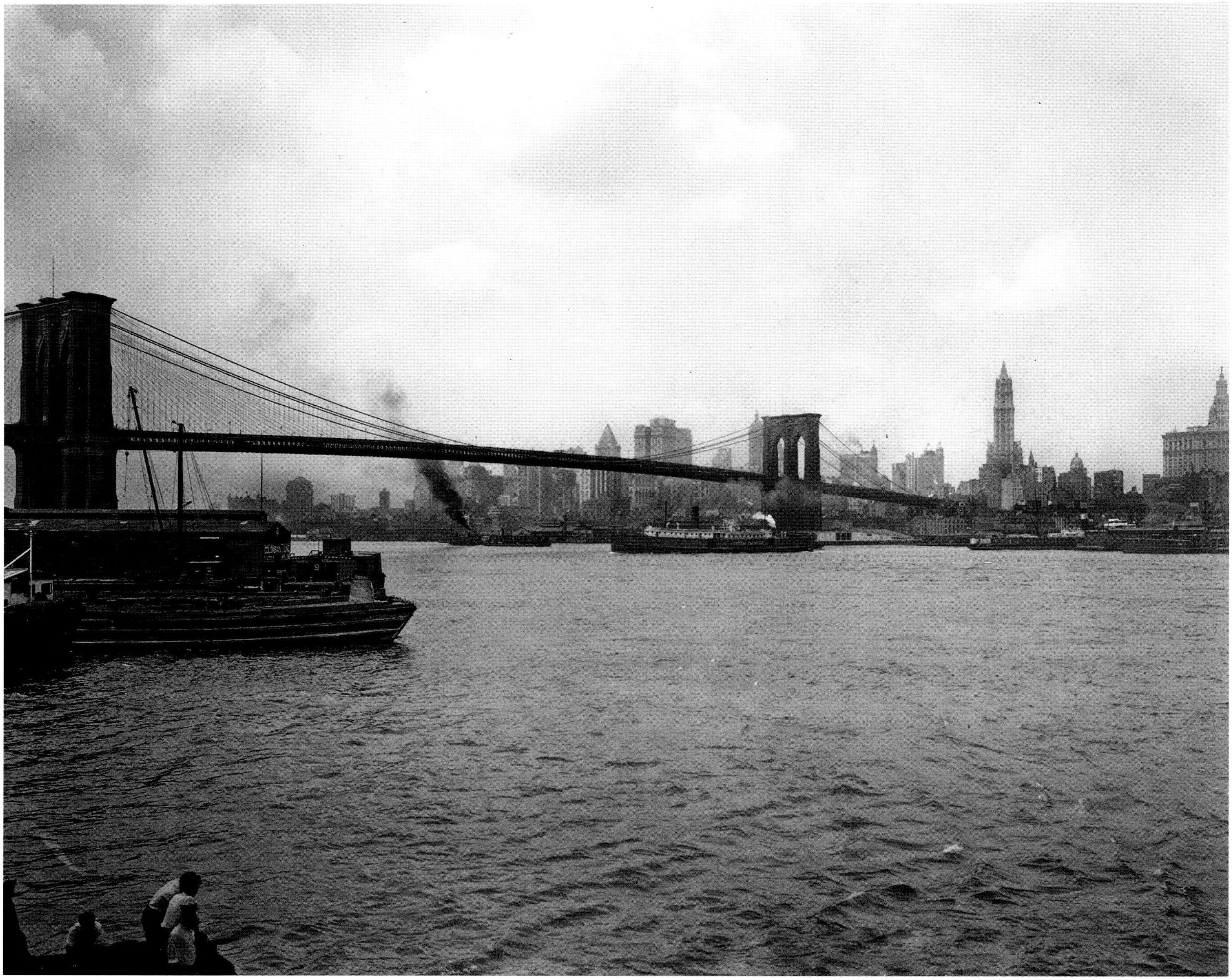

Serene majesty: Photographed from southeast of the bridge on the Brooklyn shore in 1925, the river seems placid although steamship activity can be seen on the Manhattan side.

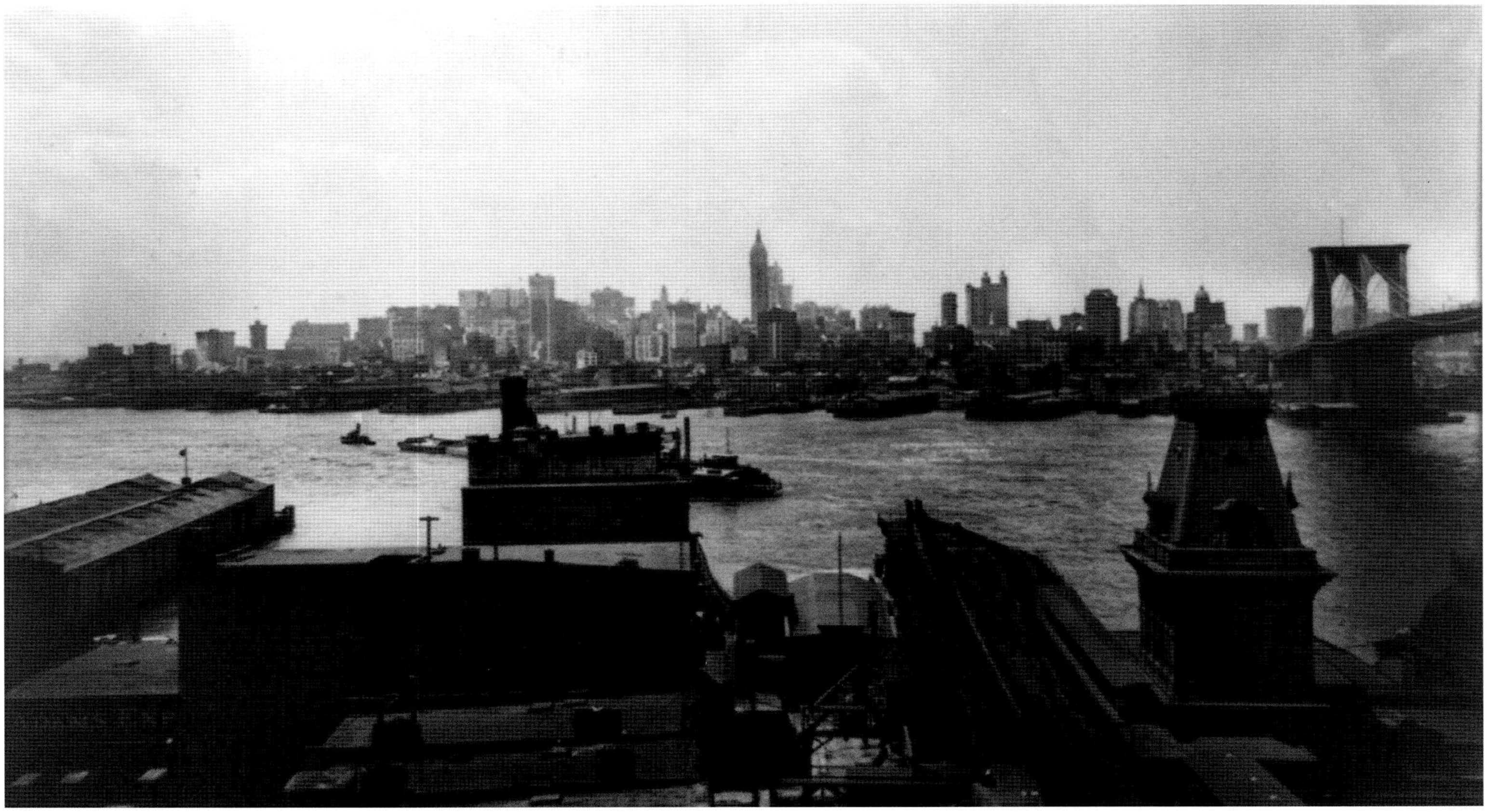

A view overlooking the Brooklyn waterfront with the Brooklyn Bridge to the right (ca. 1925). The ferry terminal and the piers would eventually be replaced by the Brooklyn Bridge Park.

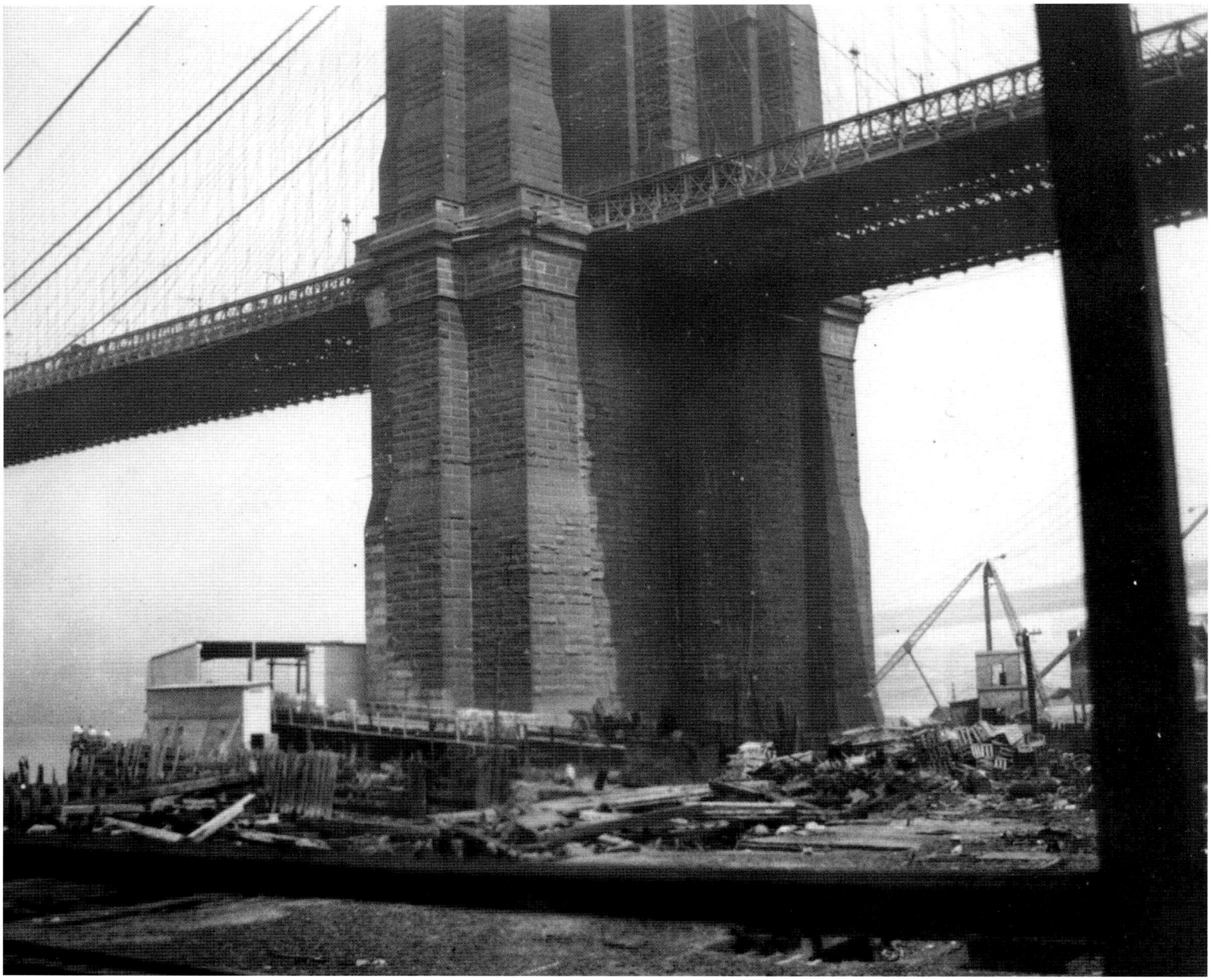

Remains of Fulton Ferry House: By 1928, only rubble of the ferry house and slip remained at the foot of the Brooklyn tower.

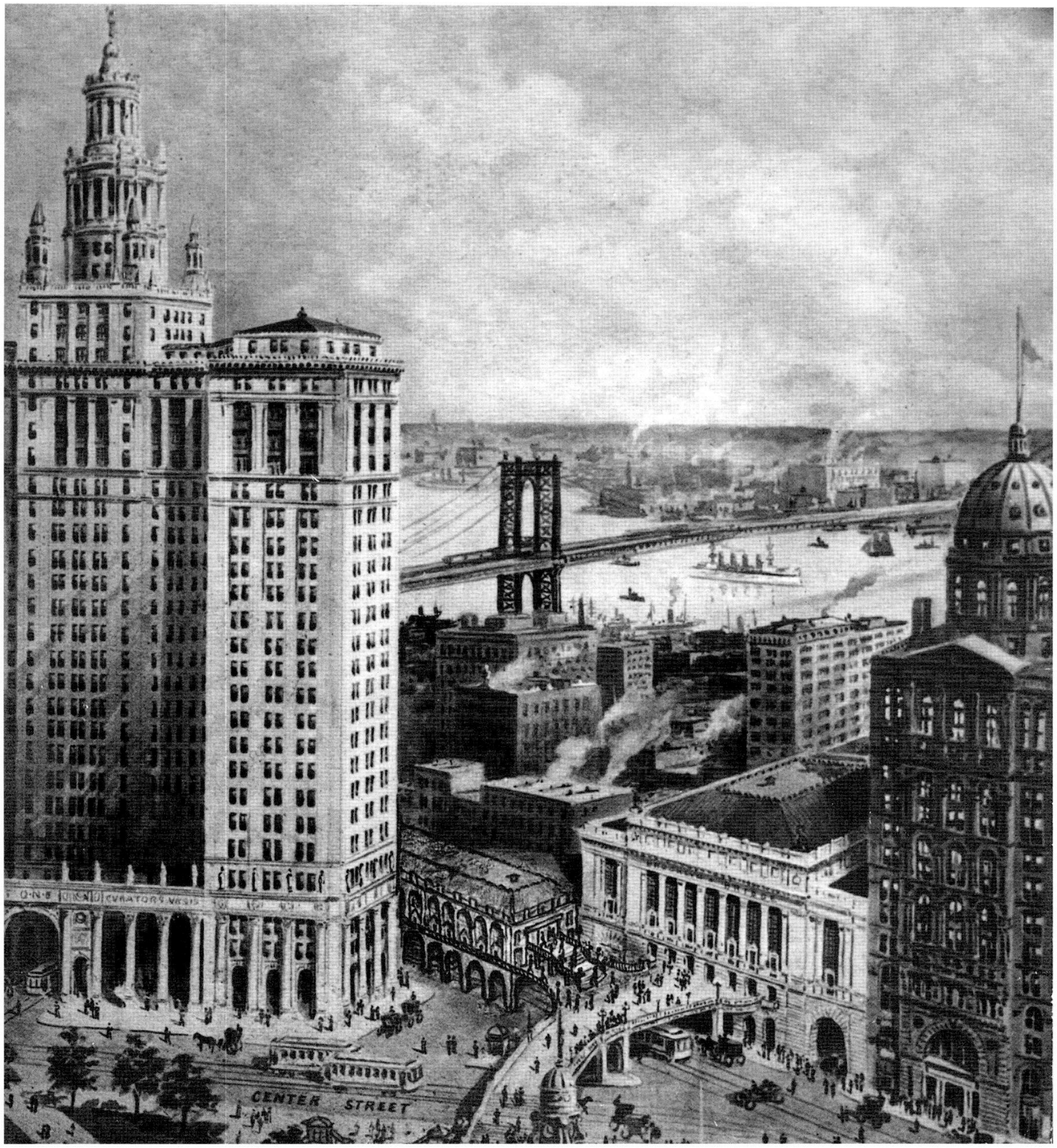

Brooklyn Bridge: A drawing of the Brooklyn Bridge, detailing the daily round in Manhattan away from the water's edge.

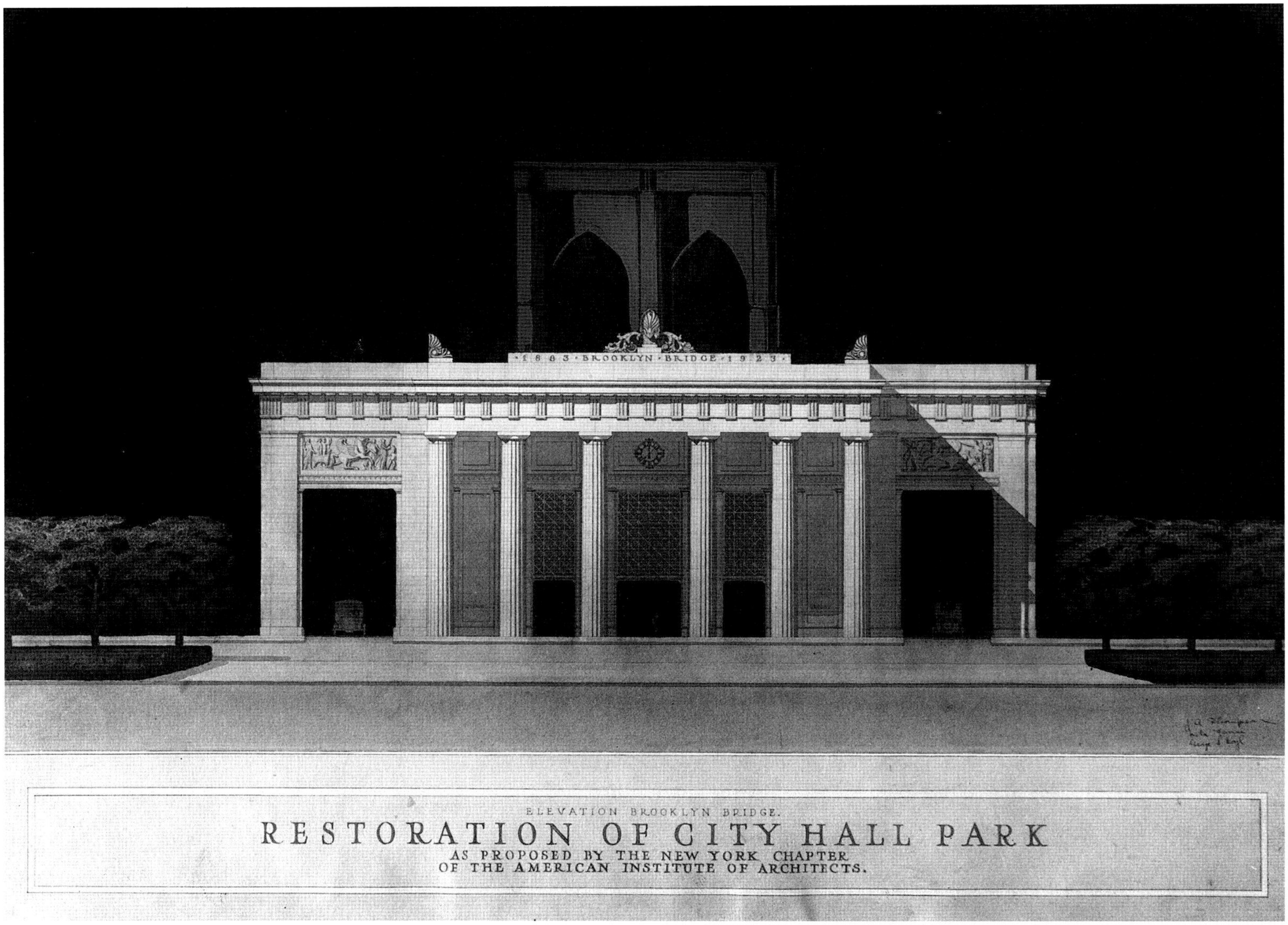

Proposed entrance: In a never-ending effort to improve the perfect bridge, city officials in 1928 proposed this entrance for the Manhattan approach to the Brooklyn Bridge. The project never reached fruition.

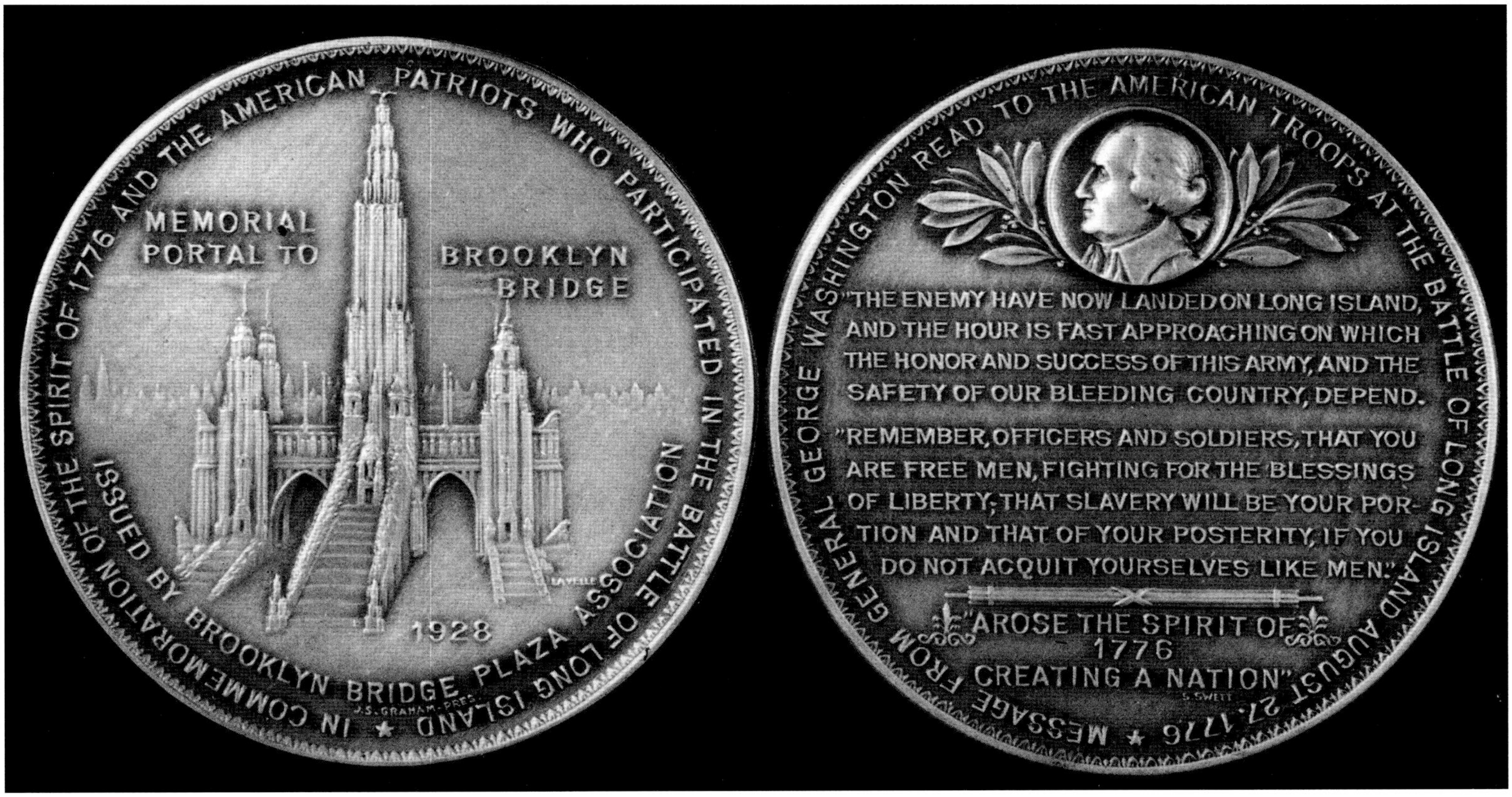

Brooklyn Bridge arch: This memorial coin was issued to honor the Brooklyn Bridge Plaza's Washington Memorial Portal, a plan that never materialized. Other Brooklyn Bridge "improvements" included a proposal for a second deck to be built on the bridge.

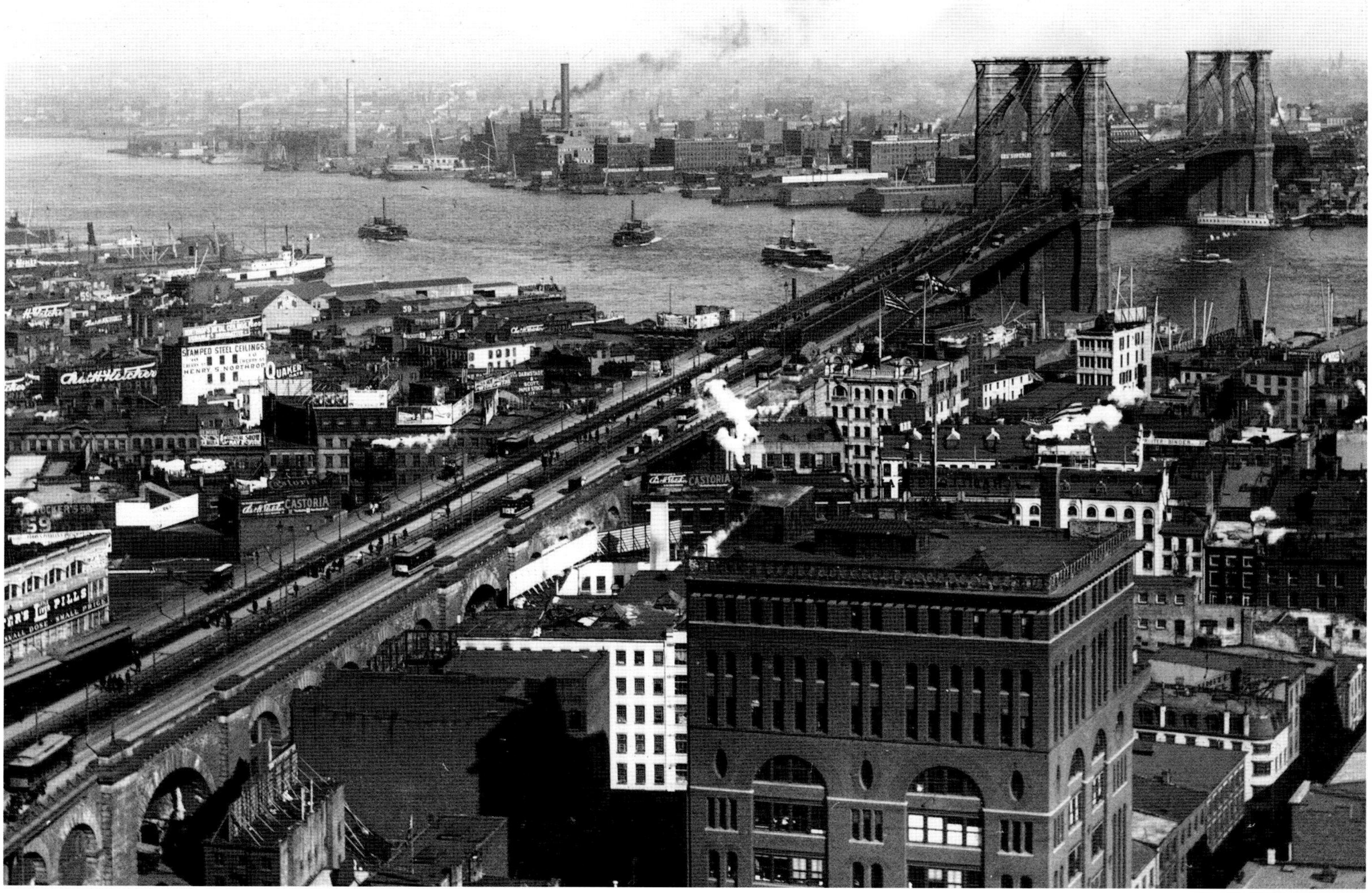

The Brooklyn Bridge from Manhattan shows trolleys and pedestrians in the 1920s.

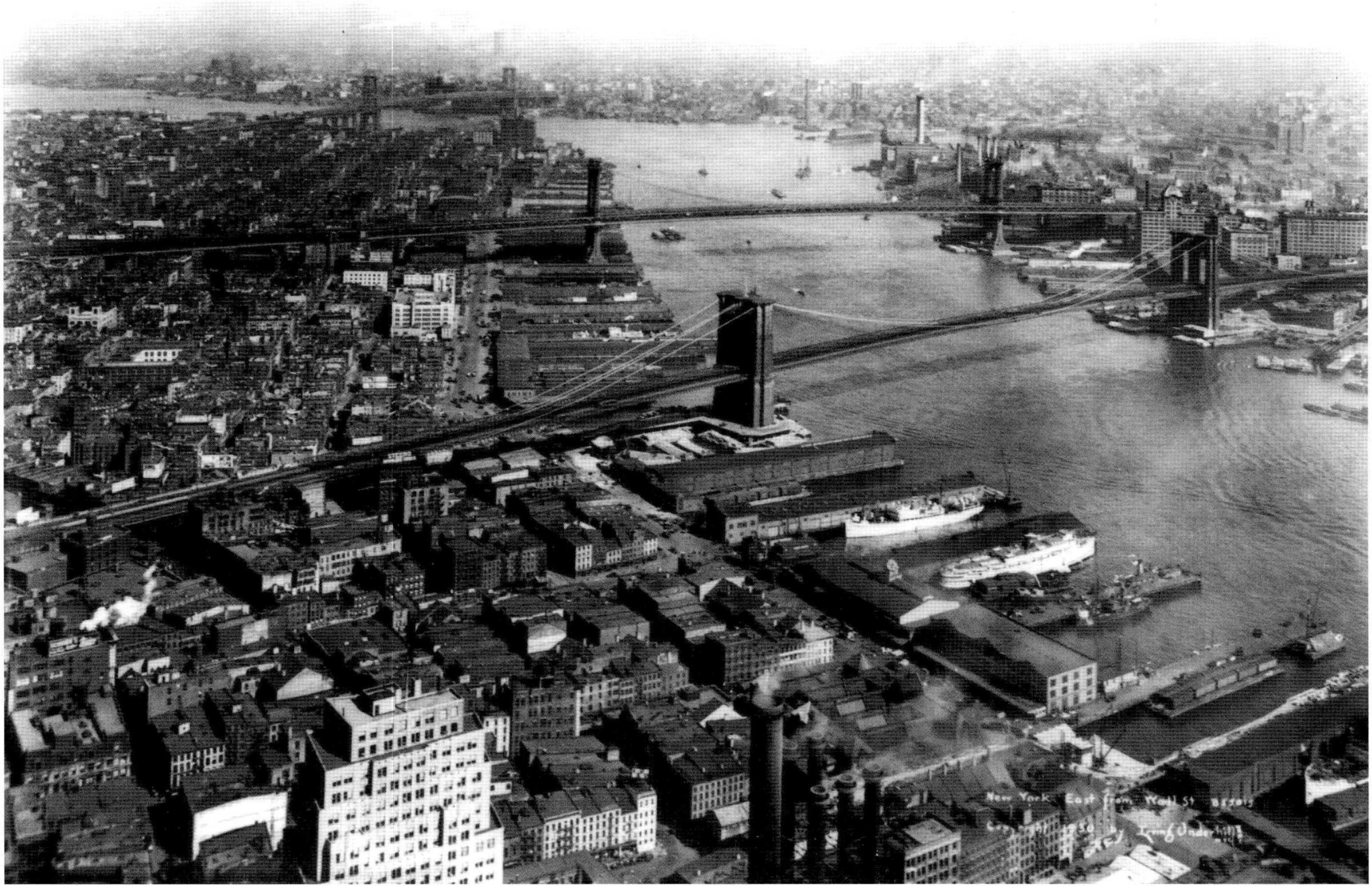

Bridge from Wall Street: Facing north from Wall Street in 1930, the photographer provided this view of the East River with all three bridges and Brooklyn in the background. The photo taken from the World Trade Center in 2001 offers a similar view.

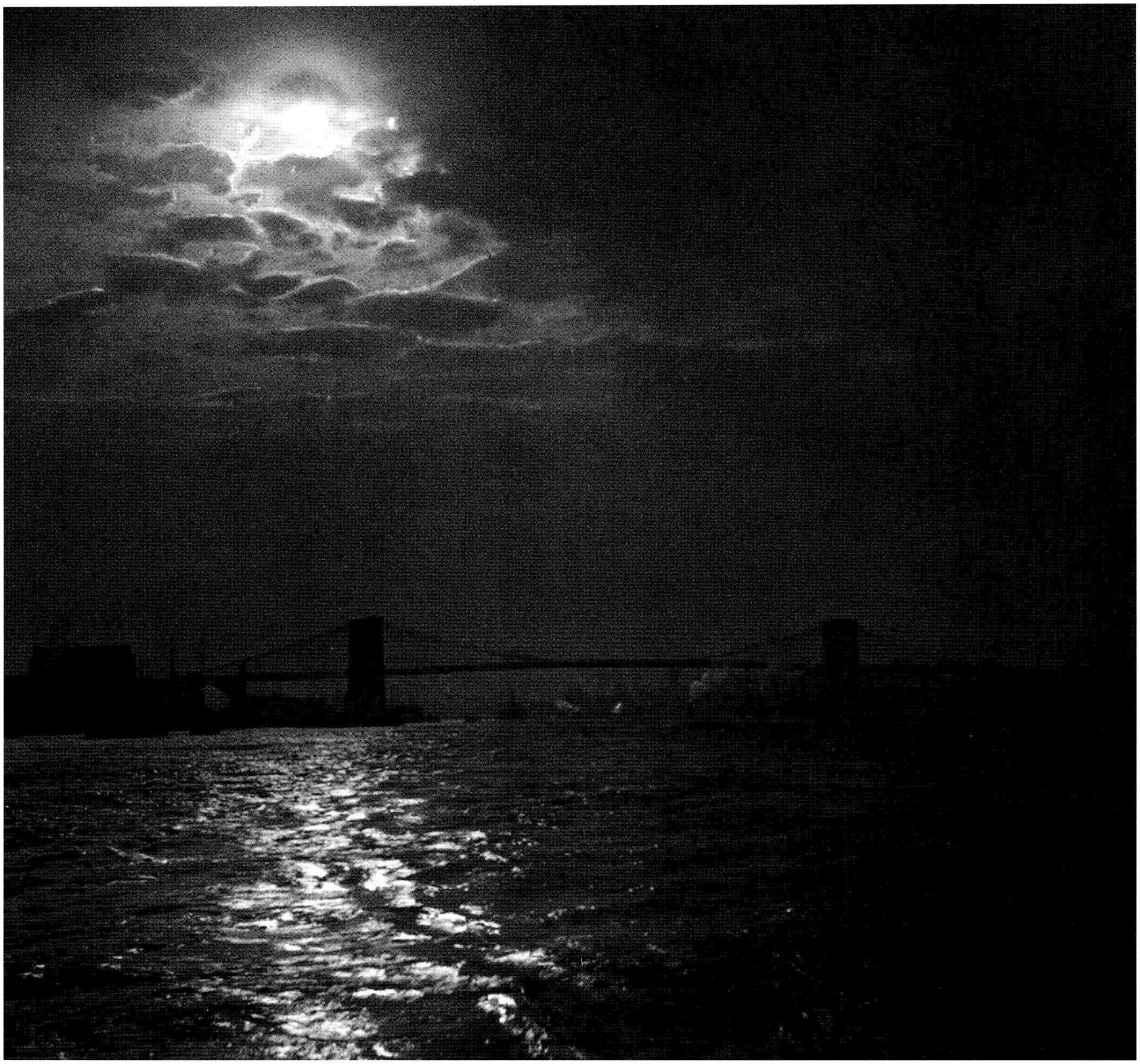

The Brooklyn Bridge at night.

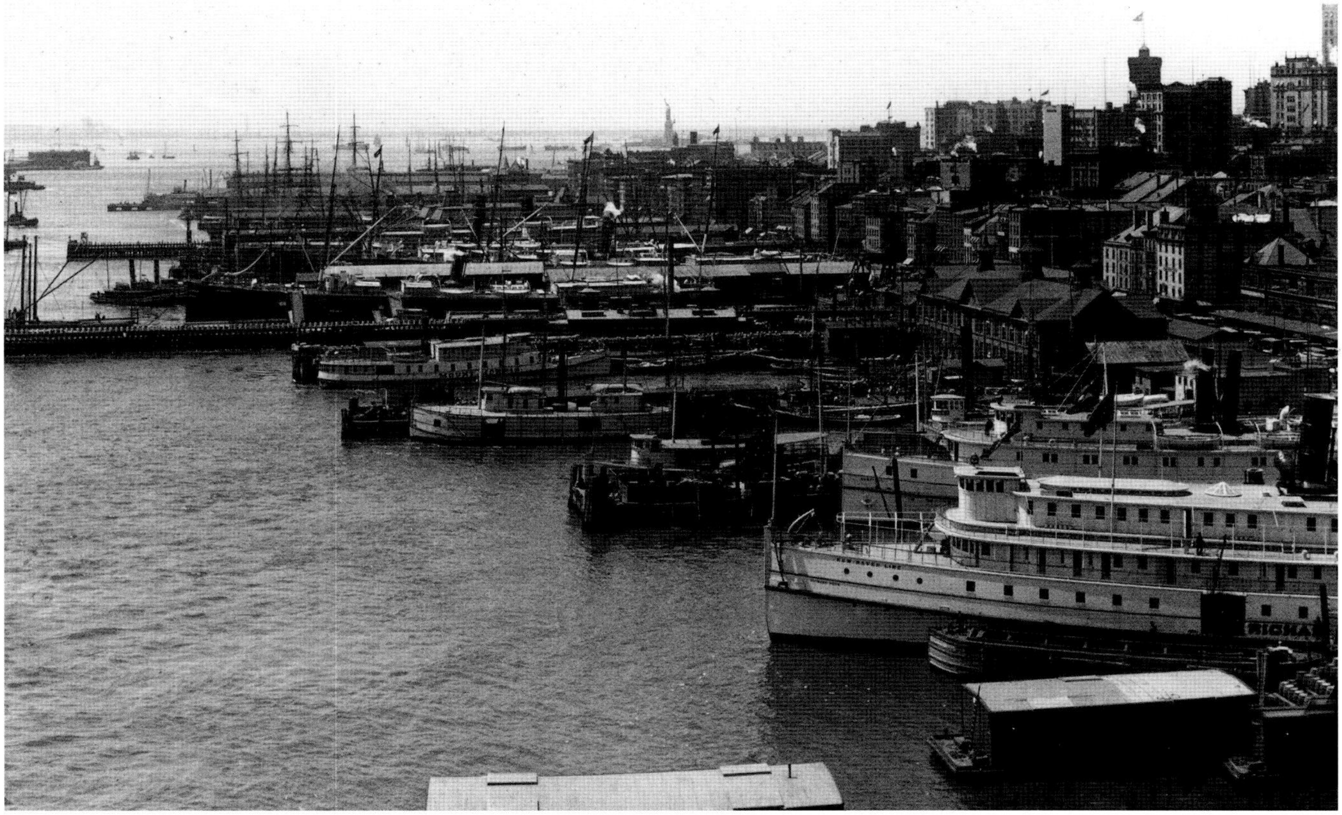

Manhattan riverfront: View of the Manhattan docks, facing south, with steam vessels docked where South Street Seaport now stands. Beyond the vessels protrude masts of sailing ships, and squat commercial buildings can be seen to the right. In the distance, the faint outline of the Statue of Liberty is in view.

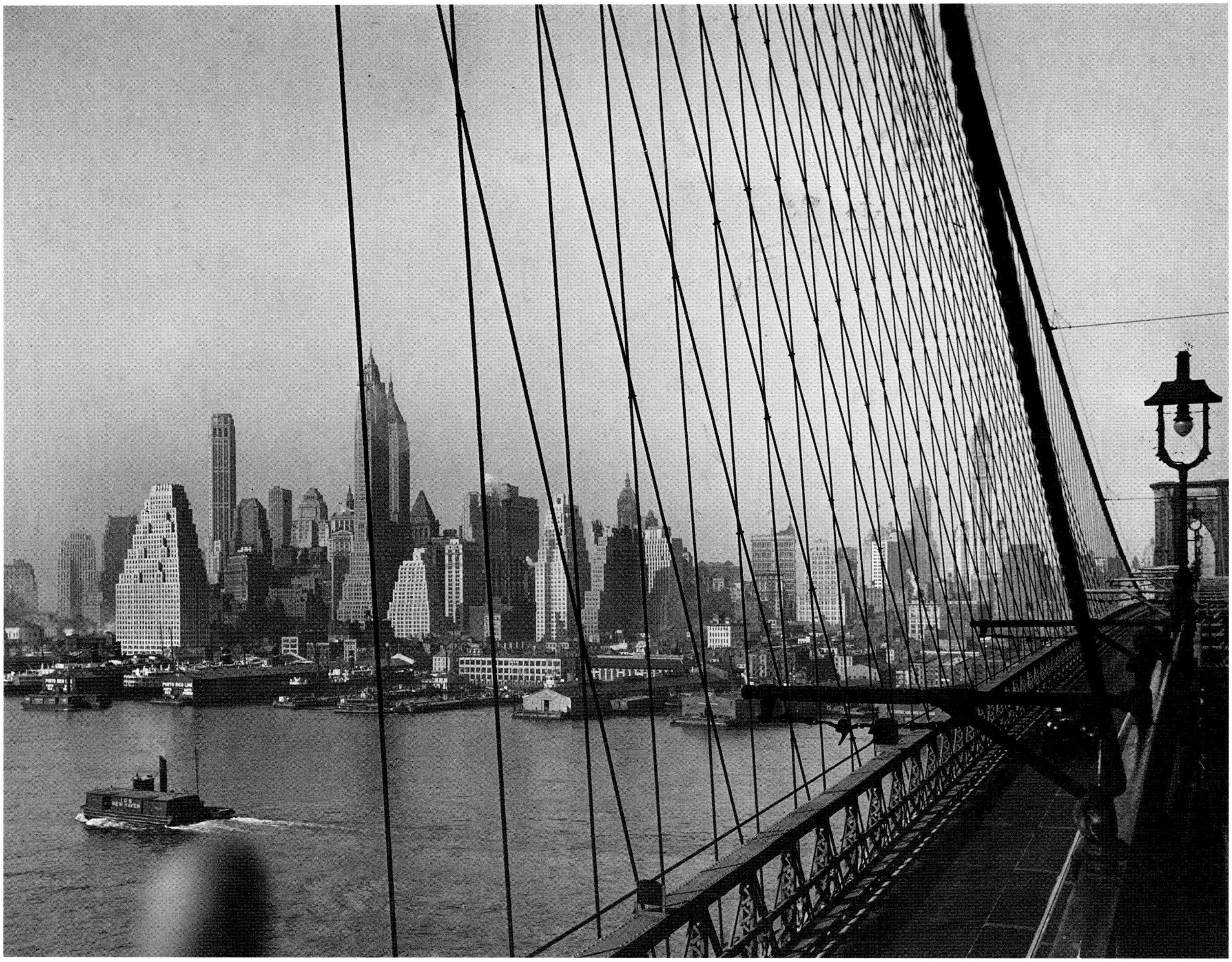

Brooklyn Bridge in motion: While a barge moves under the bridge, a car moves along the roadway above. On the right in this 1930s photograph stand a bridge light and the Manhattan tower.

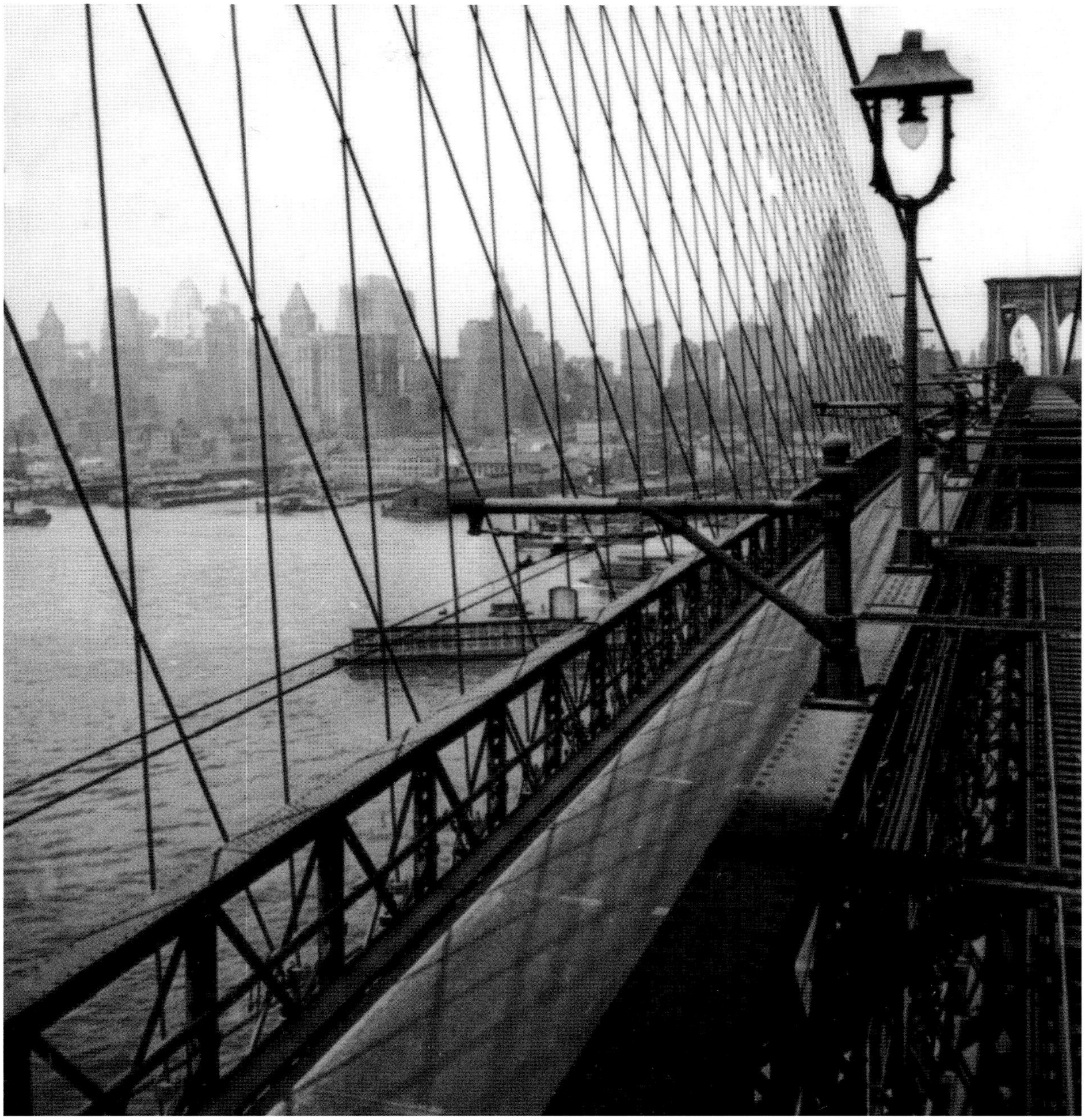

Manhattan's skyline: Through the cables of the Brooklyn Bridge, a 1930s Manhattan skyline can be seen above the bridge traffic.

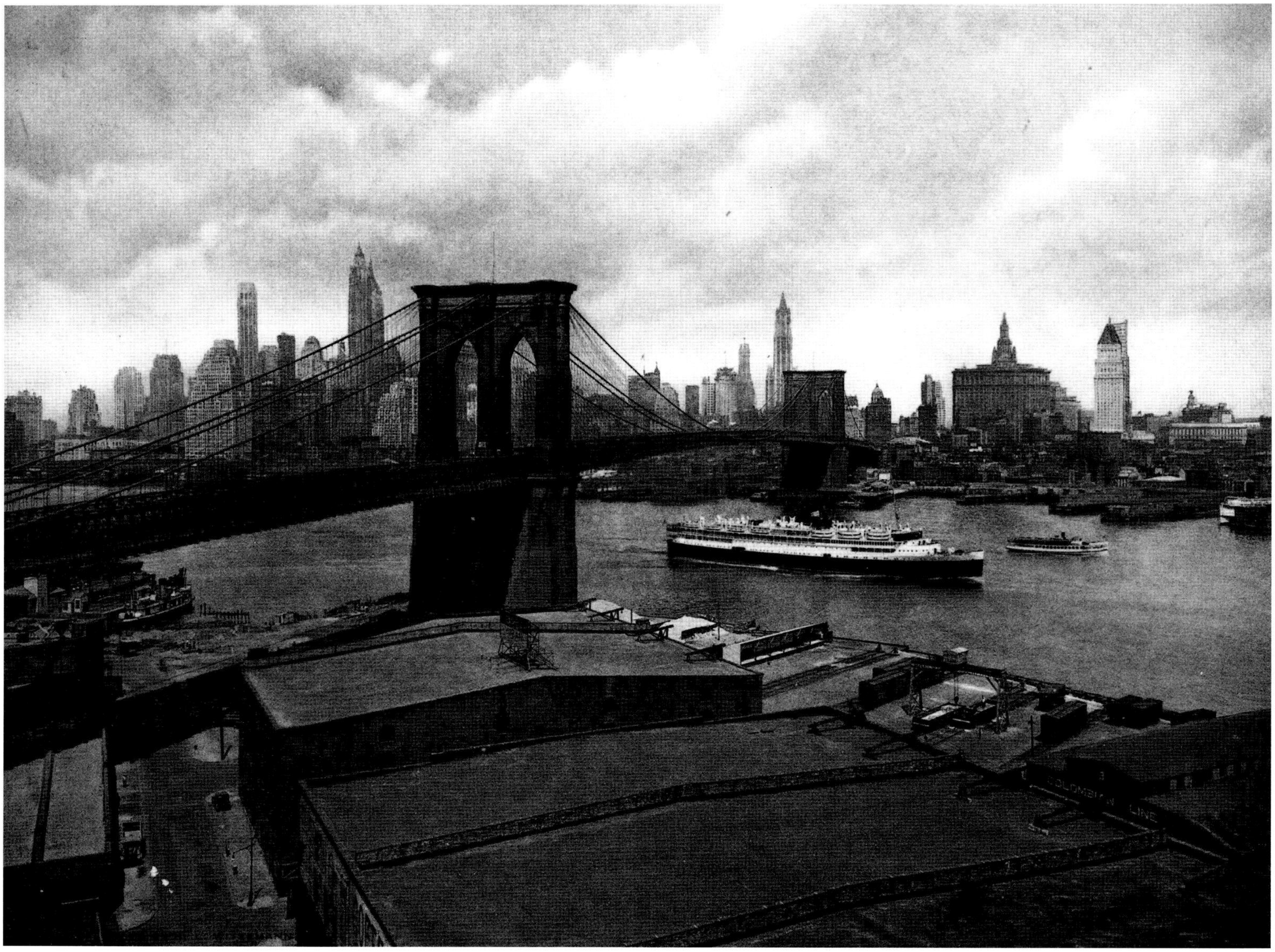

Ocean liner sailing under bridge: Taken from the DUMBO vicinity in the 1930s, this photograph shows a rare ocean liner sailing under the bridge. The Manhattan skyline appears in the distance.

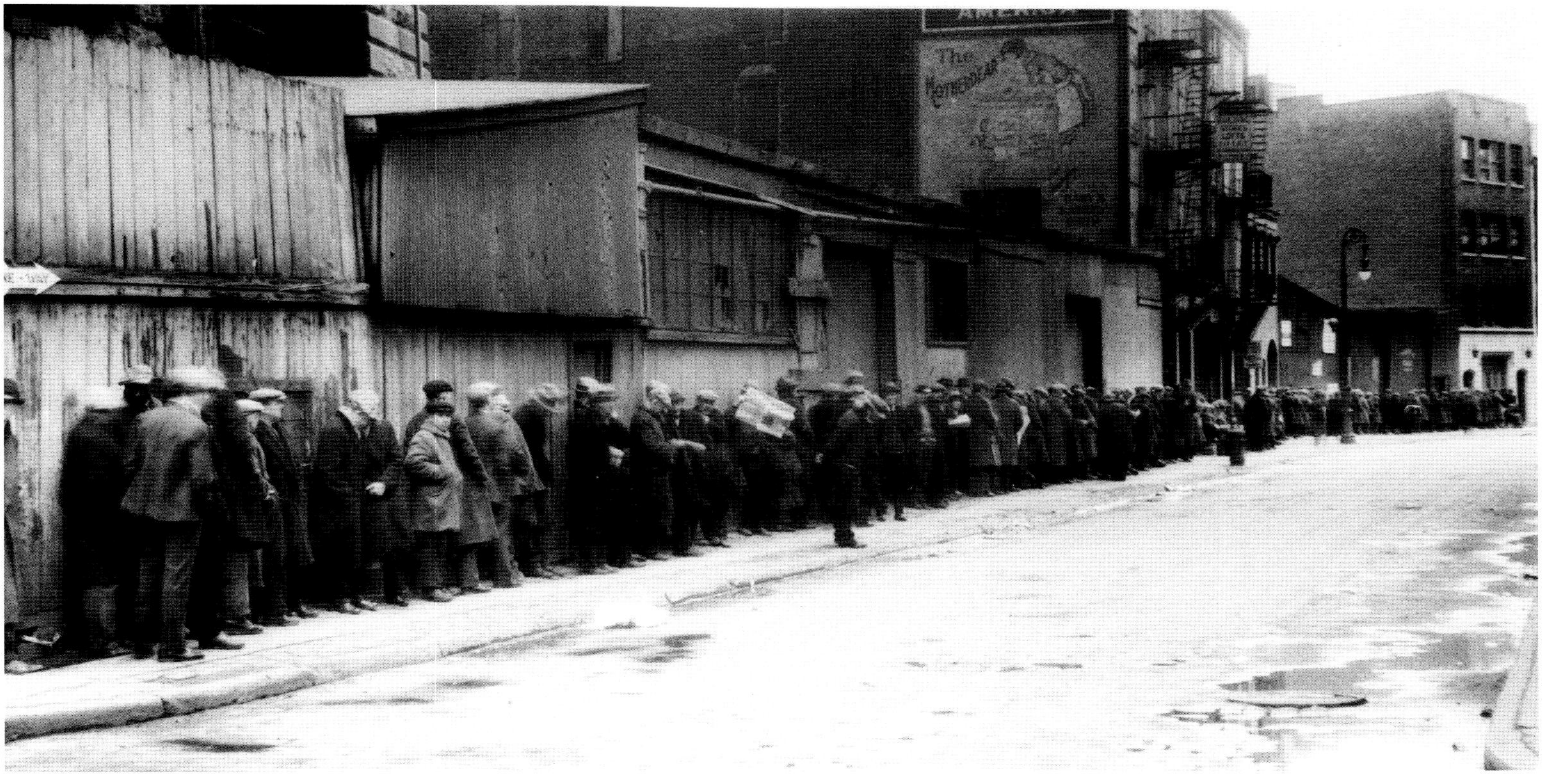

Bread line: During the Great Depression, New York residents stand in a queue at McCauley Water Street Mission near the Brooklyn Bridge, waiting for bread.

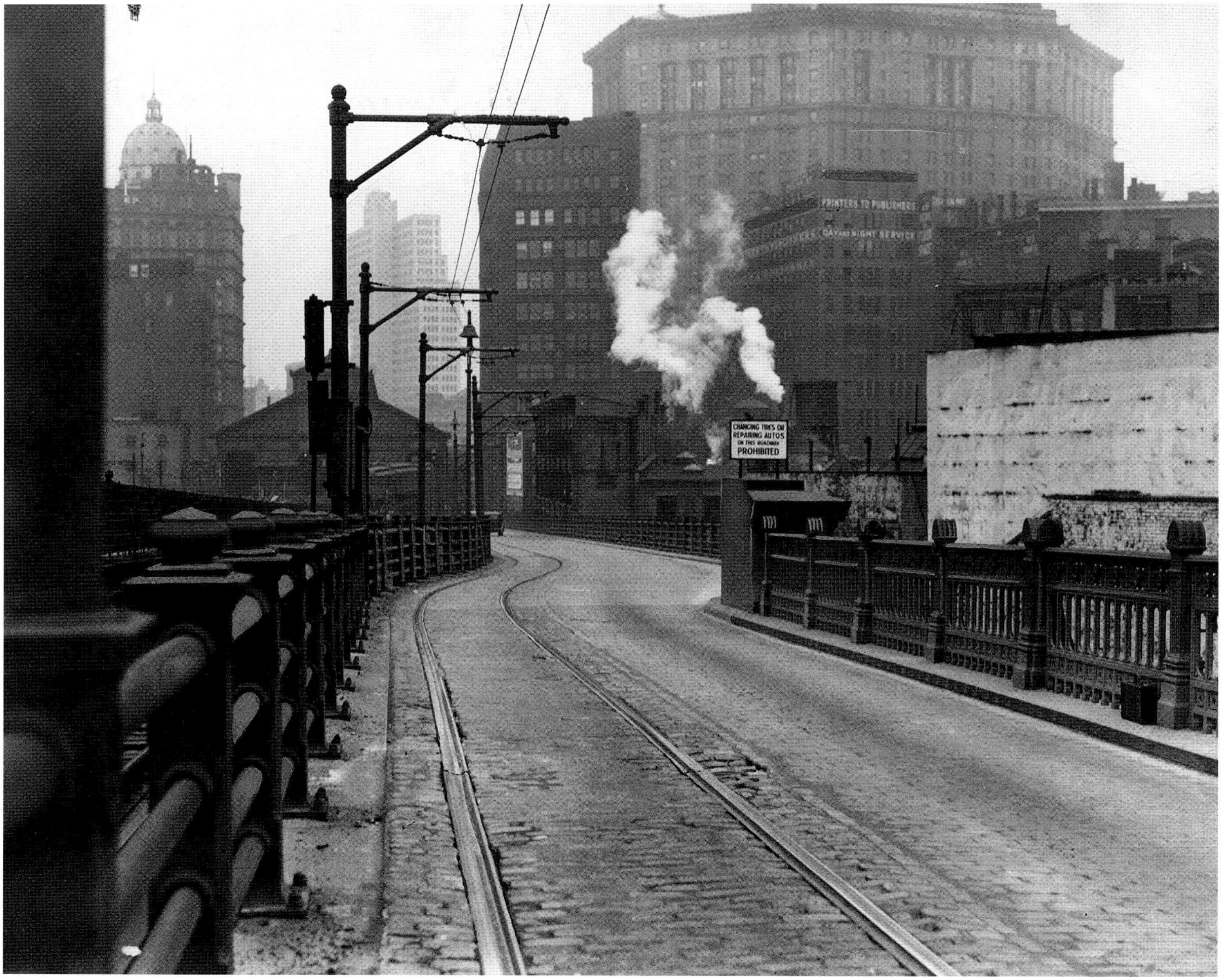

Brooklyn tracks: Trolley tracks from Brooklyn lead over a cobblestone roadway to the New York approach and then into the New York terminal. By 1932 when this picture was taken, the bridge was 49 years old; the Municipal Building at right was only 17 years old.

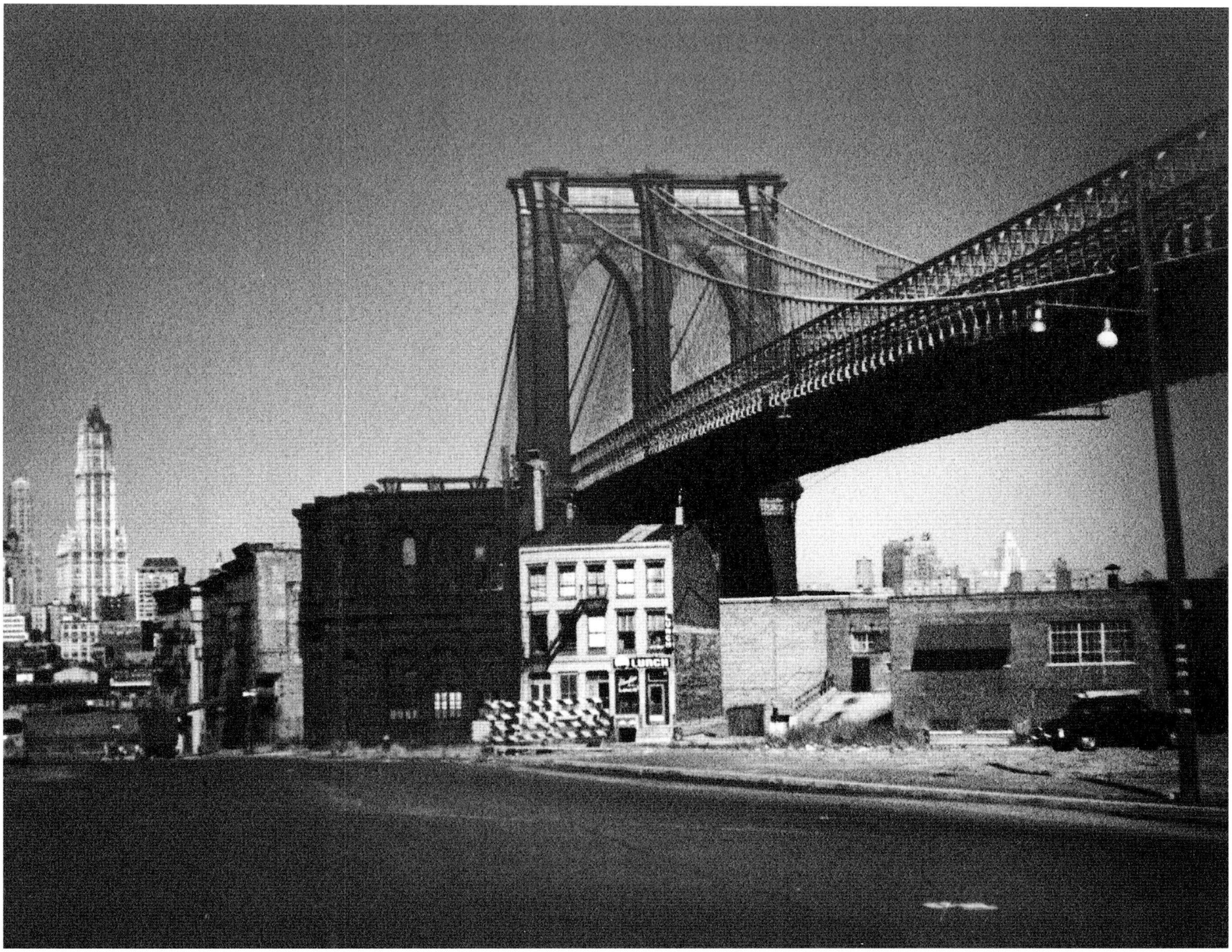

Bridge and Old Fulton Street: When the route of Fulton Street changed, the path from the former Fulton Ferry also changed to Old Fulton Street. Here in the 1930s, the commercial buildings that occupy the block beneath the bridge, also mark the entrance to DUMBO Across the East River is the Woolworth Building, which opened in 1913 as the world's tallest.

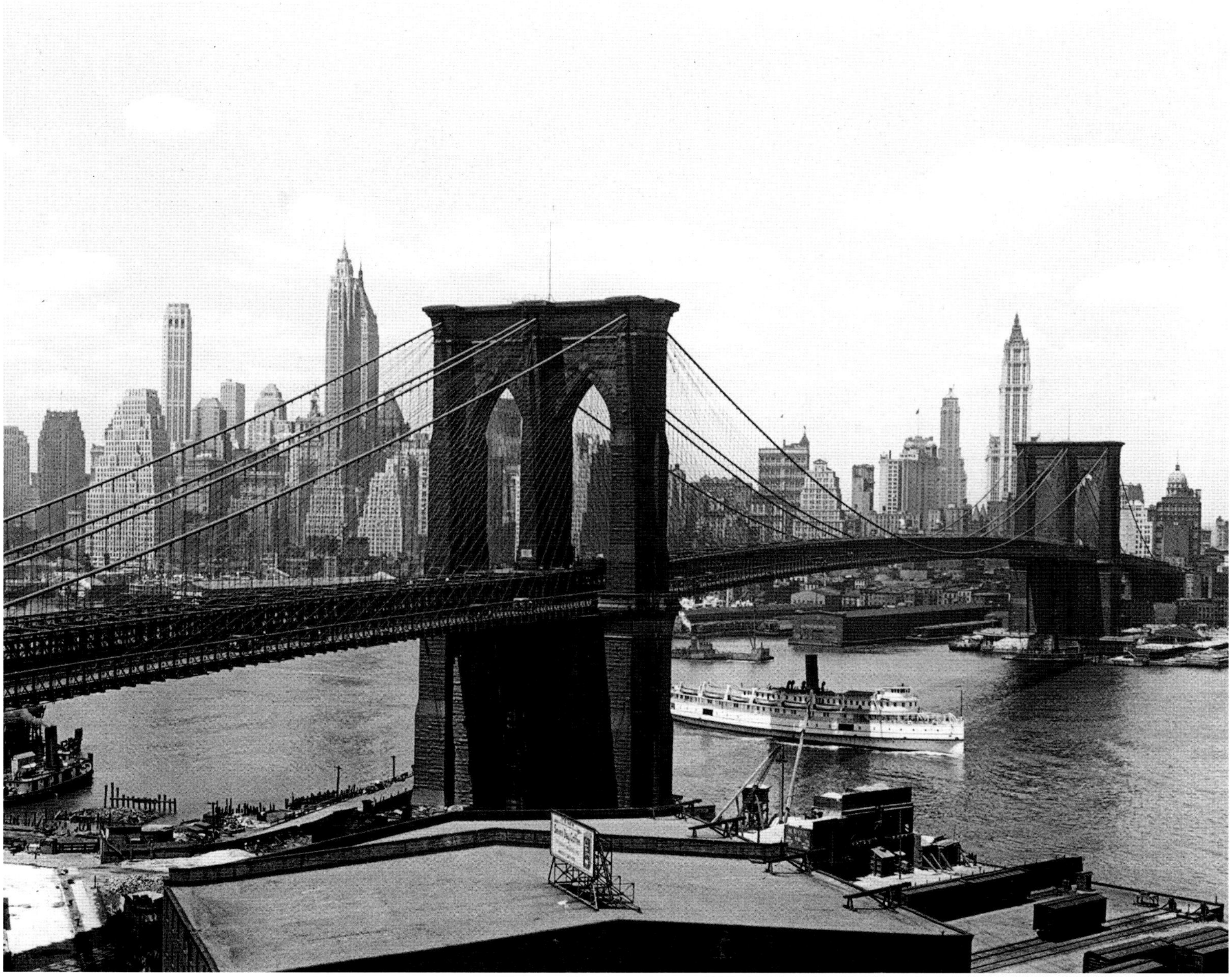

Following Spread: Brooklyn Bridge and Lower New York City: This 1932 panorama of the Brooklyn Bridge and Manhattan beyond also shows a cruise ship sailing north under the bridge.

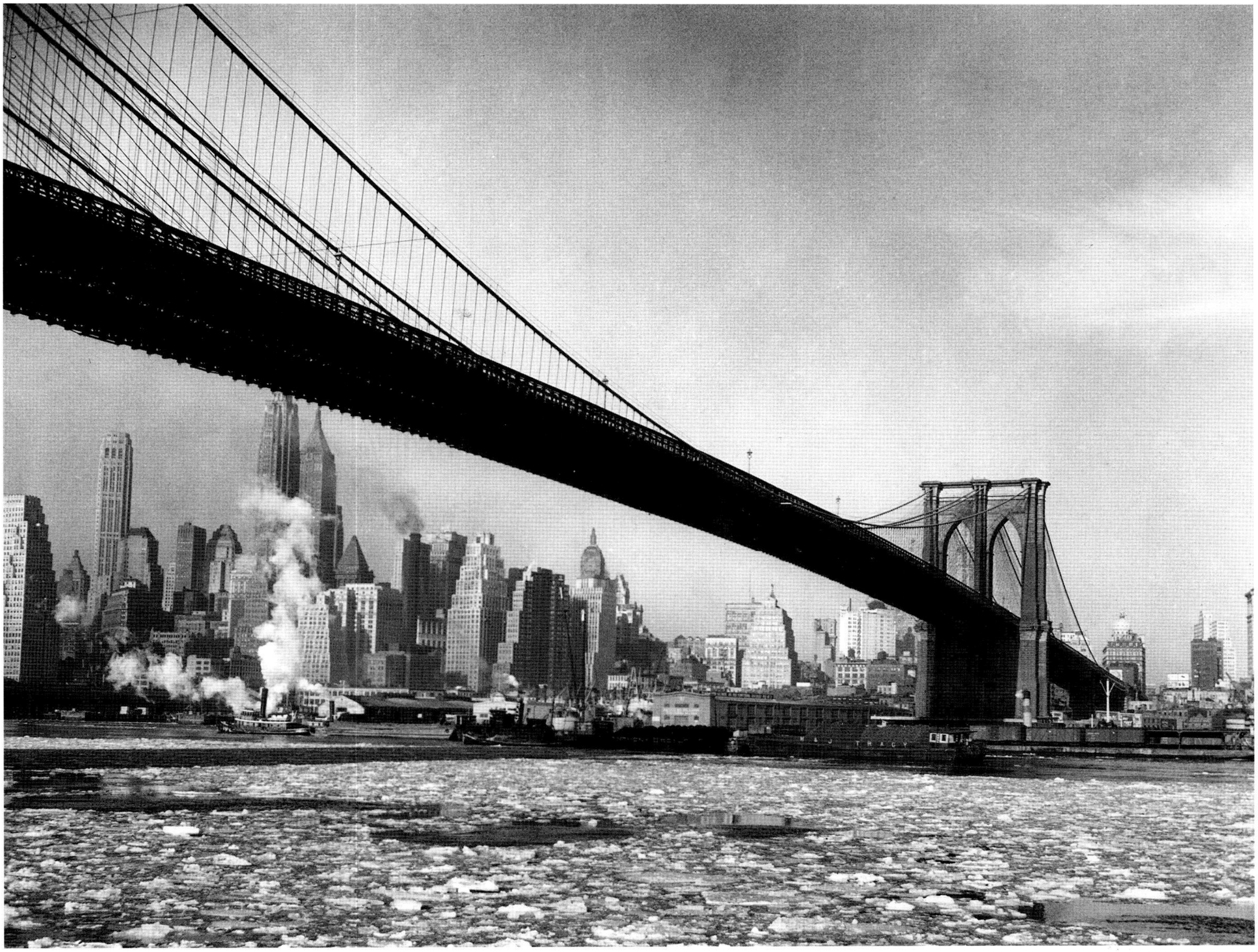

Following Spread: Ice floes: When the East River froze over and ice floes dotted the water, ferries had a difficult time navigating the channel. A similar "ice bridge" convinced the Roeblings that a suspension bridge was the most feasible answer to the ice problem. In this photo, steamboats cross the icy waters while the bridge allows for transportation overhead.

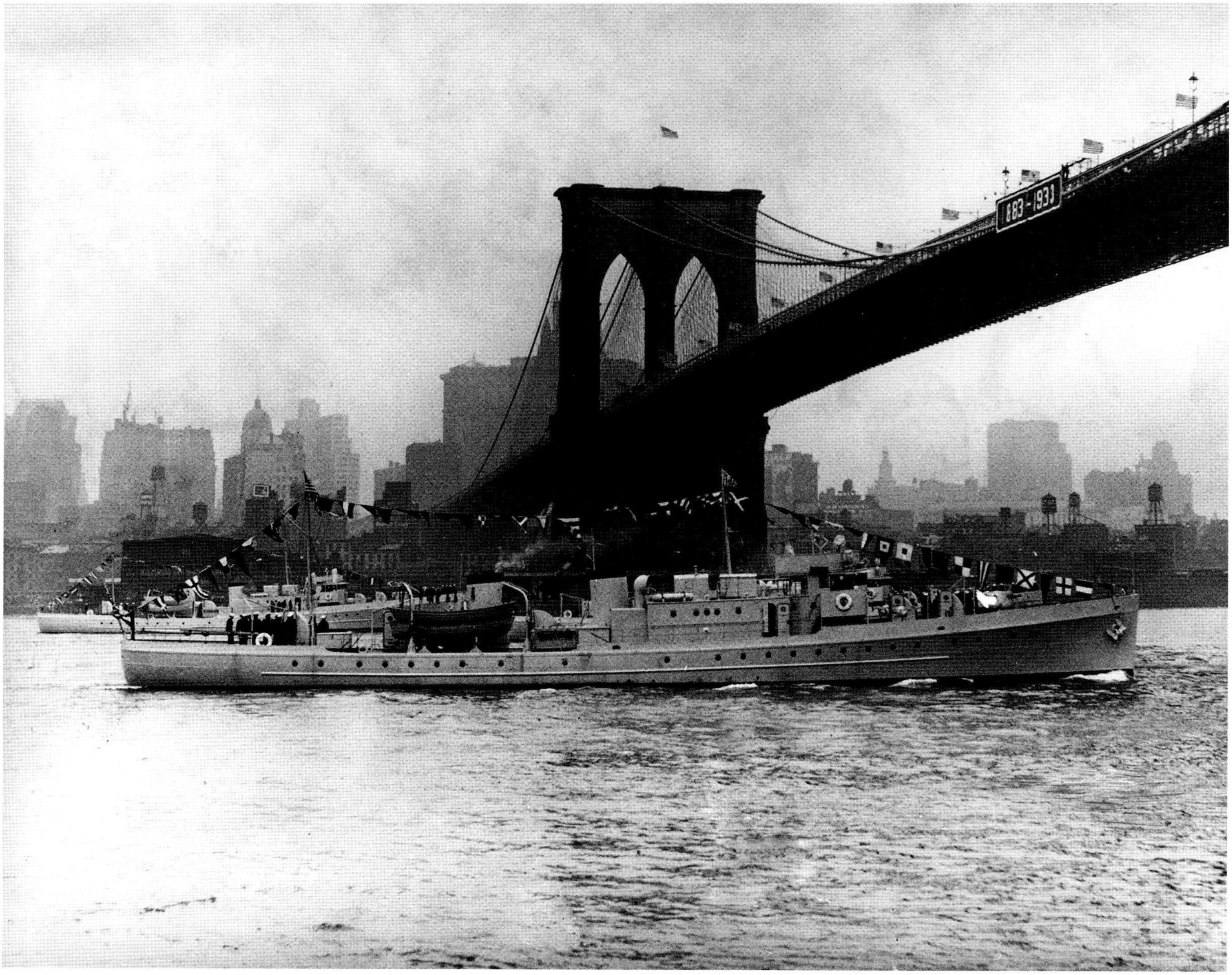

Flag be-decked ship: Celebrating the 50th anniversary of the bridge in 1933, this military ship sails under the structure.

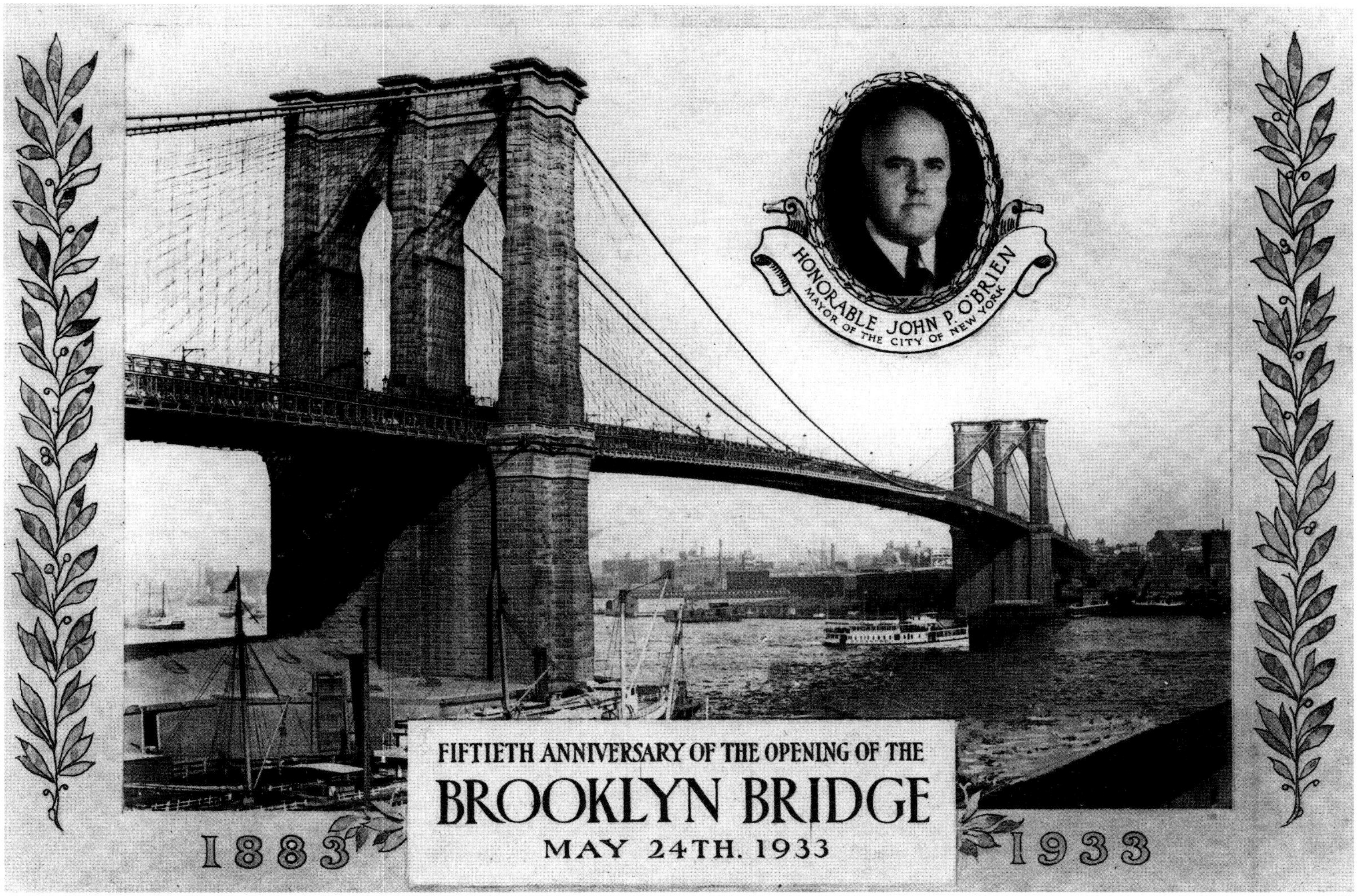

Fiftieth anniversary announcement: This program cover announcement of the 50th anniversary of the bridge includes a headshot of New York mayor John O'Brien, superimposed over an image of the Brooklyn Bridge.

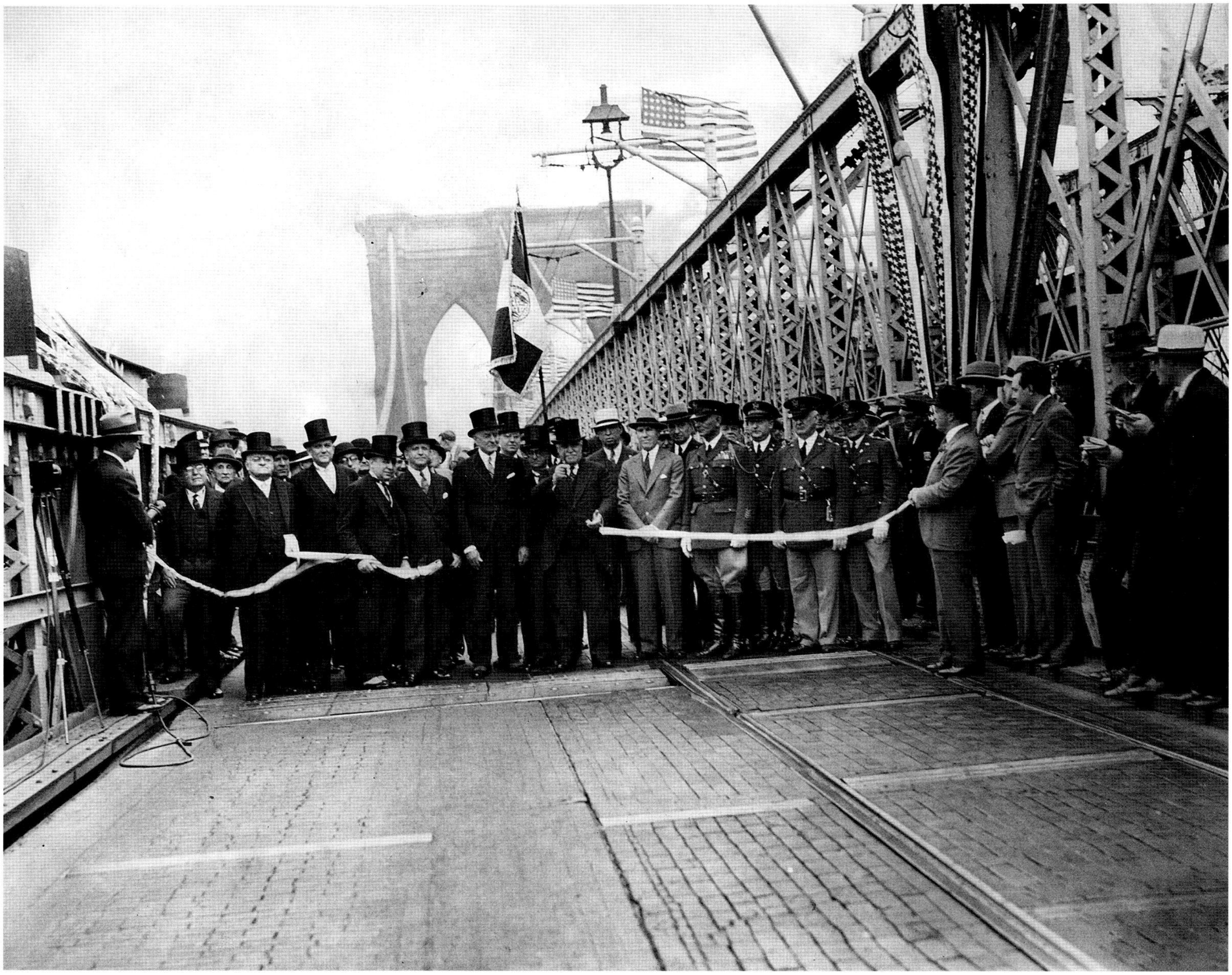

Simulated crossing: In 1933, Brooklyn and New York officials cut a ribbon on the Brooklyn Bridge to celebrate the half century of the bridge's existence.

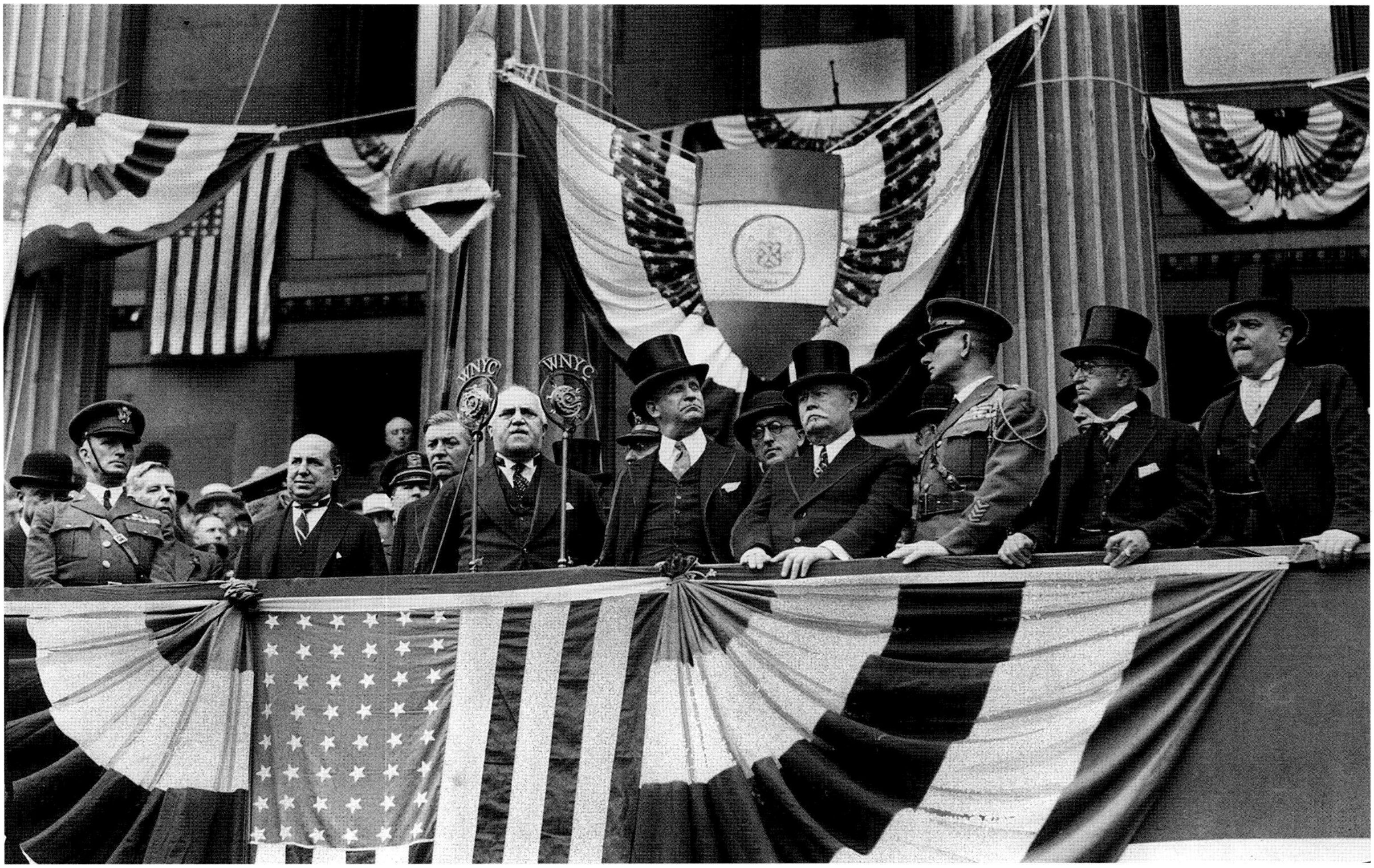

Official ceremony: On a flag-decorated platform in front of Brooklyn Borough Hall, city officials and military personnel commemorate the 50th birthday of the Brooklyn Bridge.

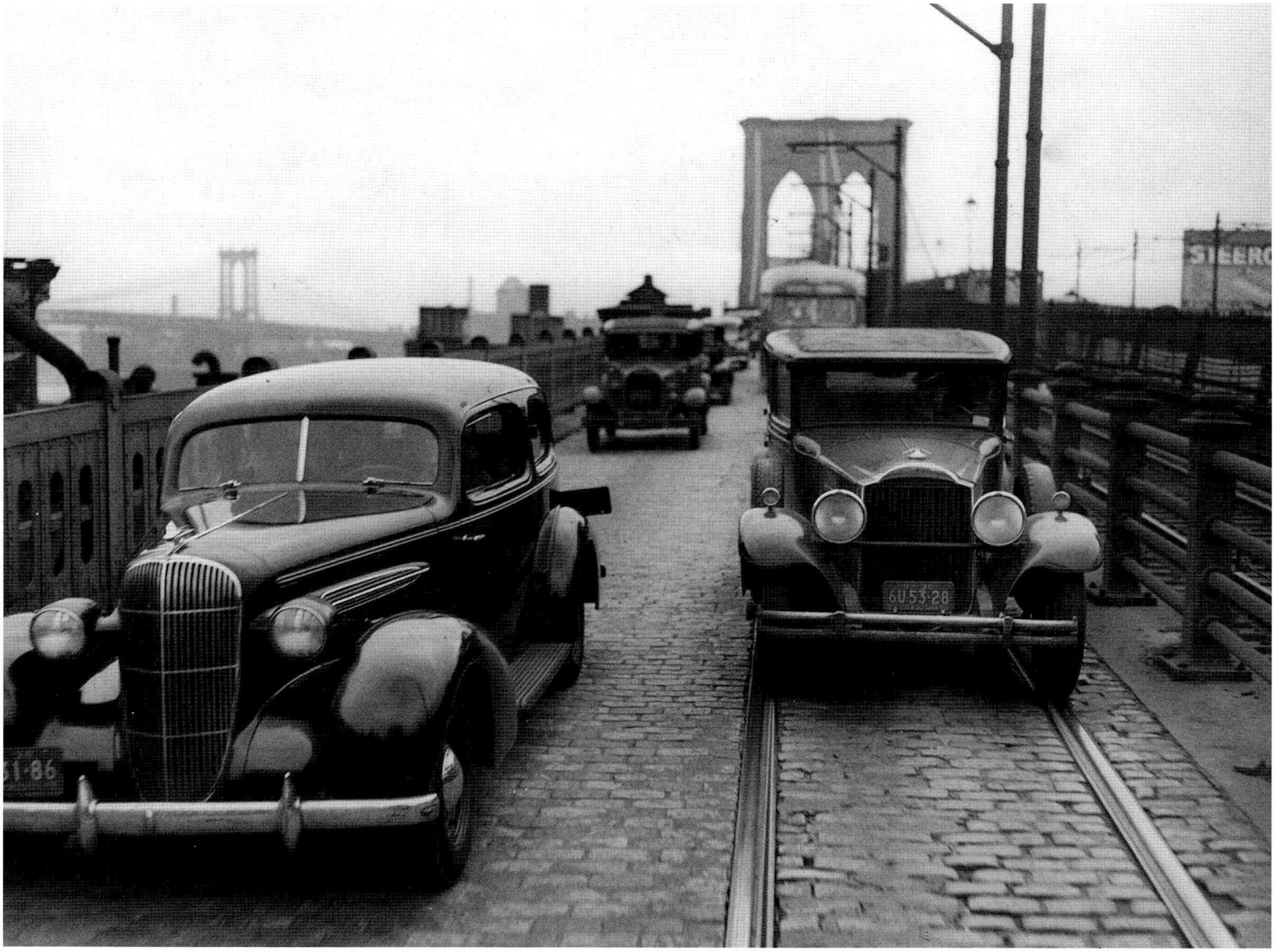

Modern vehicular traffic: Cars cross the Brooklyn Bridge in the 1930s, with trolley tracks and a trolley car behind the automobile on the right. To the left in the distance is the Manhattan Bridge.

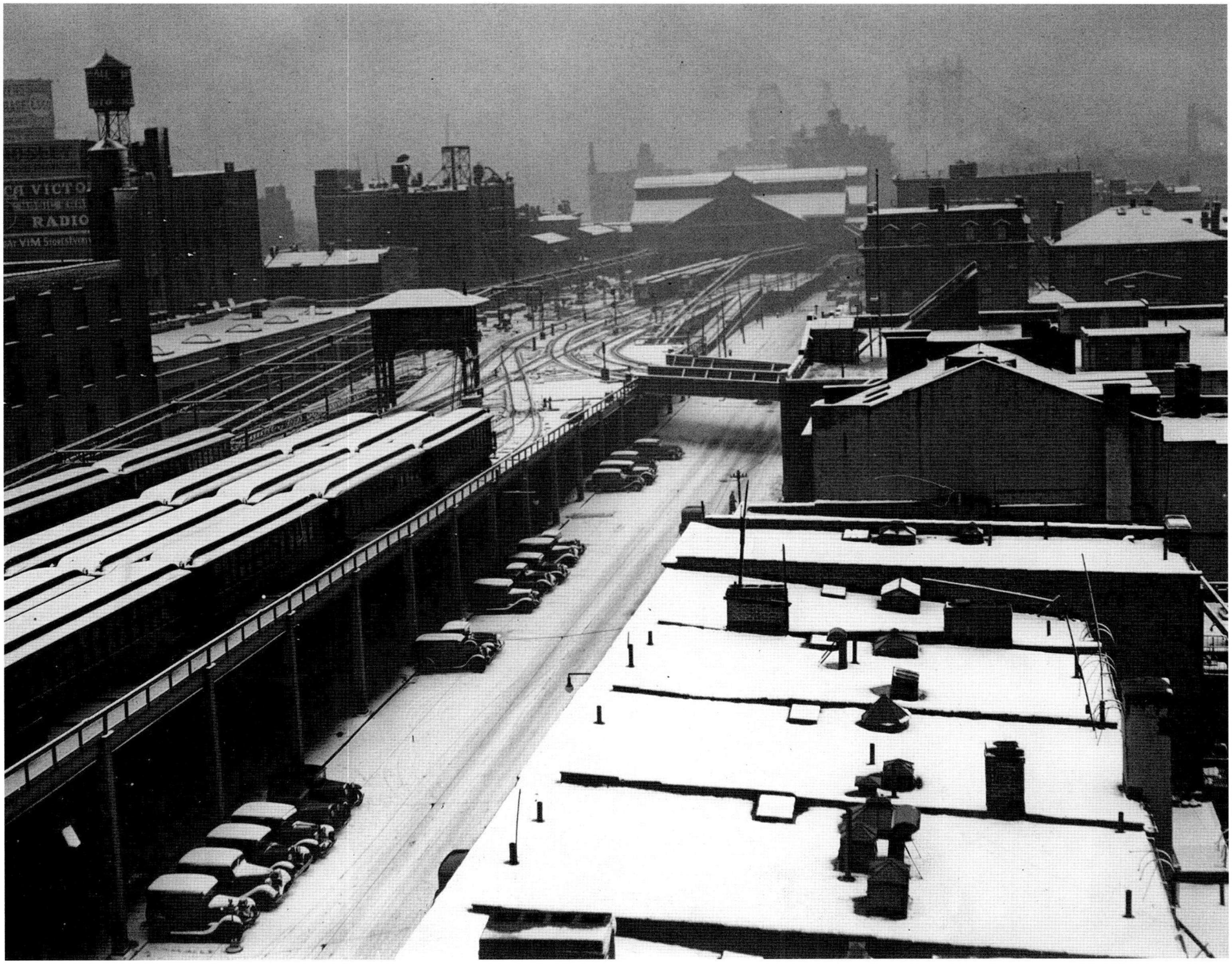

Cadman Plaza and the Brooklyn Terminal: In time, the terminals on both sides of the bridge were extended: the New York Terminal into City Hall Park and the Brooklyn Terminal down Cadman Plaza, as seen here on a snowy day in 1935.

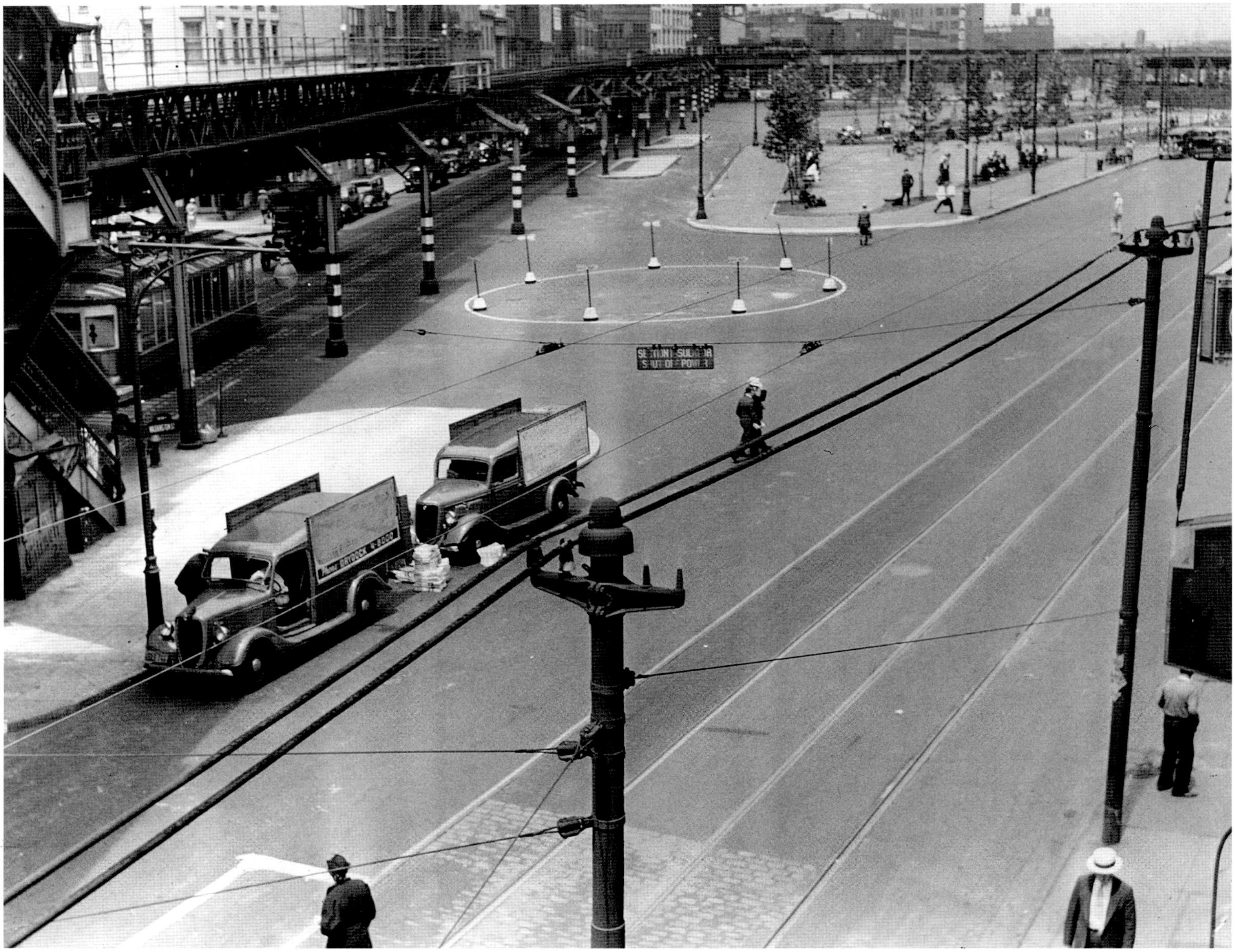

Brooklyn Bridge Plaza: In 1937, the Brooklyn Bridge area began evolving with the new plaza, shown here juxtaposed with the old elevated train.

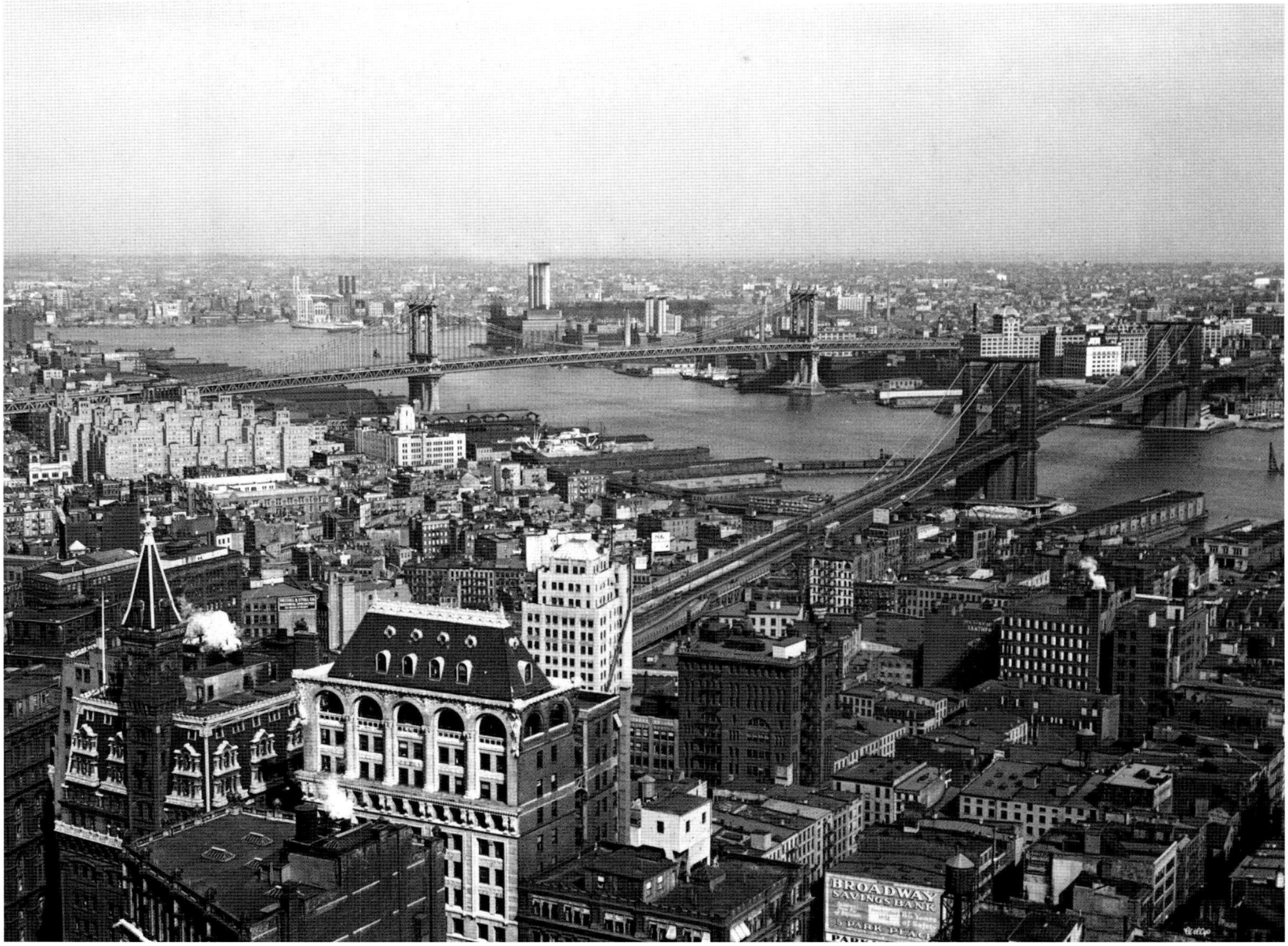

Downtown Manhattan and bridges: This view from a Manhattan skyscraper in 1938 clearly shows several official buildings as well as Knickerbocker Village, south of the Manhattan Bridge, and the Con Edison plant, with its smokestacks, to the north. The Brooklyn Bridge is is visible at right.

Marathon runners on the bridge: Pedestrians weren't the only people on foot who crossed the bridge. Marathon runners compete with trolleys and traffic in general in this 1940s photo.

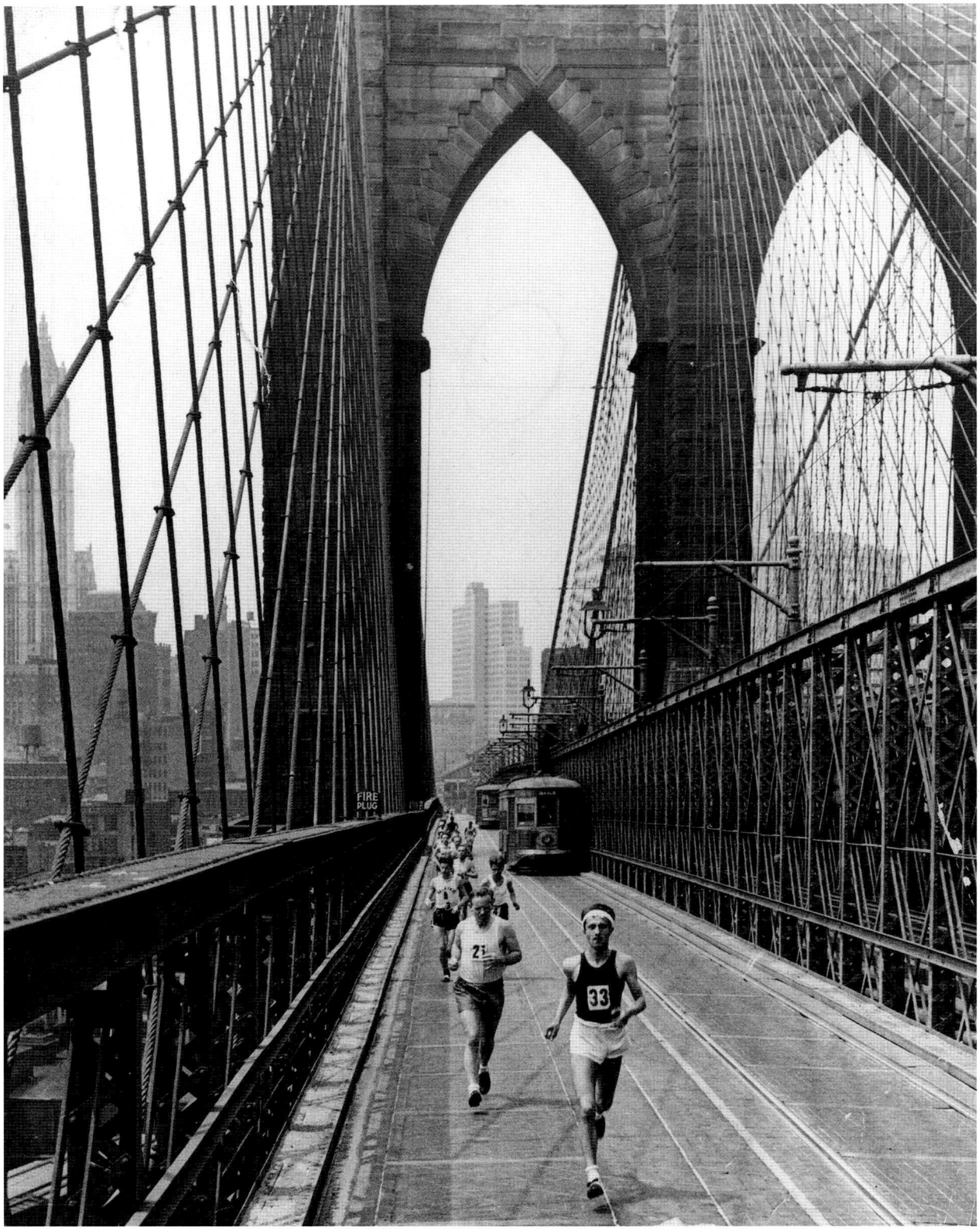

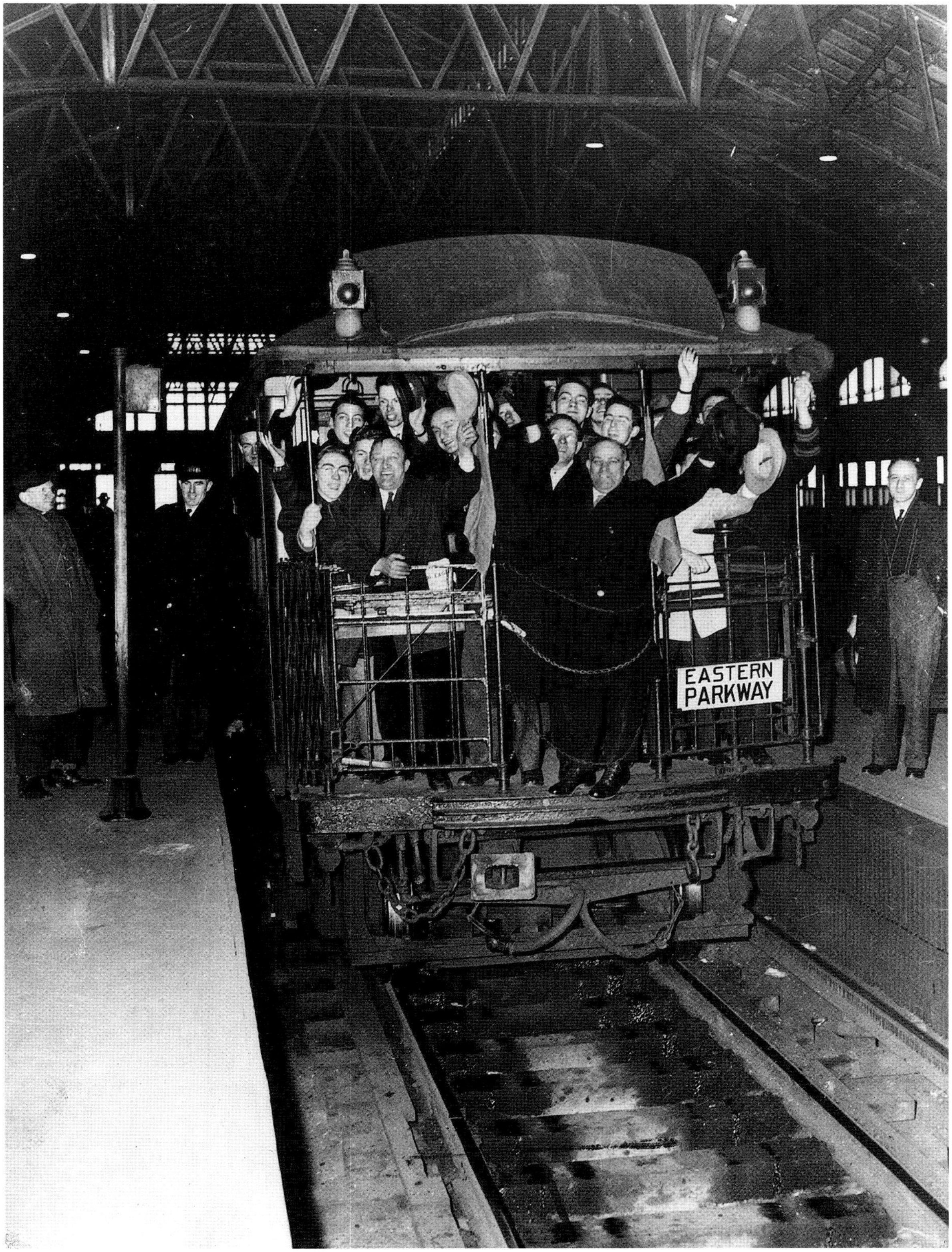

Last train across the bridge: In 1944, trains crossing the Brooklyn Bridge were discontinued. The subway had taken over the route.

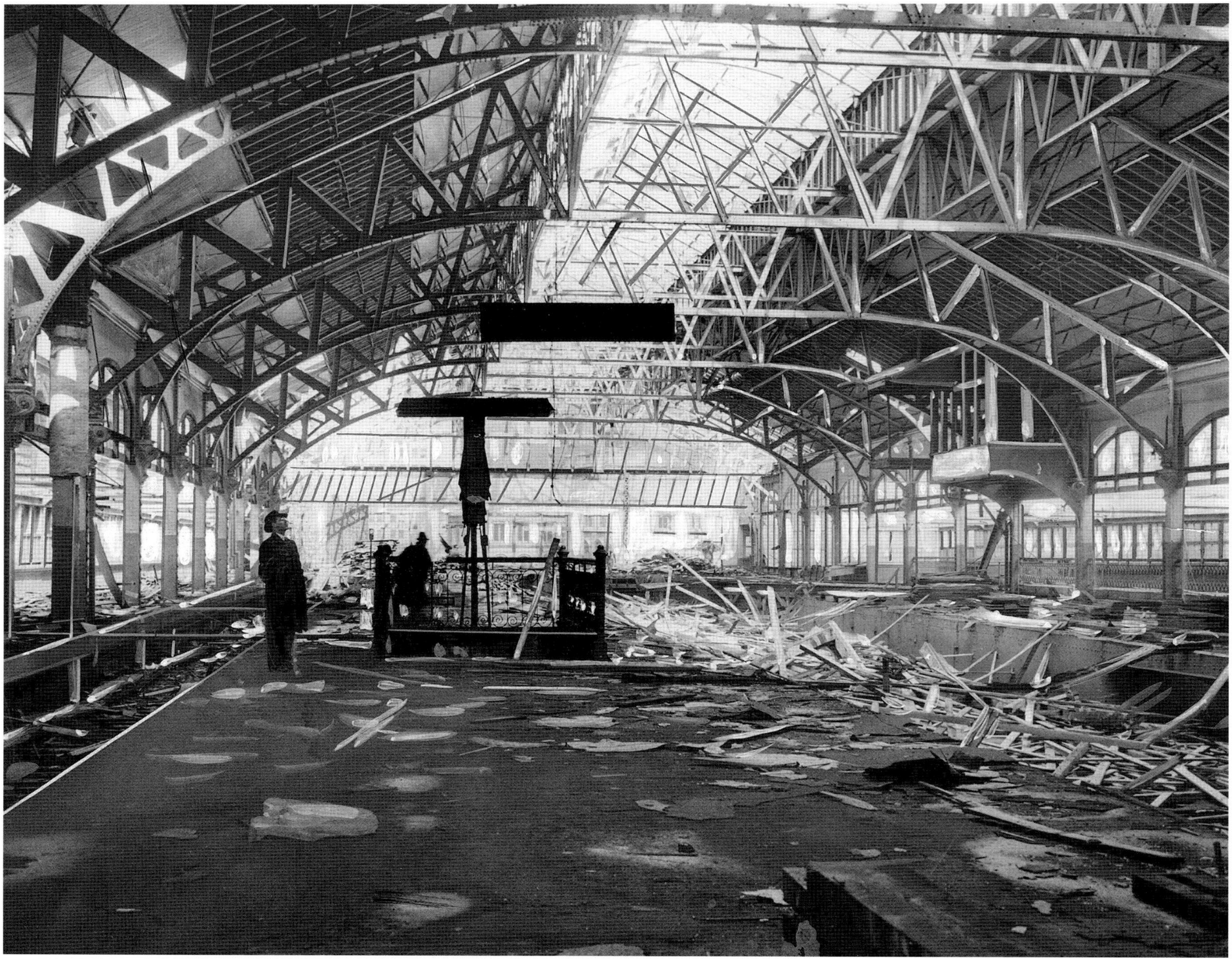

Demolition of Brooklyn terminal: With the Brooklyn terminal demolished in 1944, Cadman Plaza Park would soon be created in its place.

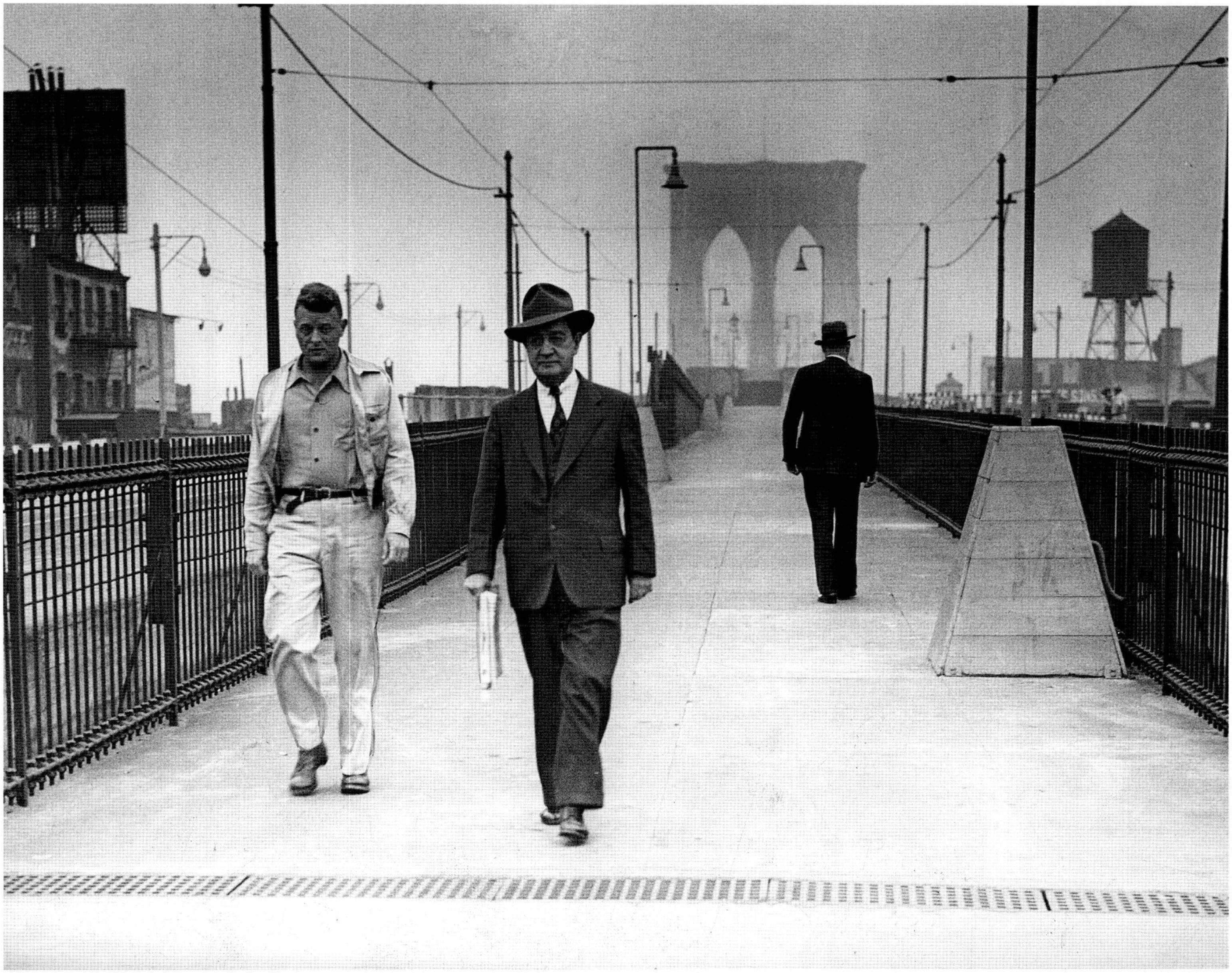

First pedestrians over modernized bridge: In 1945, photographers snapped pictures of the first pedestrians to cross the new walkway after the train tracks were removed.

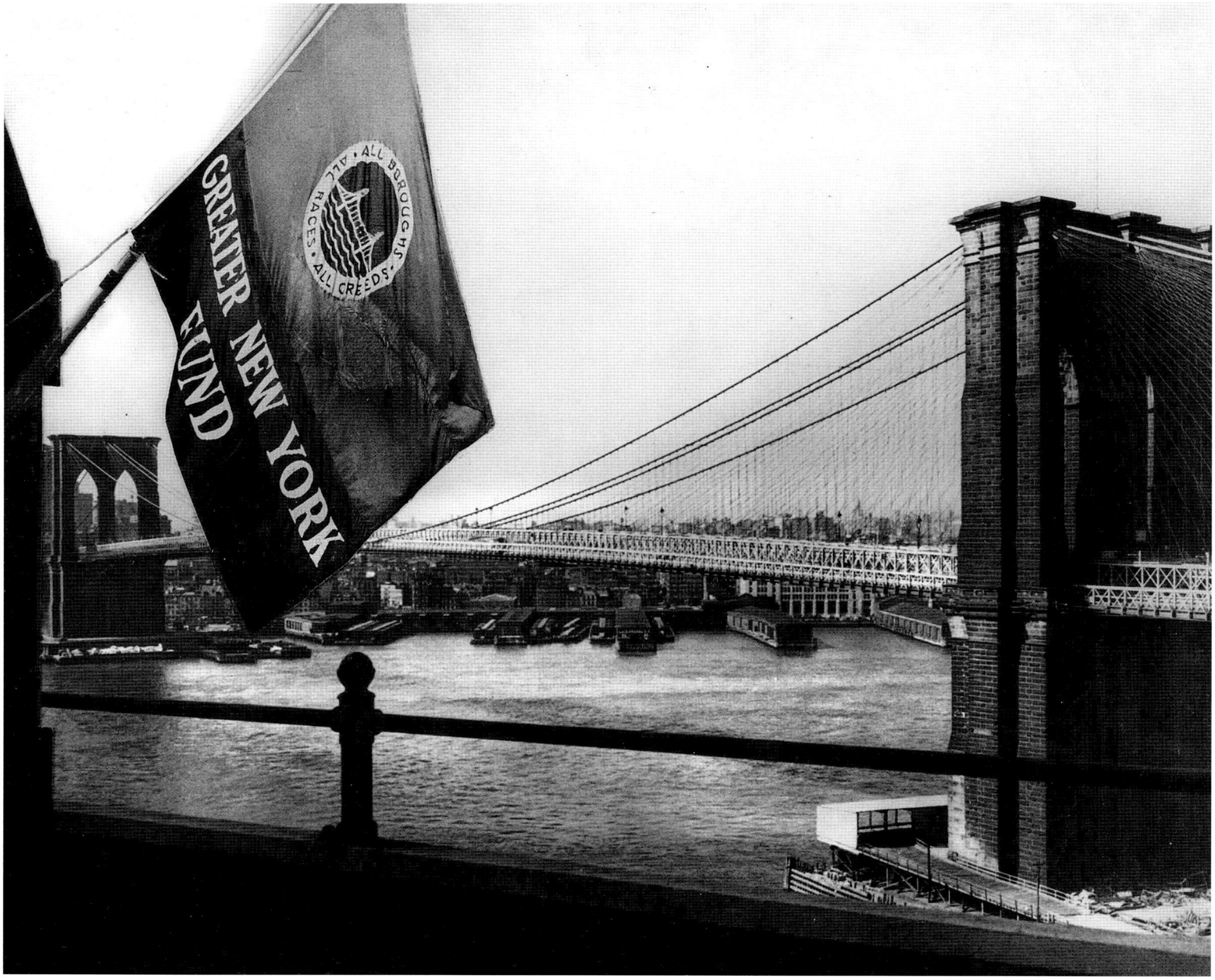

64th anniversary: The Brooklyn Bridge's 64th anniversary was celebrated in 1947.

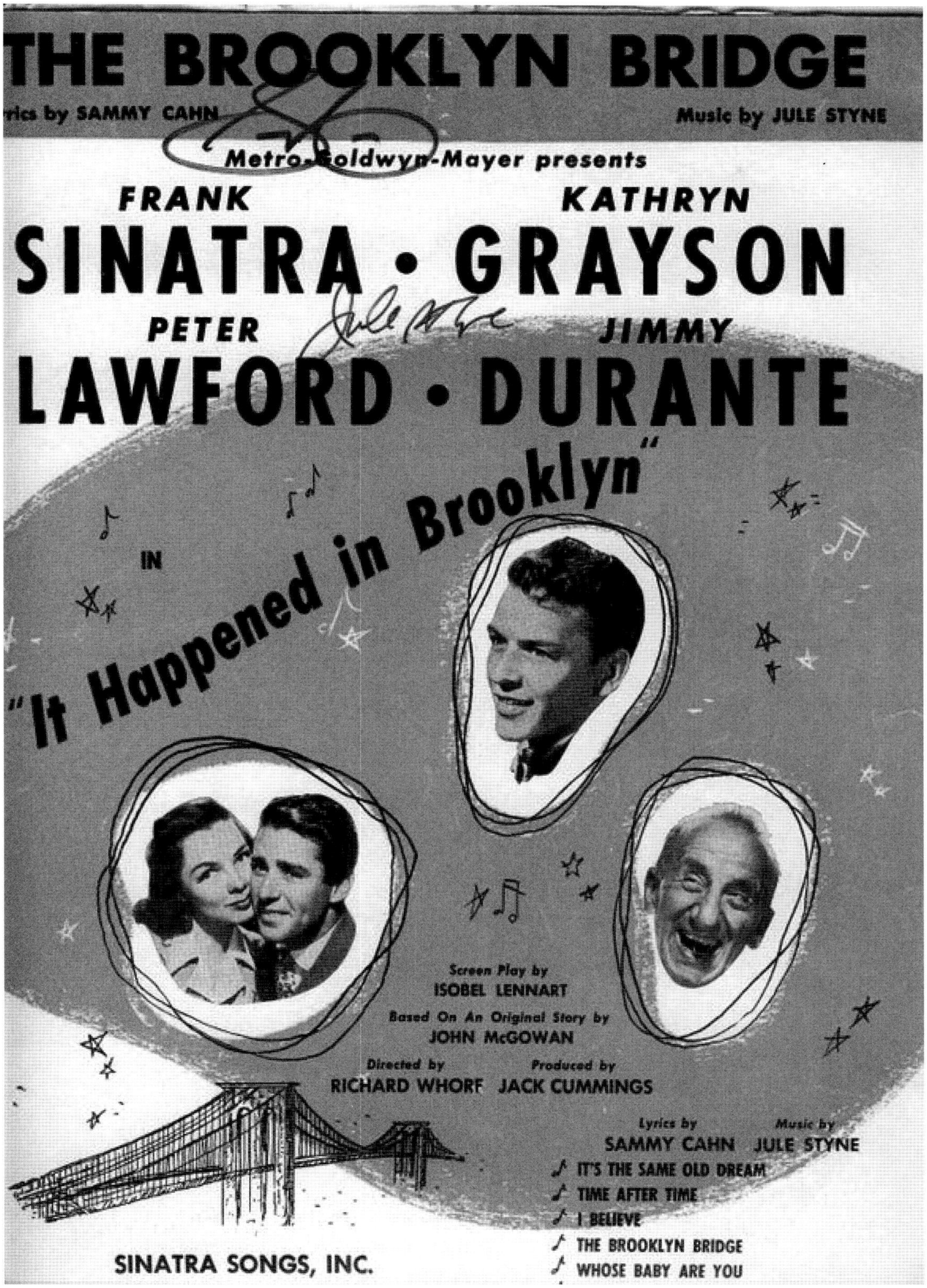

"The Brooklyn Bridge": This song, made famous by Frank Sinatra in the 1947 movie *It Happened in Brooklyn,* is just one of more than 15 songs written in tribute to the bridge. This copy is signed by the composers Sammy Cahn and Jule Styne.

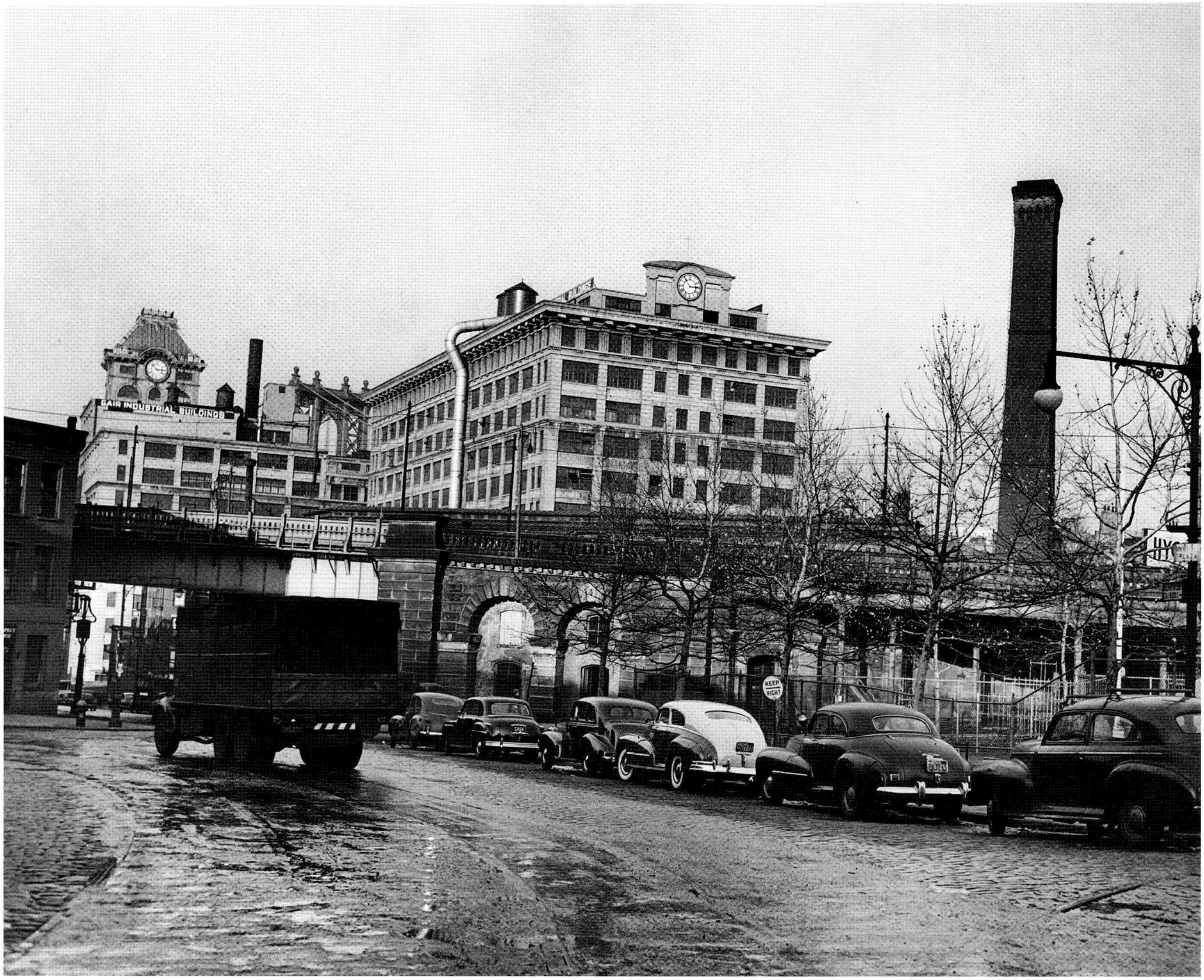

New vehicular approach: With the construction of the Brooklyn-Queens Expressway in 1949, new approaches led to the Brooklyn Bridge.

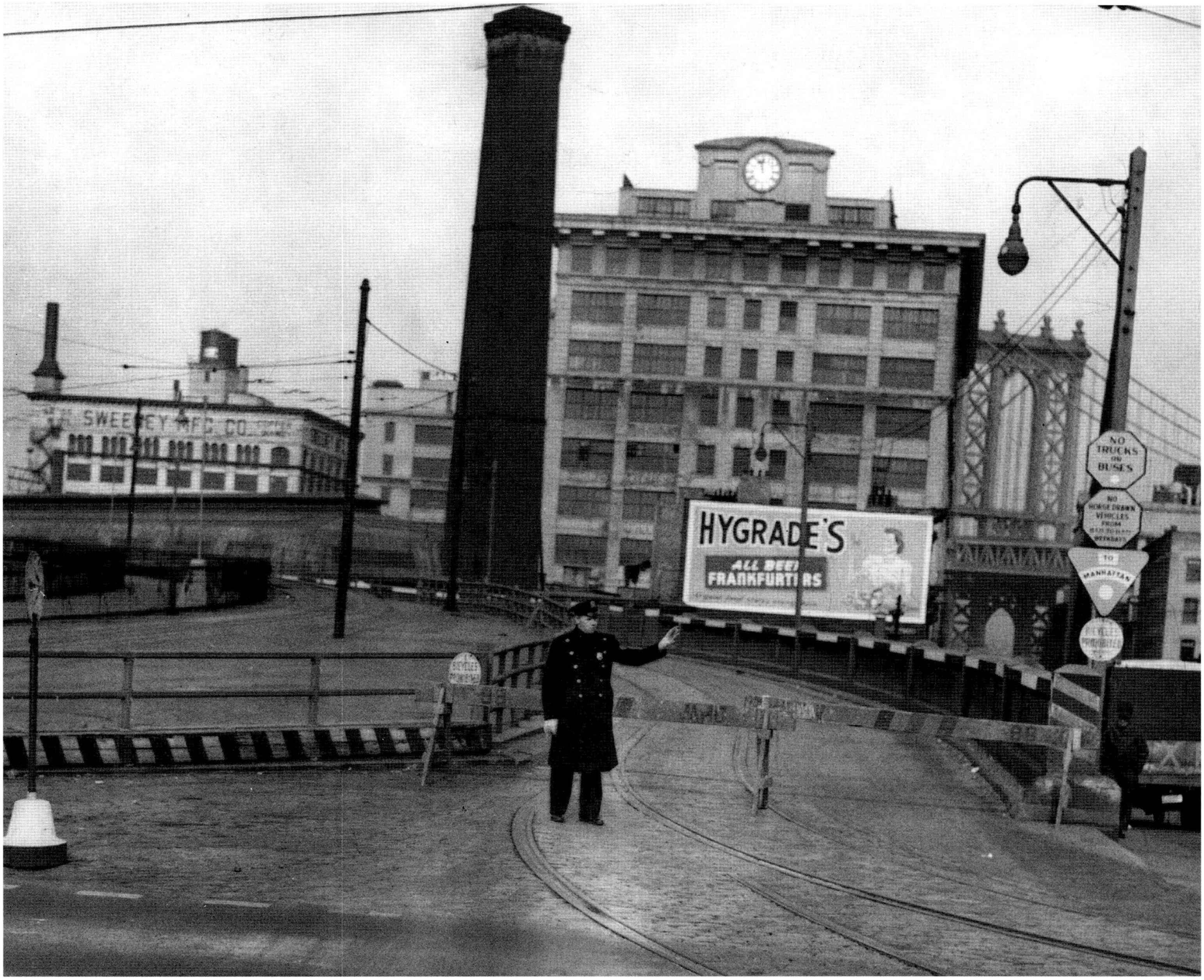

Closed: Signs close Brooklyn Bridge to traffic for repairs and modernization. The Manhattan Bridge is visible to the right.

Bridge painter: This bridge painter takes a lunch break during routine maintenance in 1951.

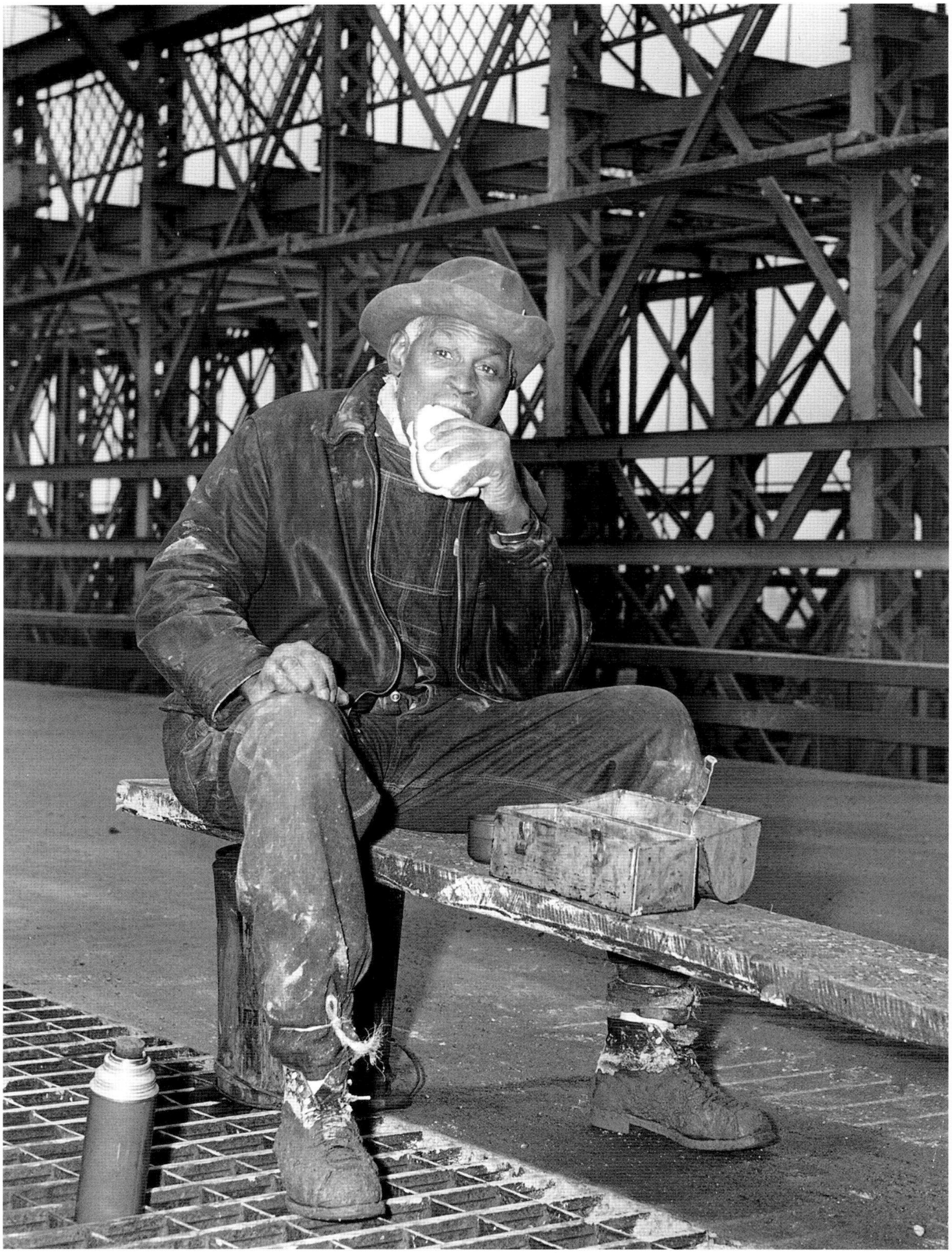

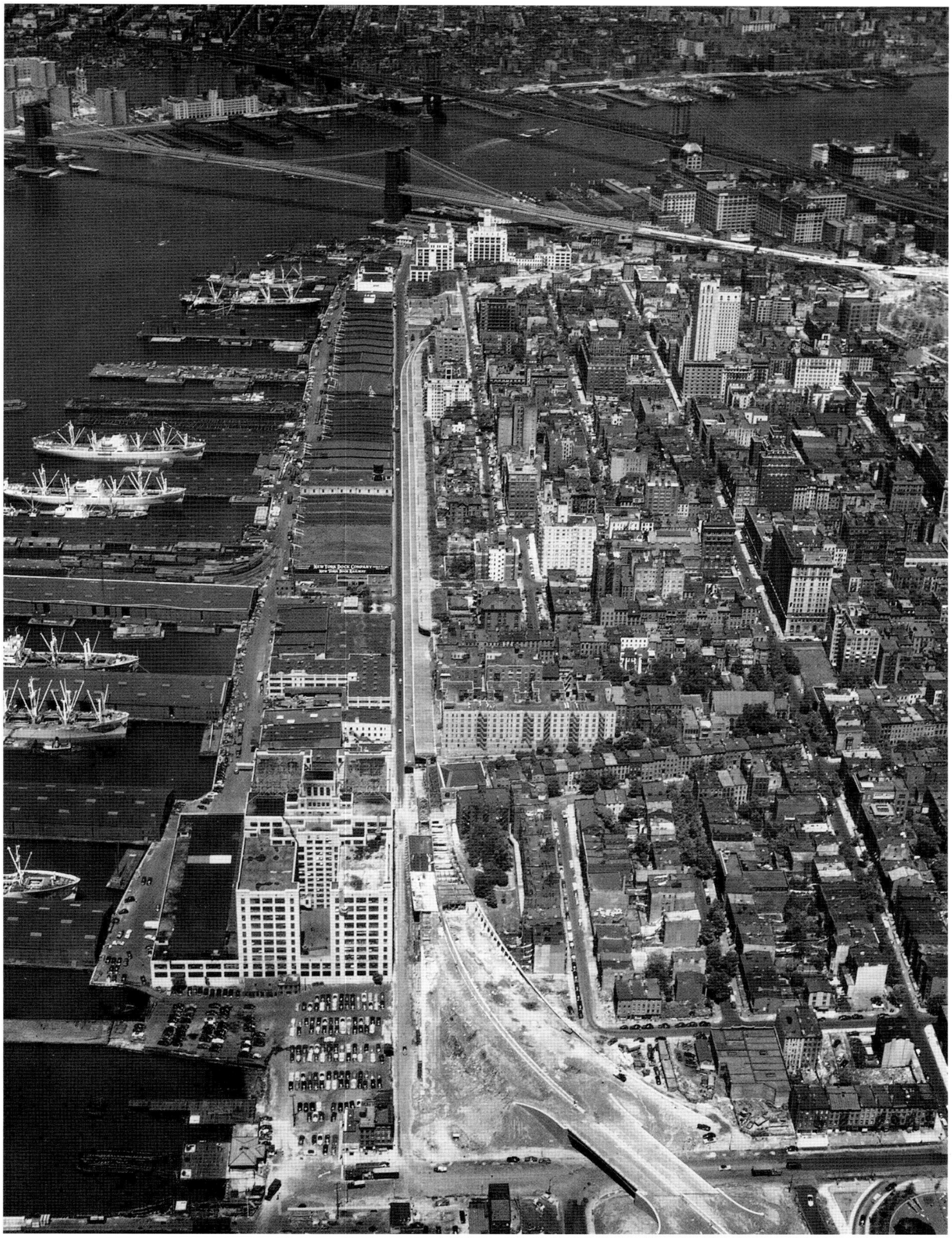

New expressway: Finishing touches to the Brooklyn-Queens Expressway in 1953 permitted automobiles to make a smooth transition from the expressway to the bridge.

Following Spread: Inspecting the bridge: During renovation in 1950, inspectors check the Brooklyn Bridge for significant structural flaws. They didn't find any.

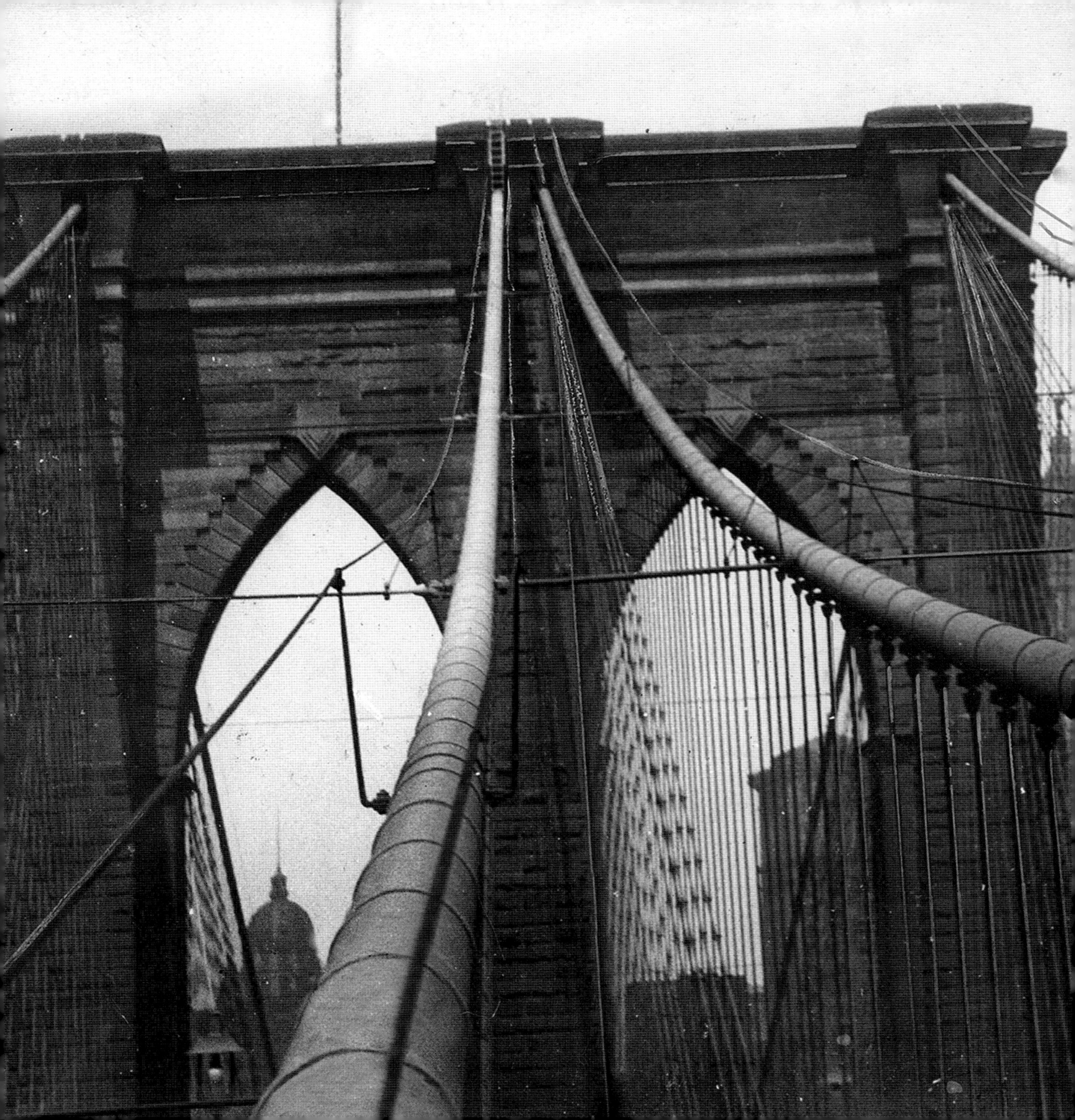

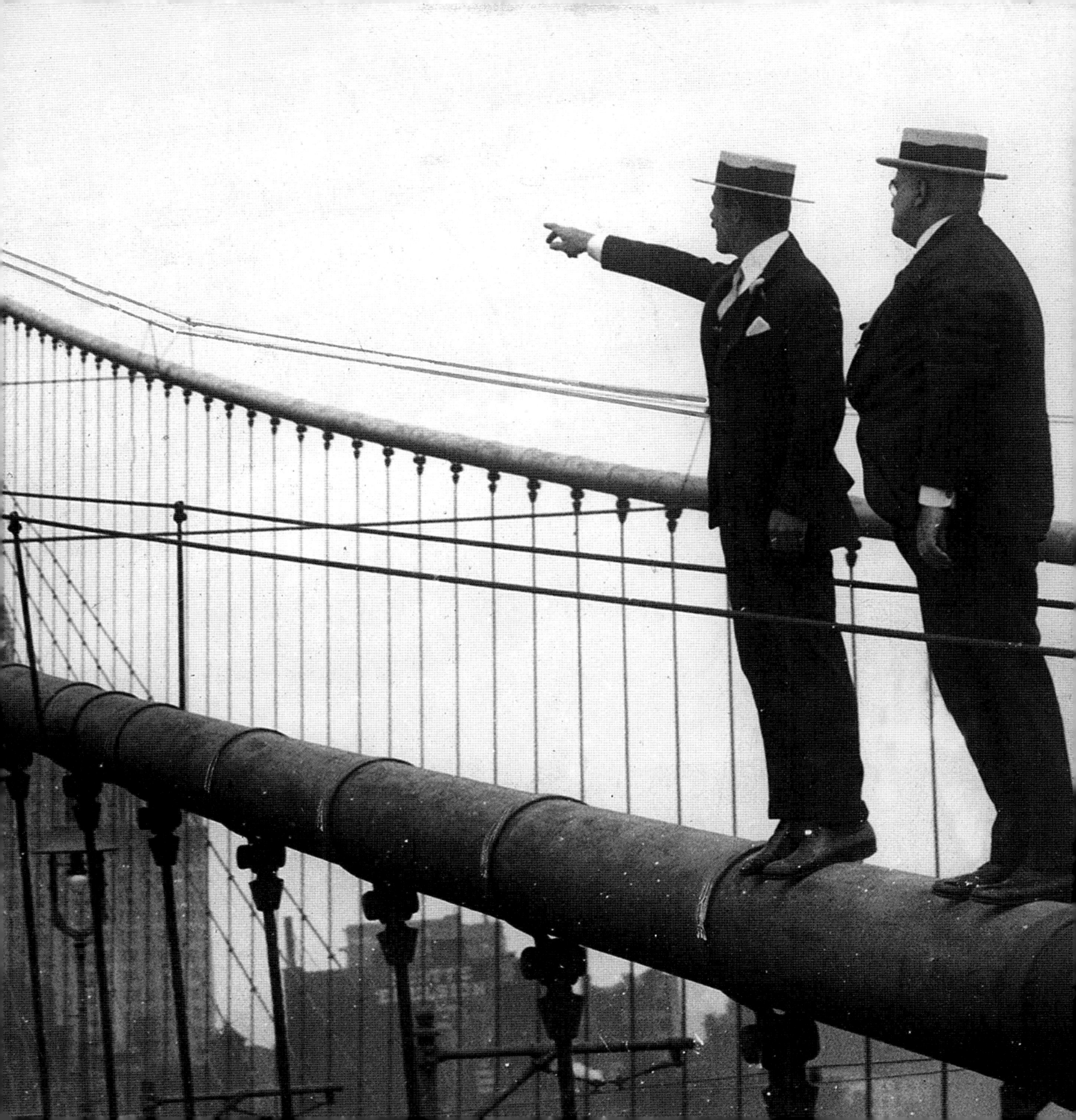

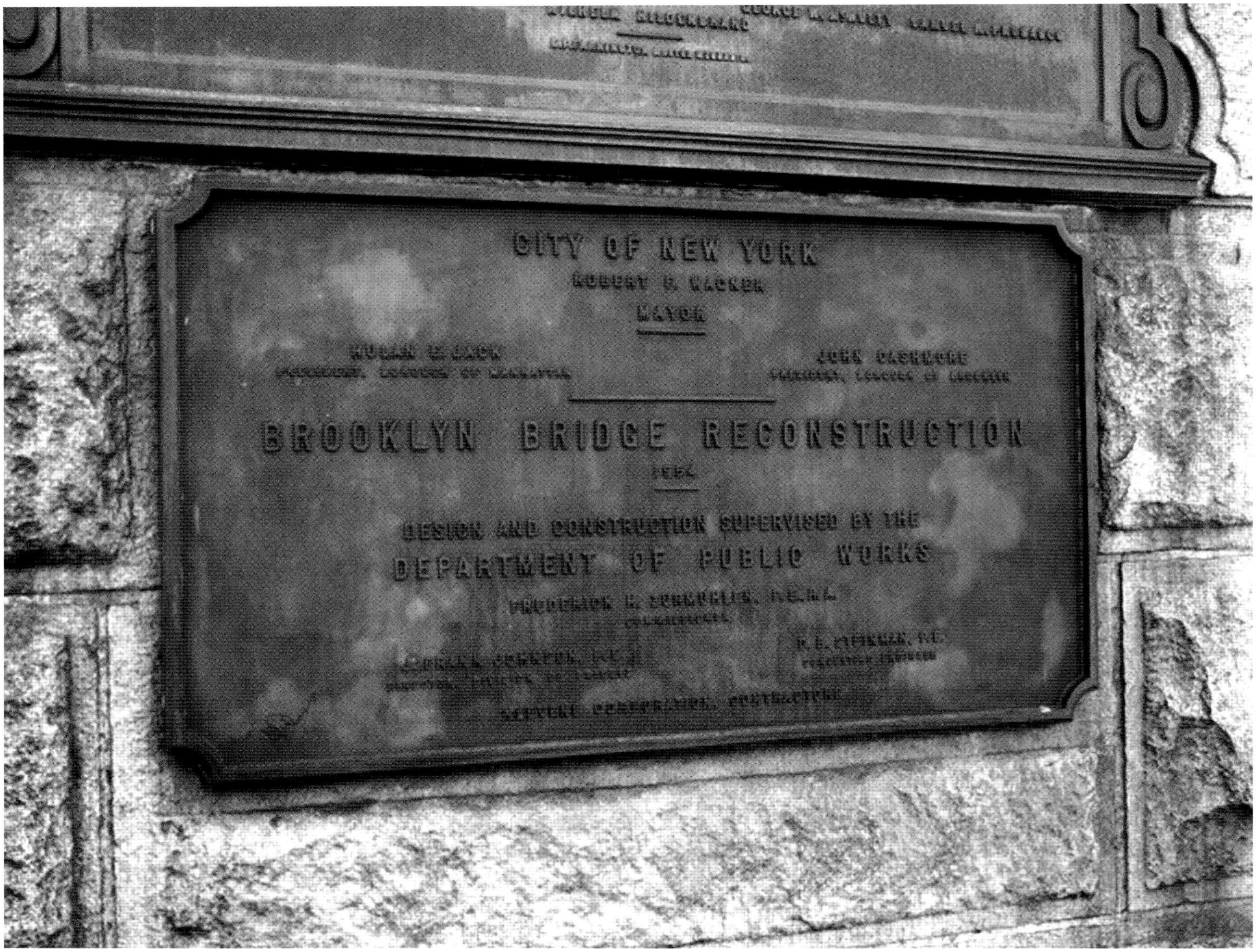

Restoration plaque: This plaque pays tribute to the mid–twentieth century restoration of the Brooklyn Bridge.

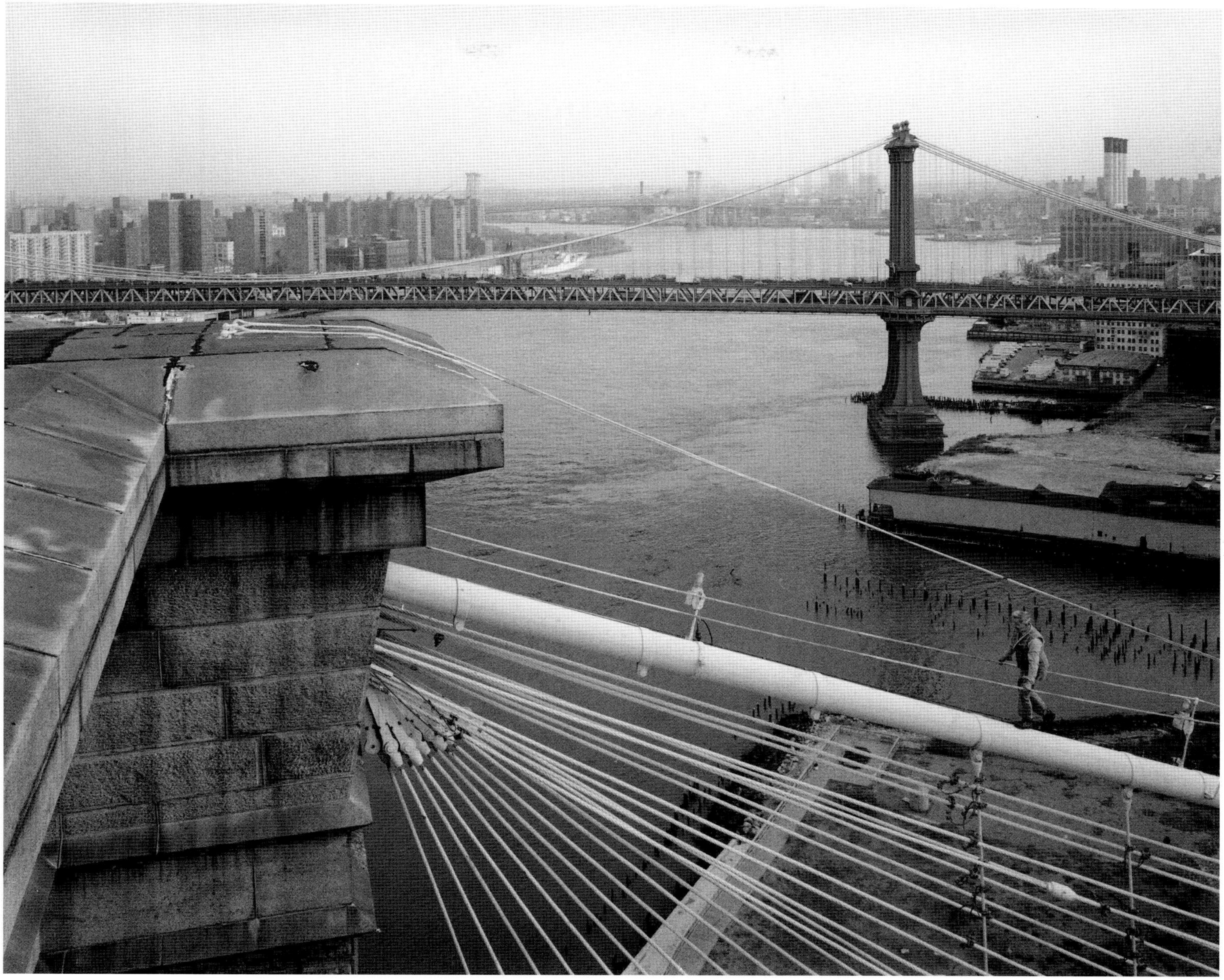

This close-up of a workman climbing the Brooklyn tower cables offers a view of the Manhattan Bridge (completed in 1909). To the right behind the Manhattan Bridge is the Brooklyn Navy Yard, and in the distance to the north stands the Williamsburg Bridge (completed in 1903). Pilings that once supported a ferry pier are visible in the water behind the workman.

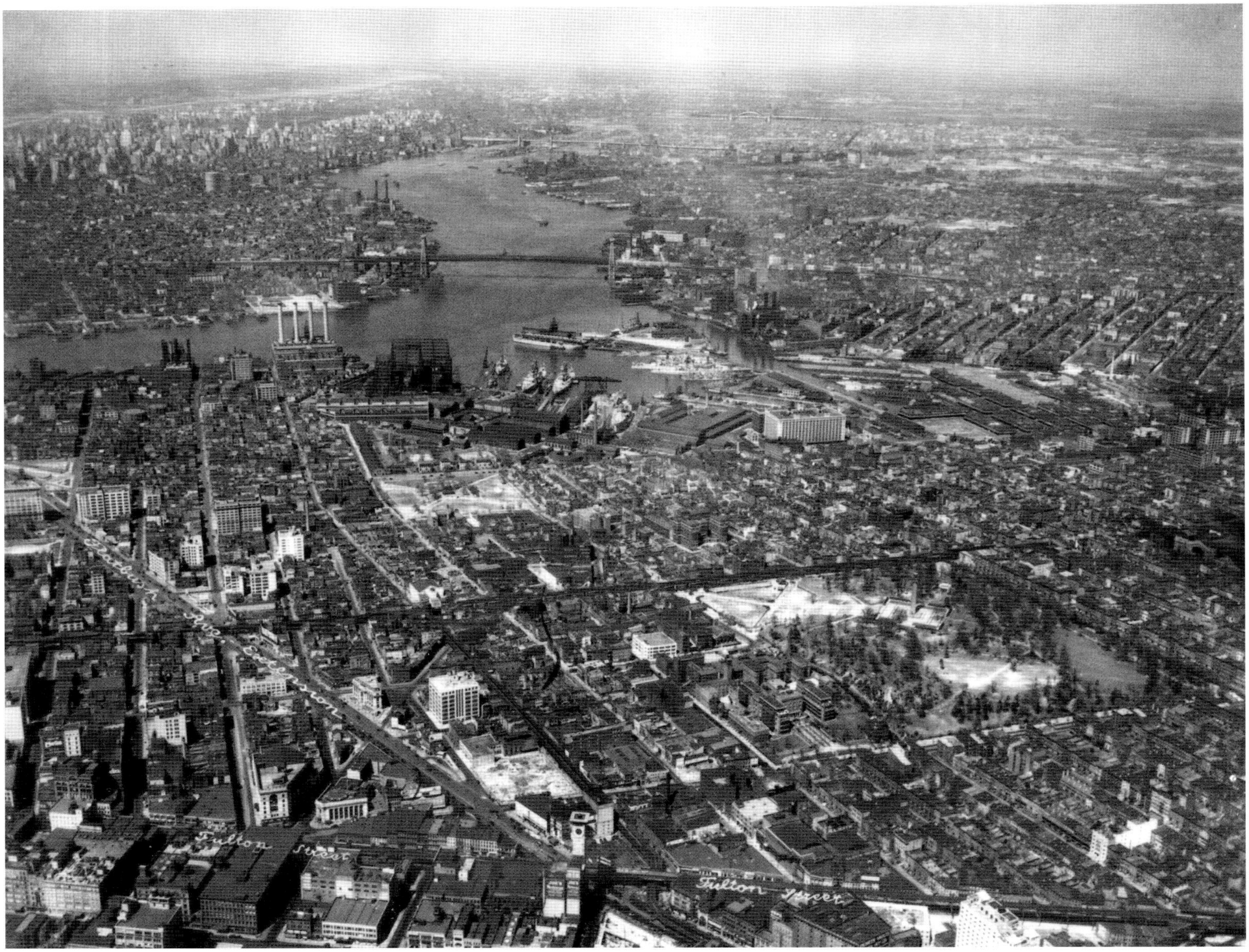

Bridges and Navy Yard: This 1970s aerial view encompasses the area surrounding the Williamsburg Bridge, built 20 years after the Brooklyn Bridge opened. Also visible are the Navy Yard and the scope of downtown Brooklyn.

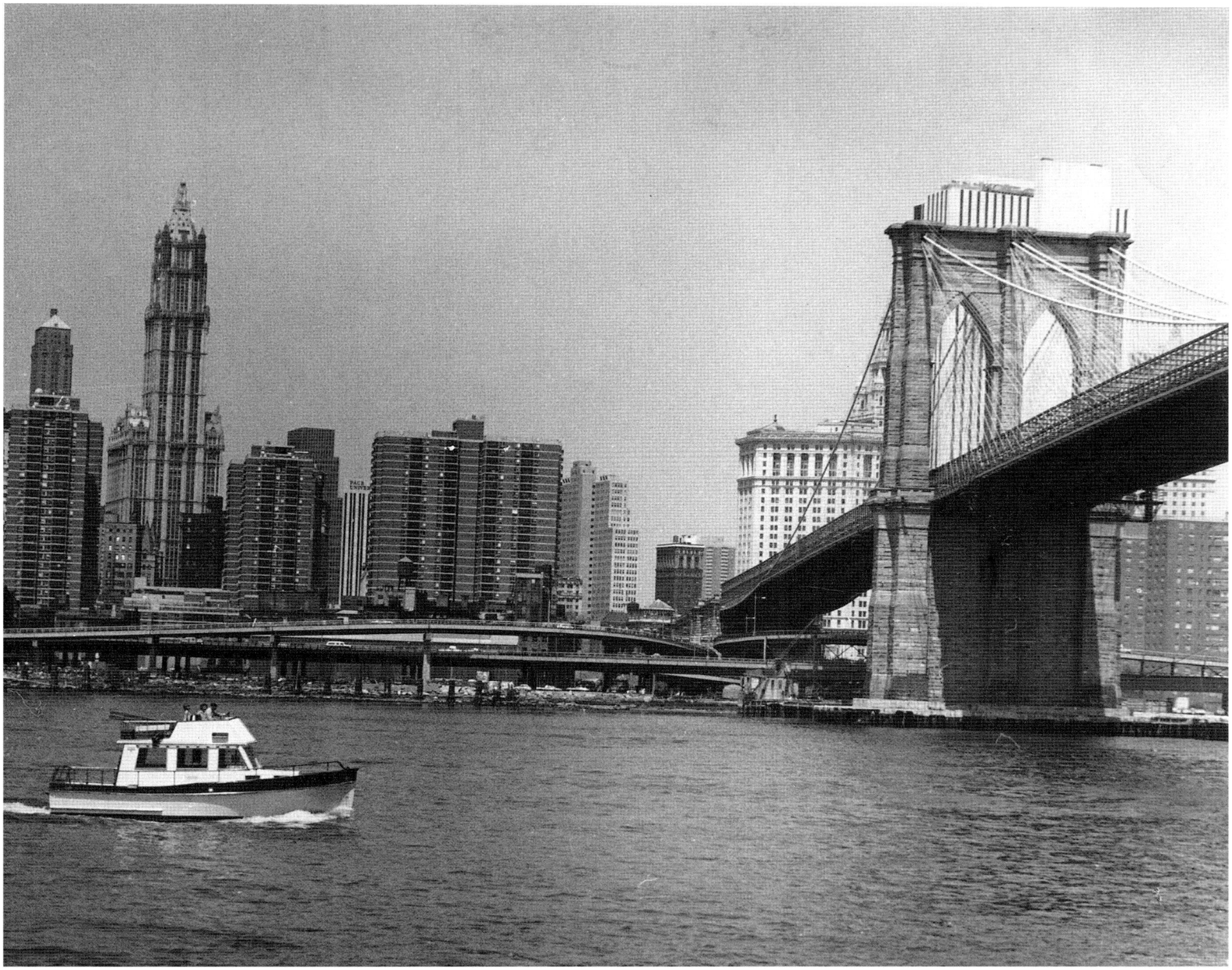

Speedboat: As the Brooklyn Bridge maintains its stately silence, a private speedboat prepares to pass underneath.

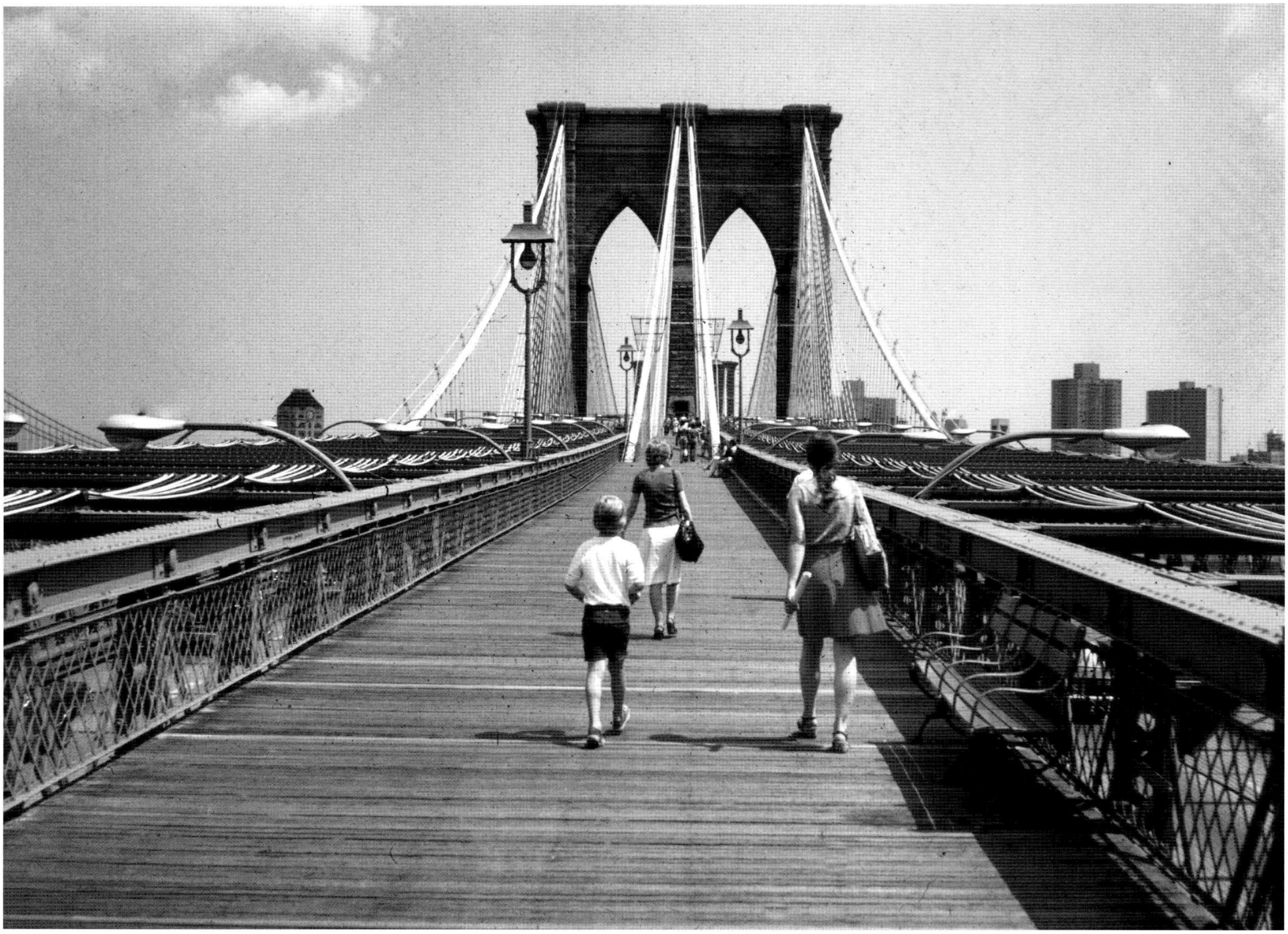

Crossing the bridge in 1976: Pedestrians cross the Brooklyn Bridge during America's bicentennial celebration.

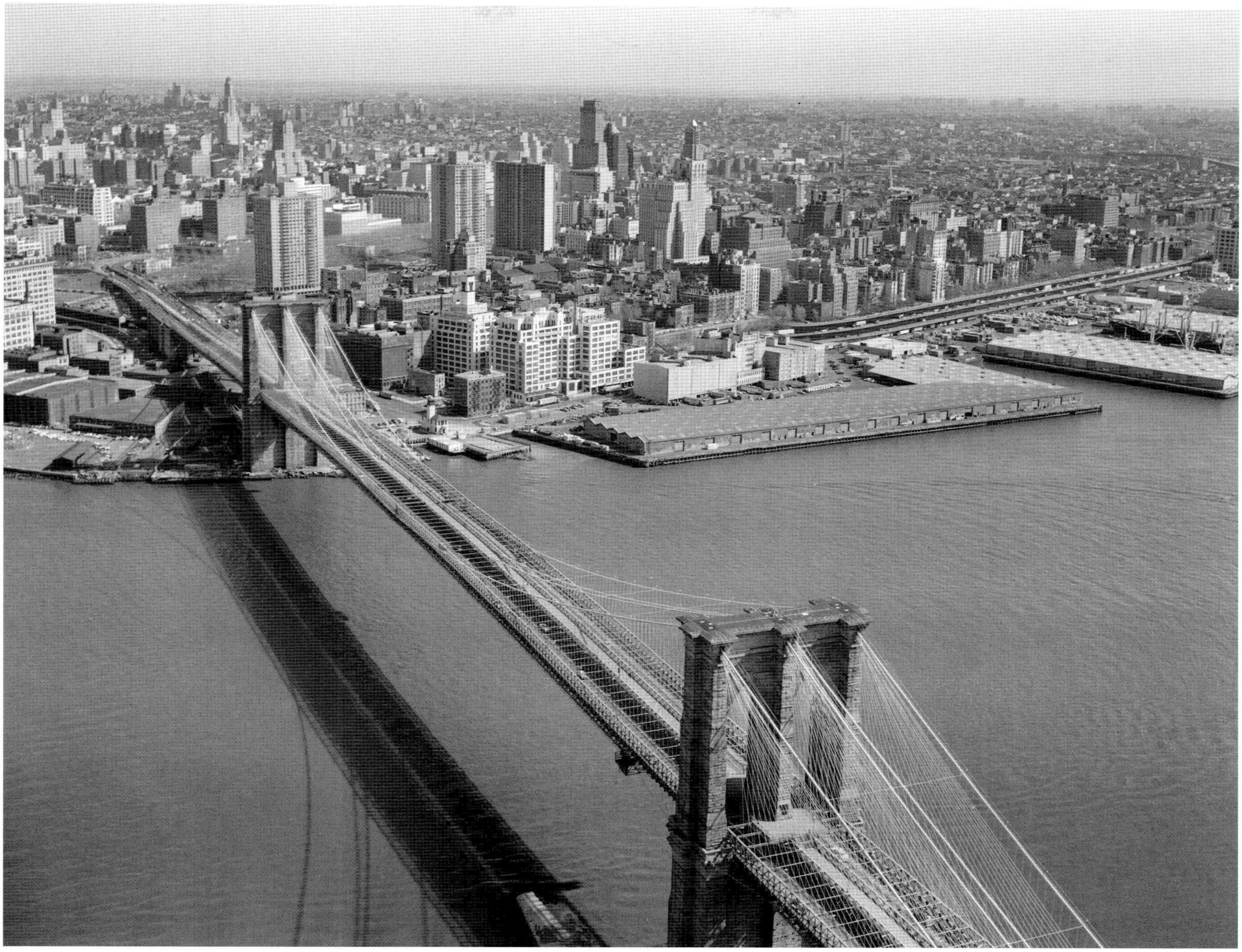

A 1978 view of the Brooklyn Bridge leading to the Brooklyn waterfront and Brooklyn Heights. Captured in the photograph is the Squibb Building, now headquarters for Jehovah's Witnesses, and, to the right, the path of the Brooklyn-Queens Expressway. Below is the Fulton Ferry Pier with a firehouse, and to the left are Civil War–era warehouses, now part of DUMBO and Brooklyn Bridge Park.

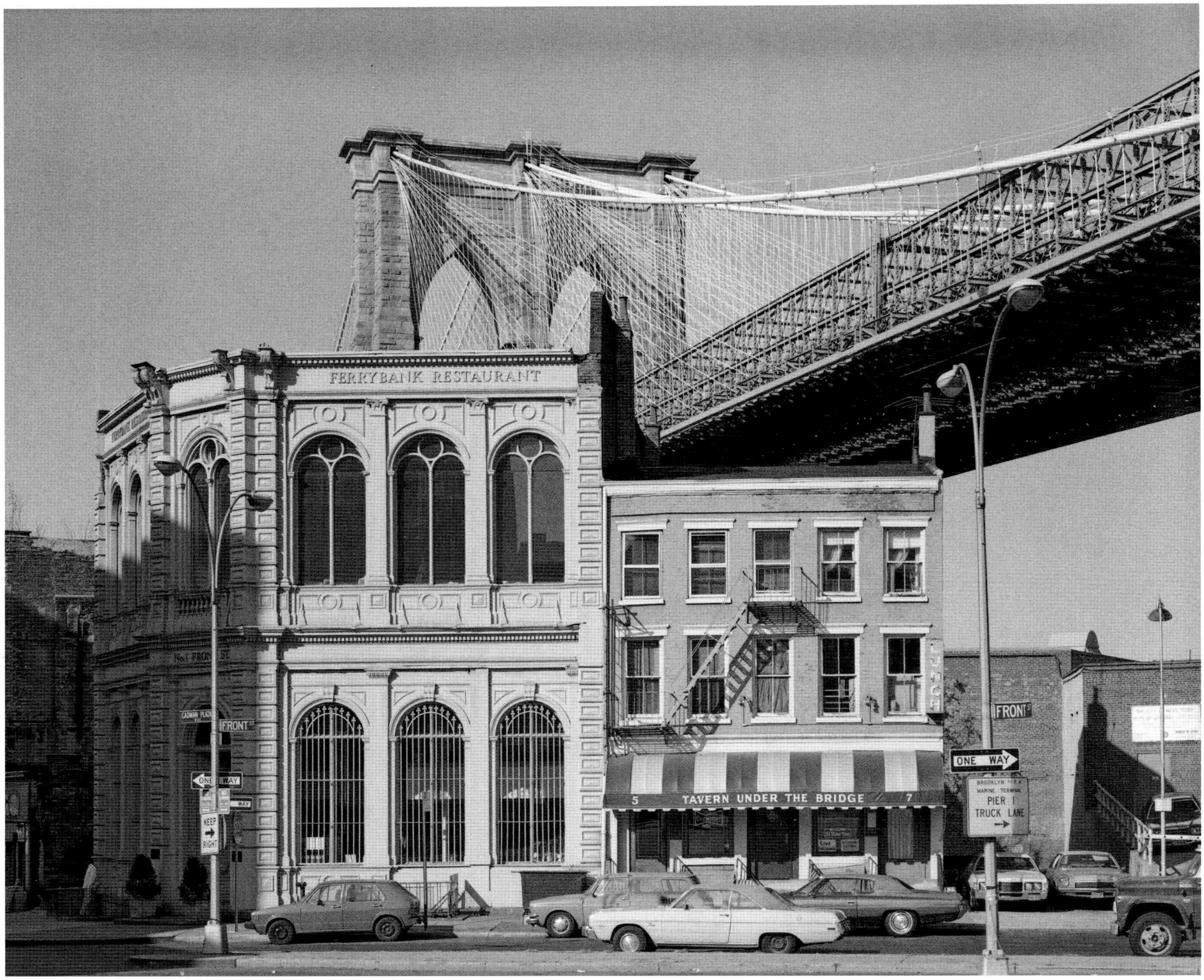

A street entering DUMBO at Front Street and Cadman Plaza (now Old Fulton Street). This 1982 photo of the Fulton Ferry Historical District shows the Brooklyn Bridge hovering over the former Long Island Safe Deposit Company building, at the time the Ferrybank Restaurant.

A Bridge for All Time: Magnificence (1983–2001)

As the years have passed, the Brooklyn Bridge has grown in stature while the city surrounding it has spread. As the skyline grew, the bridge redistributed its transportation. Over time, proposals to add or alter the bridge and its approaches have been effected or forgotten. As traffic moved from public conveyances such as omnibuses, trains, trolleys, and horse-drawn vehicles to private automobiles, the bridge adjusted to the new patterns. The one constant, however, has been the pedestrian walkway, or the Promenade.

Throughout this evolution, the bridge remained "more impressive than the skyscrapers," declared science-fiction futurist H. G. Wells. The bridge withstood two major world wars as well as several other conflicts, even providing an escape route for ash-covered refugees from the World Trade Towers' collapse. It became a National Historic Landmark and is honored whenever public officials see a reason to do so. Its history can be discovered on the bridge itself, recounted in plaques around the towers and in memorials affixed to the granite proclaiming its honors.

The people of New York have celebrated this impressive monument at its 25-year anniversary, its half-century anniversary, its 75th anniversary, and finally, its 100th birthday in 1983. During its centennial celebration, President Ronald Reagan traced President Arthur's original path—except he didn't have a mayor of Brooklyn to congratulate. And now the city's largest park, Brooklyn Bridge Park, will honor it for years to come.

The Brooklyn Bridge has been lauded, admired, and honored not least because it is a hand-built bridge. John Roebling, when he conceived the plan, predicted that the "towers will serve as landmarks." He was right. The towers are unique to the city's skyline and its lore. The bridge has been worth every cent of its almost $16 million construction price.

In all its splendor, the Brooklyn Bridge dominates. An 1883 sign appearing in a Brooklyn department store window read, "Babylon had her hanging gardens; Egypt her pyramids; Athens her Acropolis; Rome her Athenaeum; so Brooklyn has her bridge."

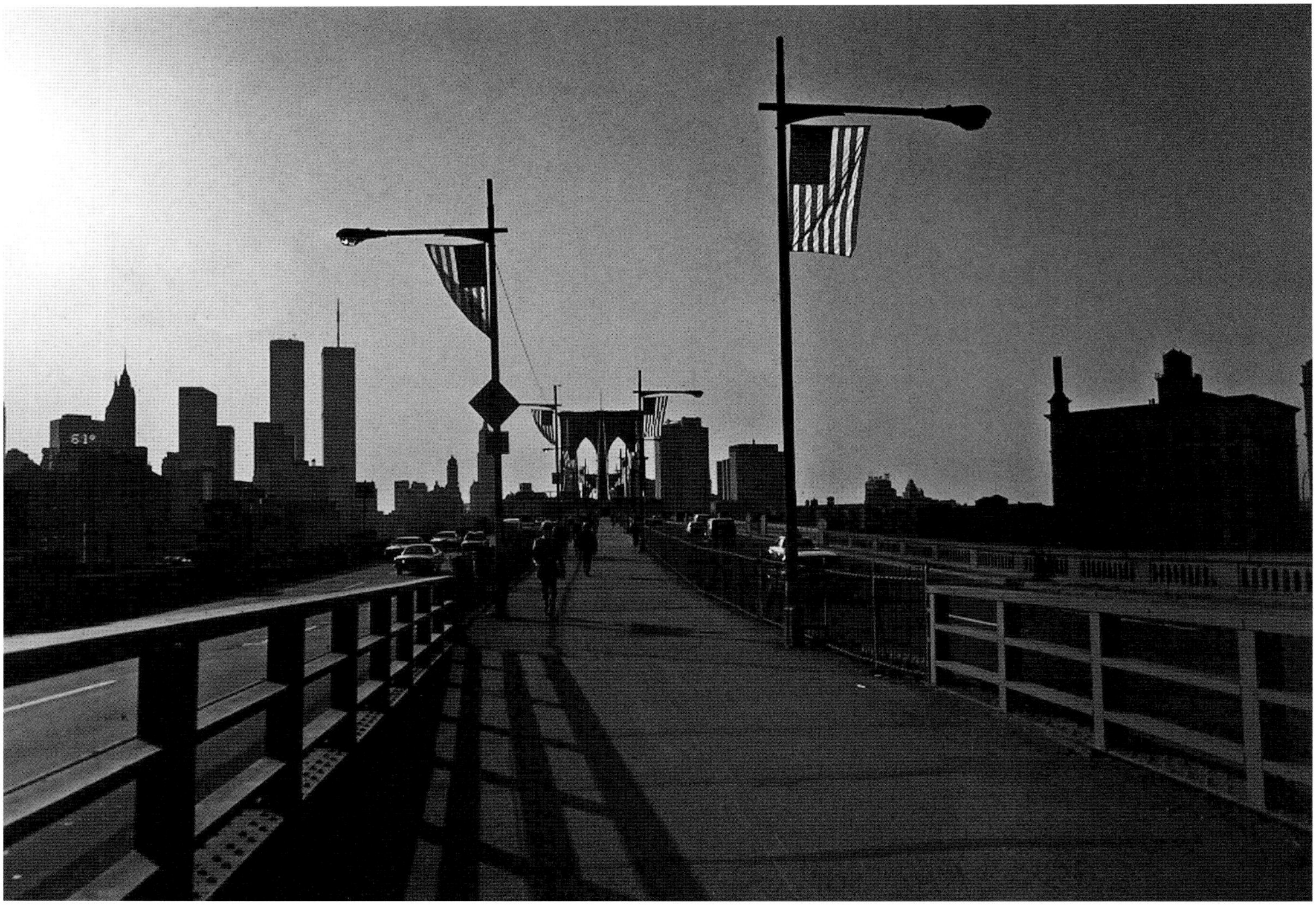

Flags aloft: Flags fly on the Brooklyn Bridge for its centennial celebration in 1983.

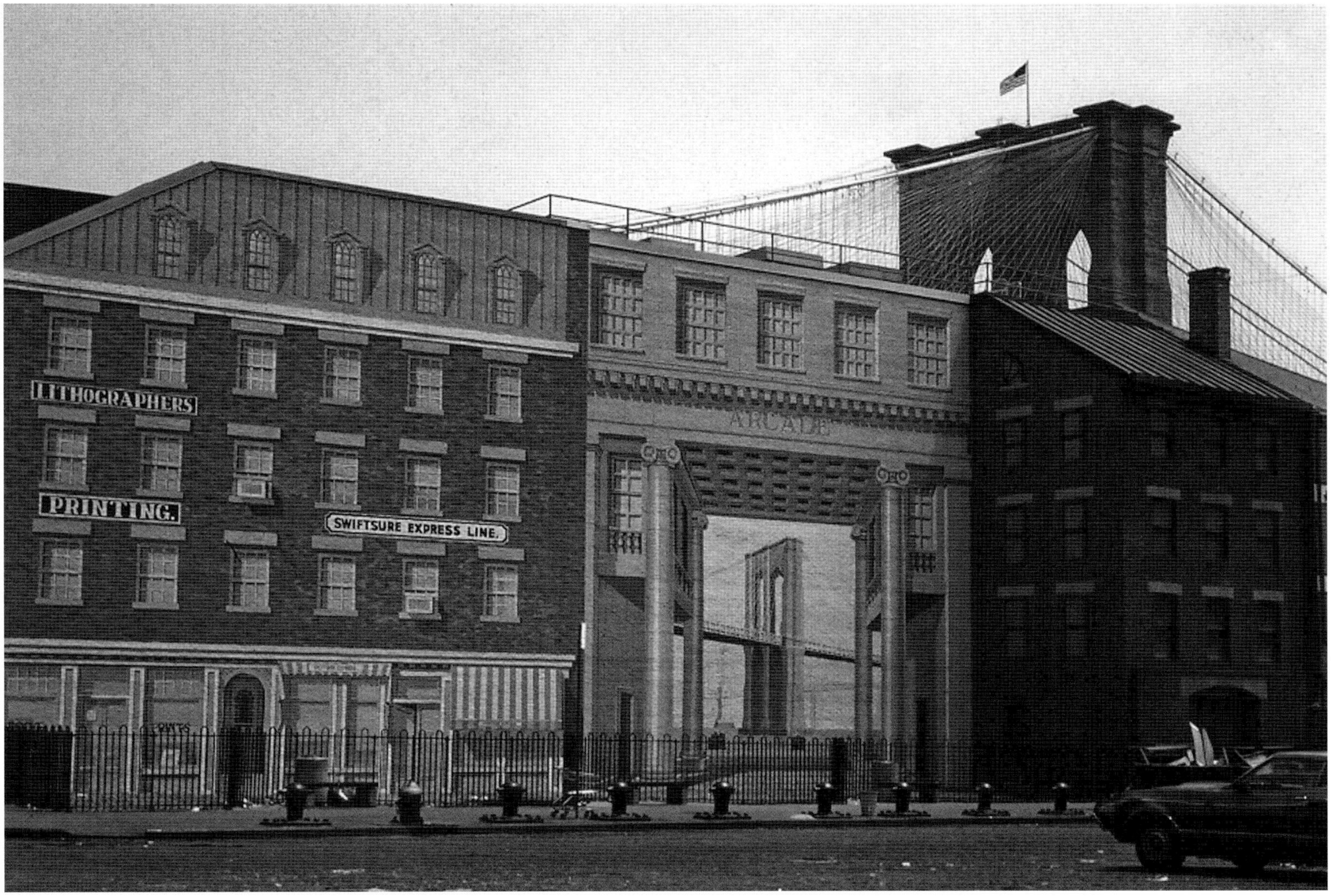

Bridge mural: A wall mural at South Street Seaport in 1983 was painted to resemble stores as well as the bridge, visible through an arch. The New York tower of the real bridge displays itself above the building.

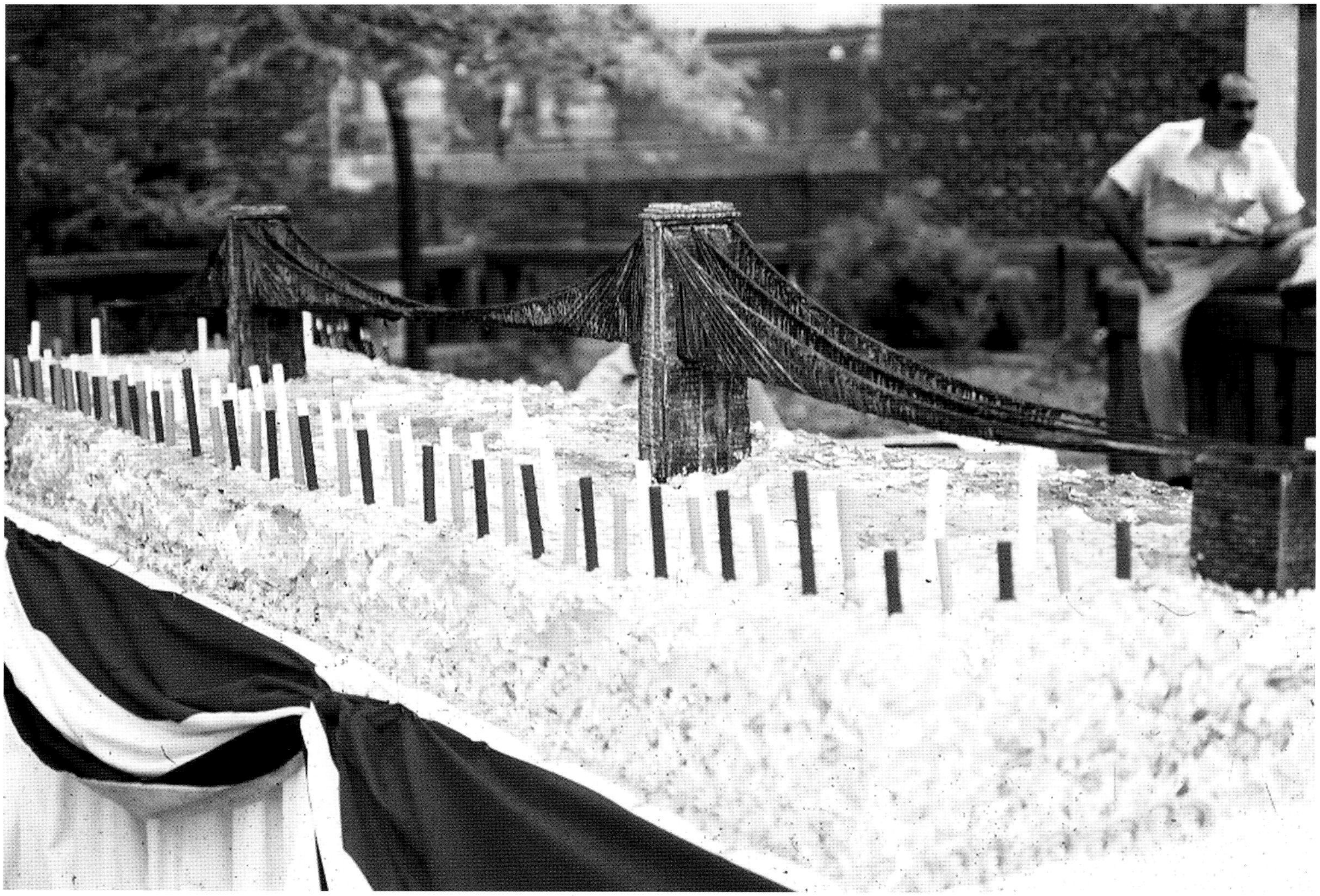

Birthday bash: In honor of the bridge's centennial celebration, this birthday cake was created with a candy Brooklyn Bridge on top.

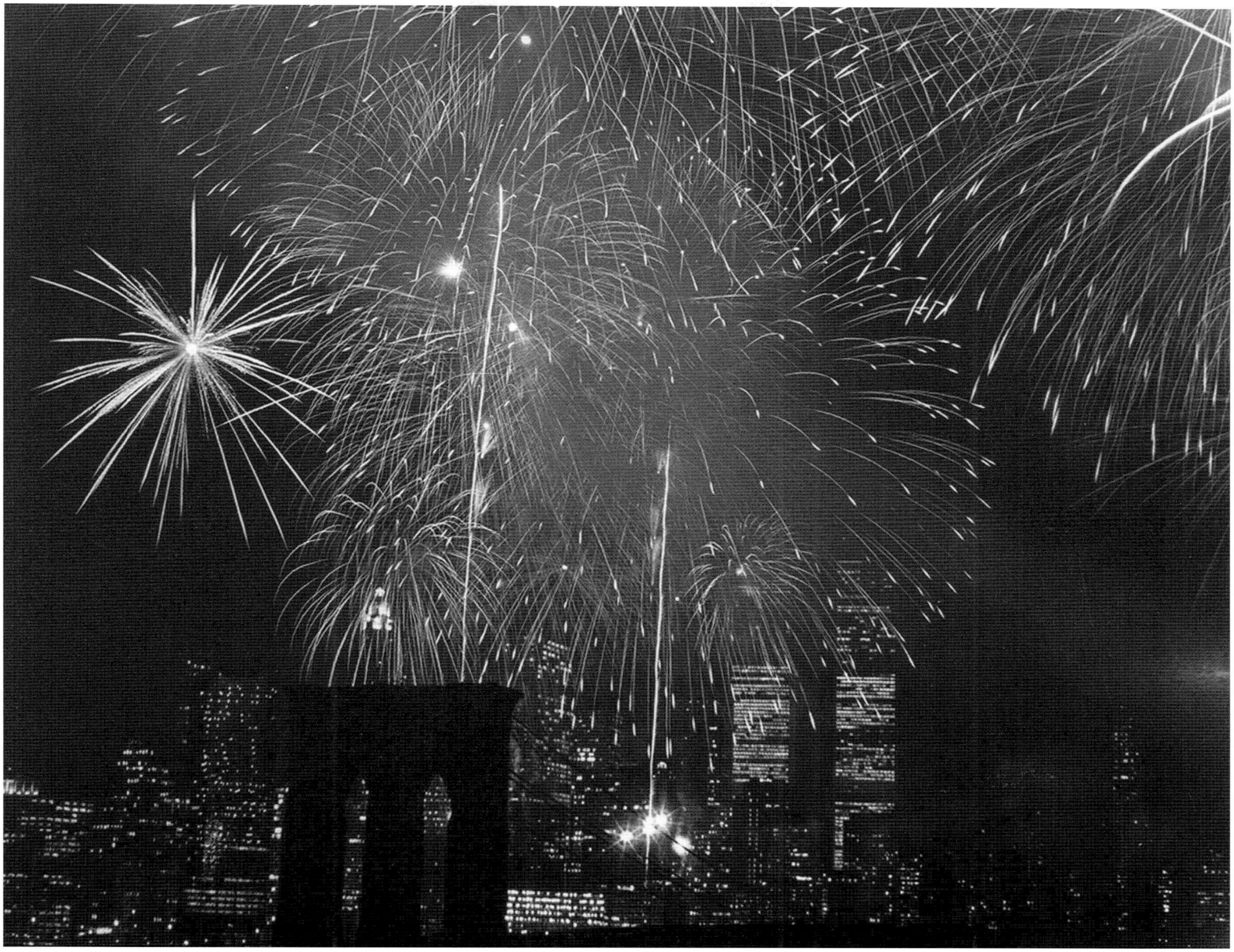

Fireworks: In an attempt to imitate the original celebration, fireworks are set off from the Brooklyn Bridge for its 100th anniversary. Note the illuminated twin towers of the World Trade Center in the distance.

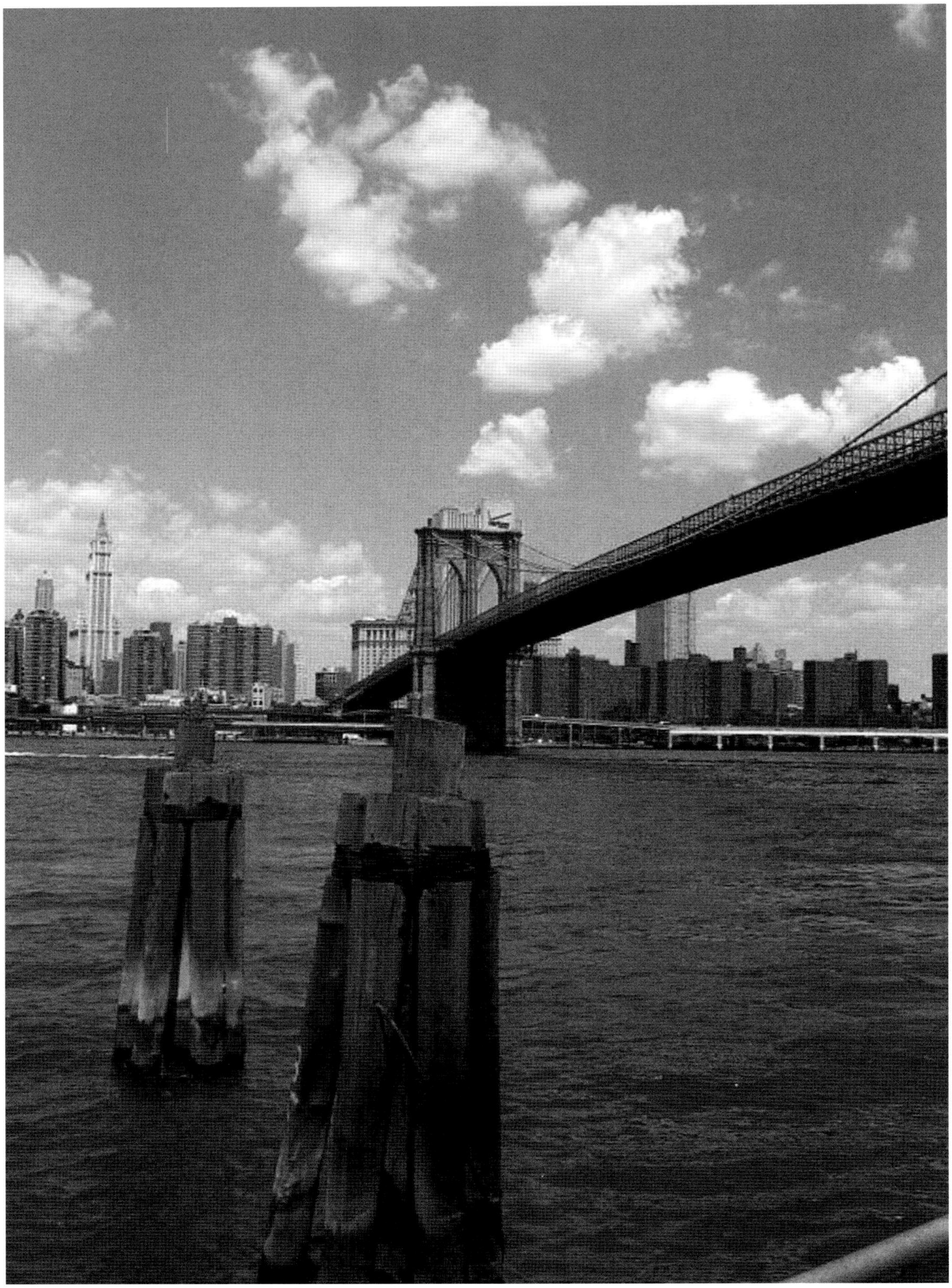

Brooklyn Bridge from Fulton Ferry Pier: With old pilings in the foreground, the Brooklyn Bridge stretches to Manhattan under a clear sky.

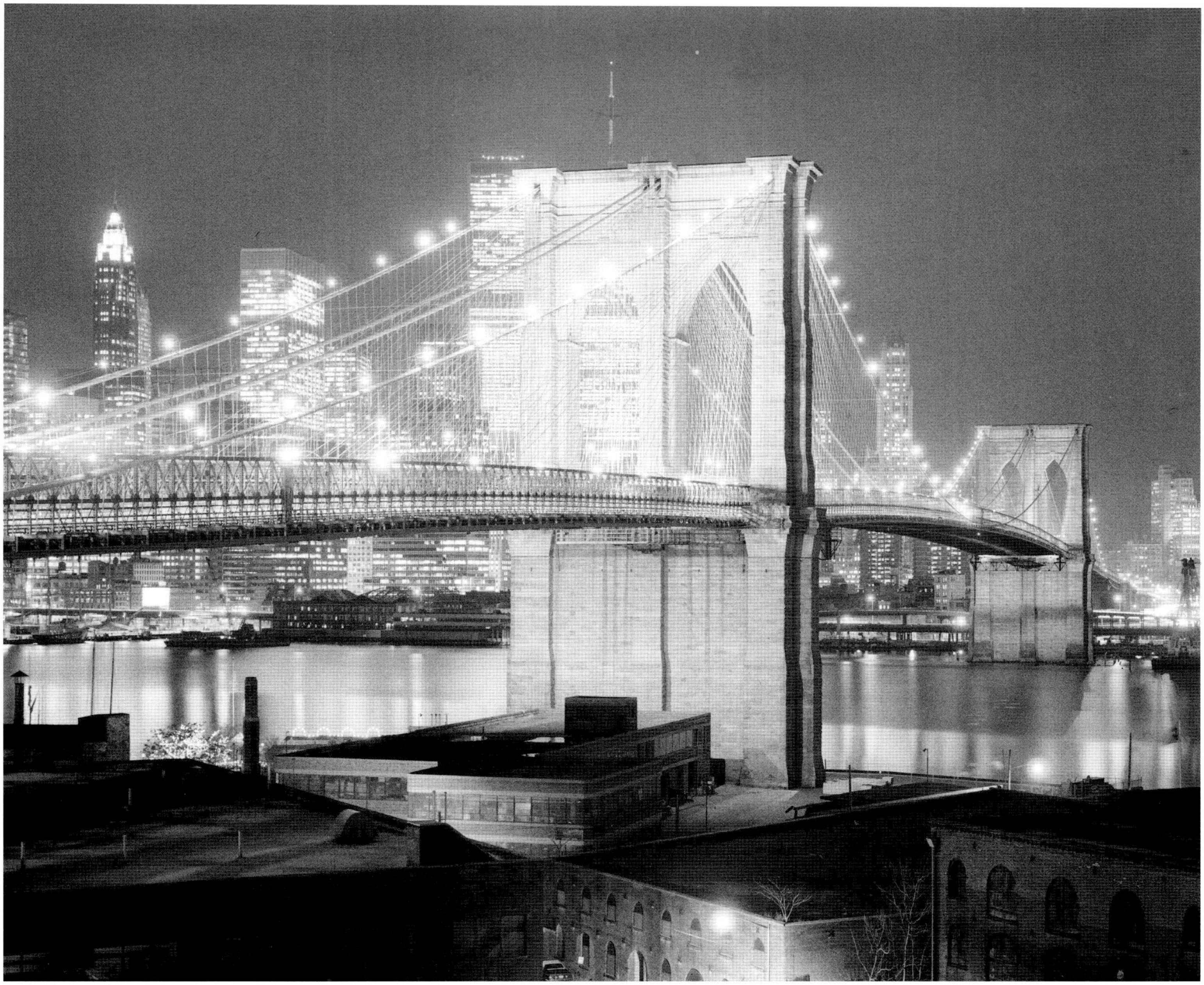

A night view of the Brooklyn Bridge with its "necklace of lights": In the foreground are buildings of DUMBO's New York State Empire Park, including the Tobacco Warehouse and the now demolished Purchase Building from the 1930s.

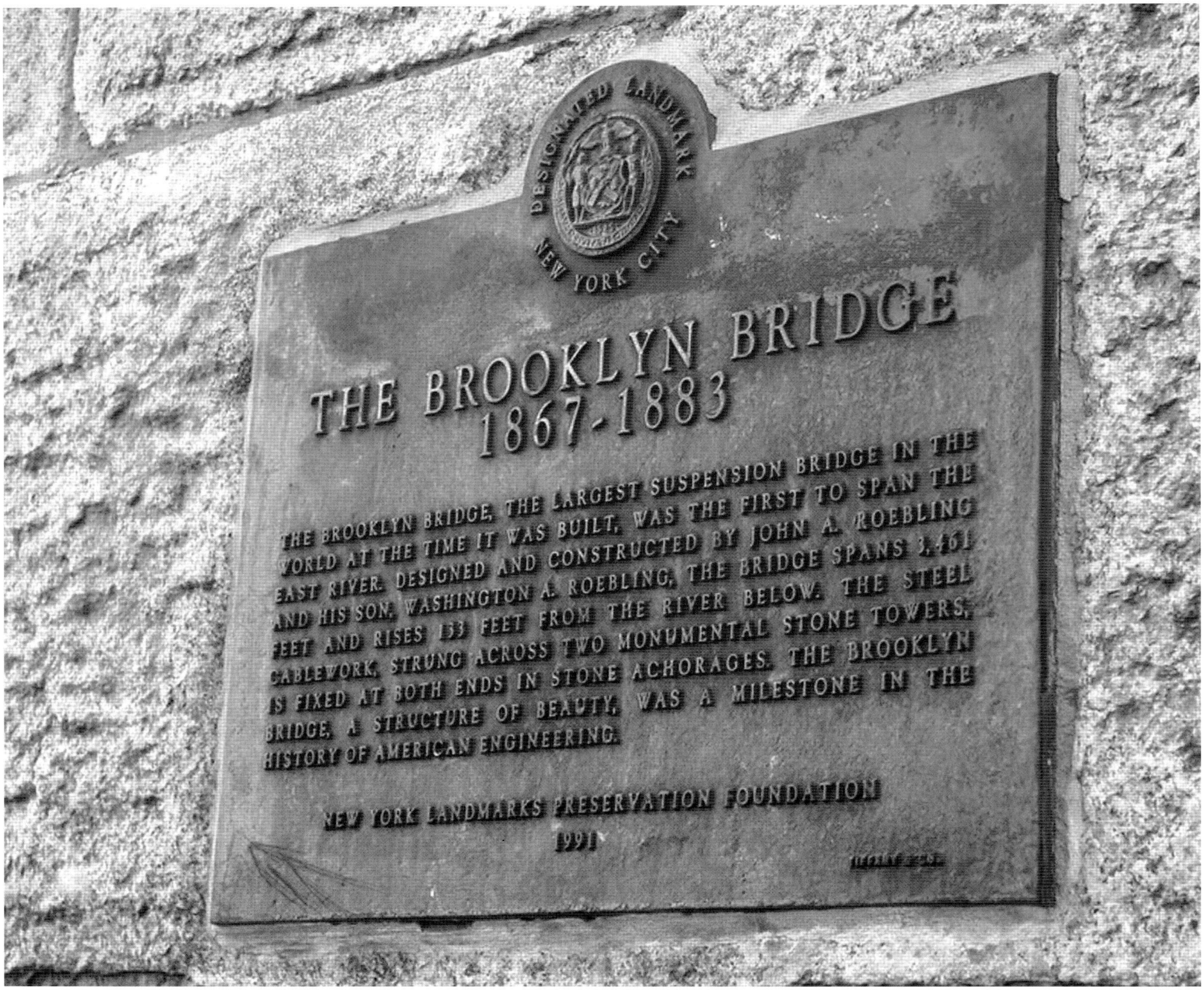

Landmark plaque: This plaque was mounted on the Brooklyn Bridge in 1991, naming it a National Historic Landmark.

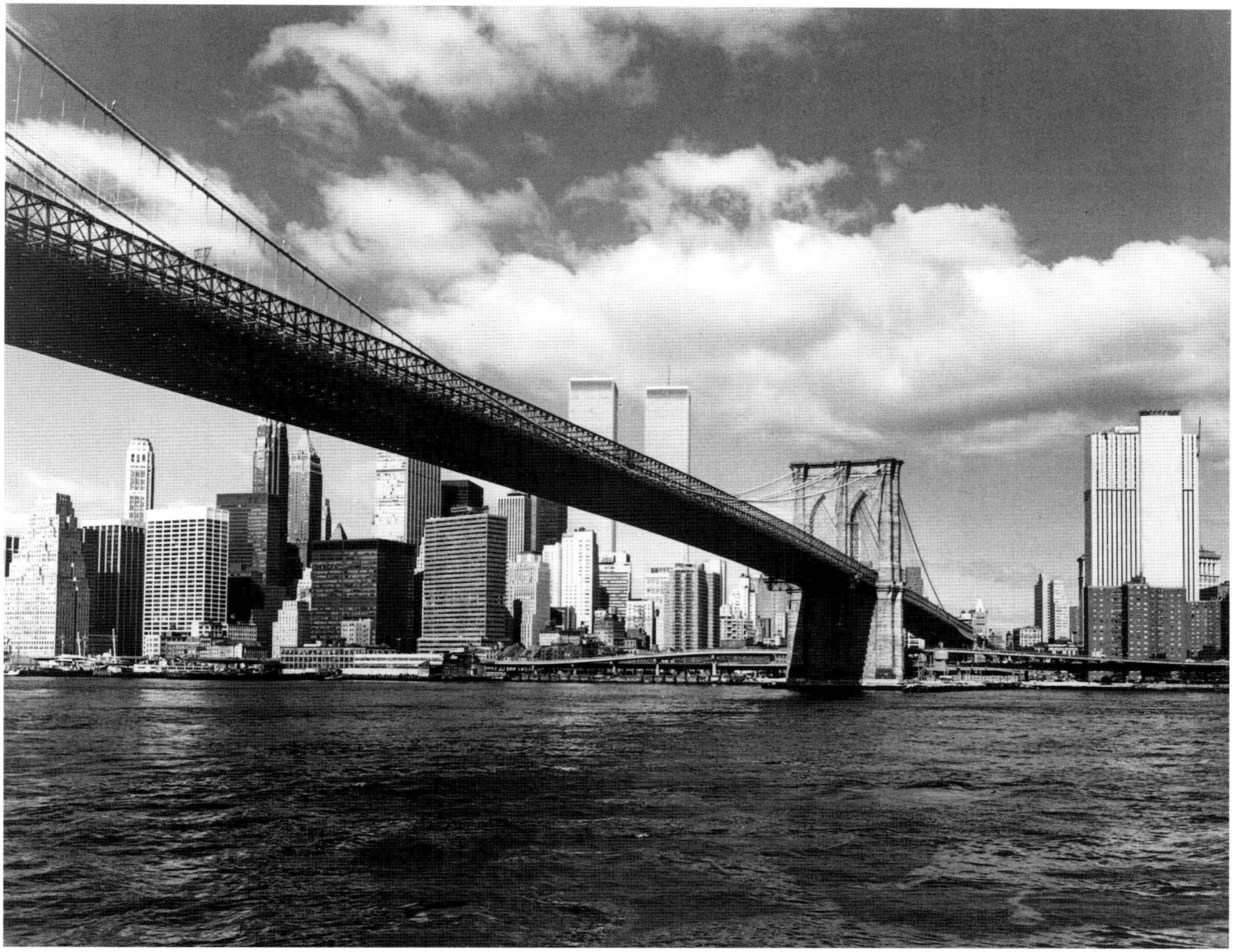

Brooklyn Bridge from afar: Panorama of the Brooklyn Bridge as it appeared in recent times. The twin towers of the World Trade Center hover in the background.

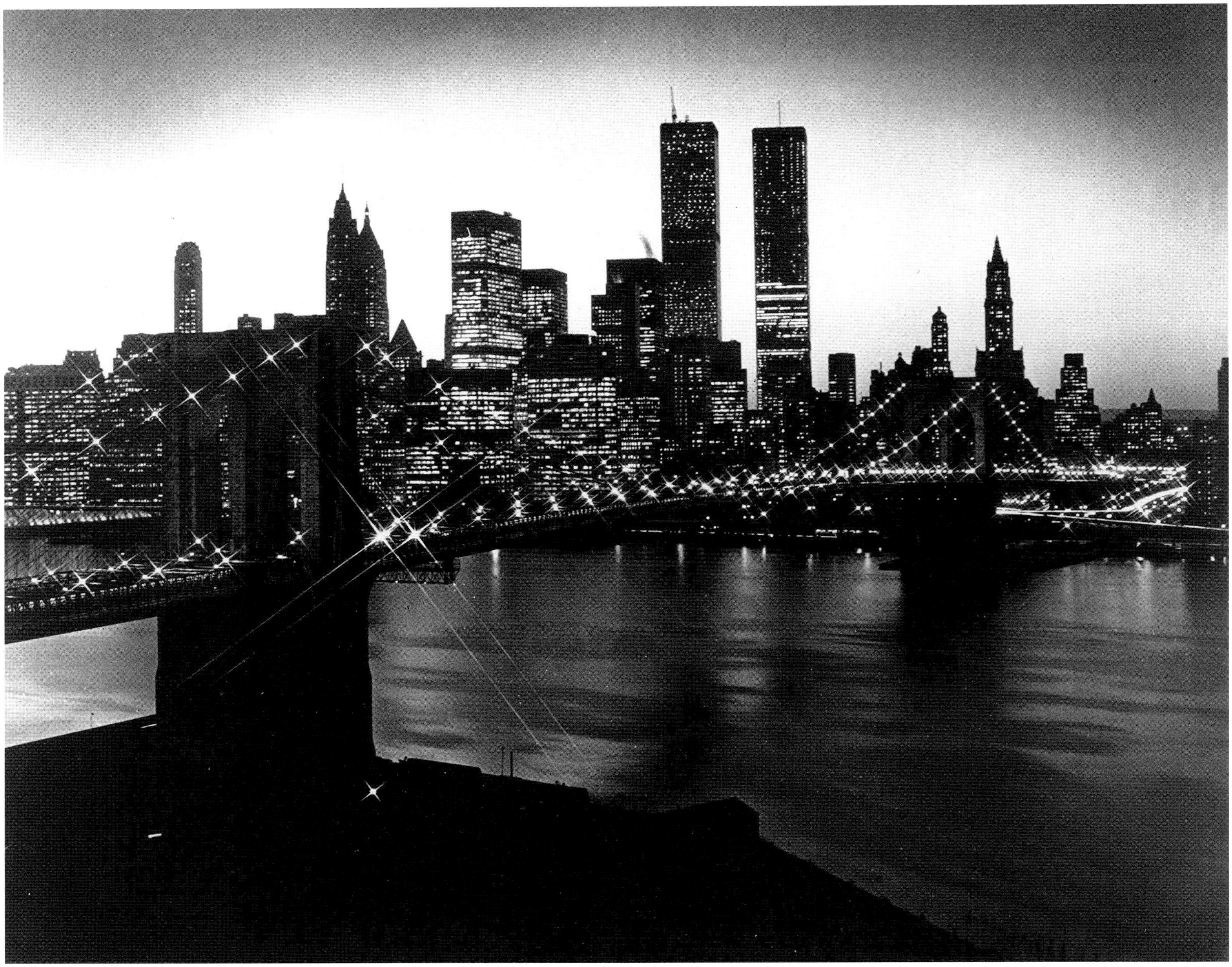

Bridge and towers at evening: Here the Brooklyn Bridge stands illuminated along with the twin towers of the World Trade Center and other surrounding buildings in Manhattan.

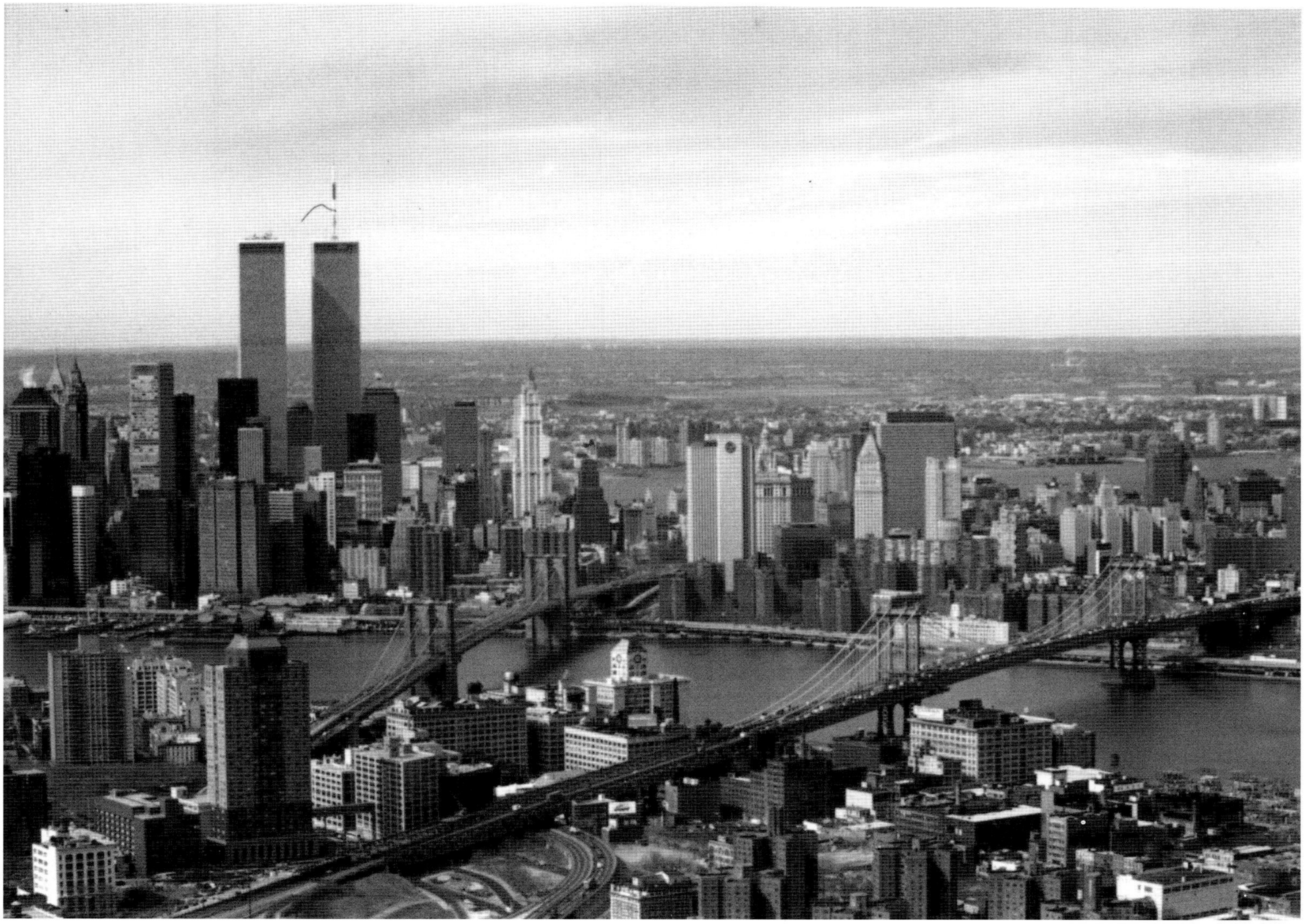

Bridges and towers: Shown here in 1995 are the Brooklyn Bridge and the World Trade Center, its two towers dominating the skyline.

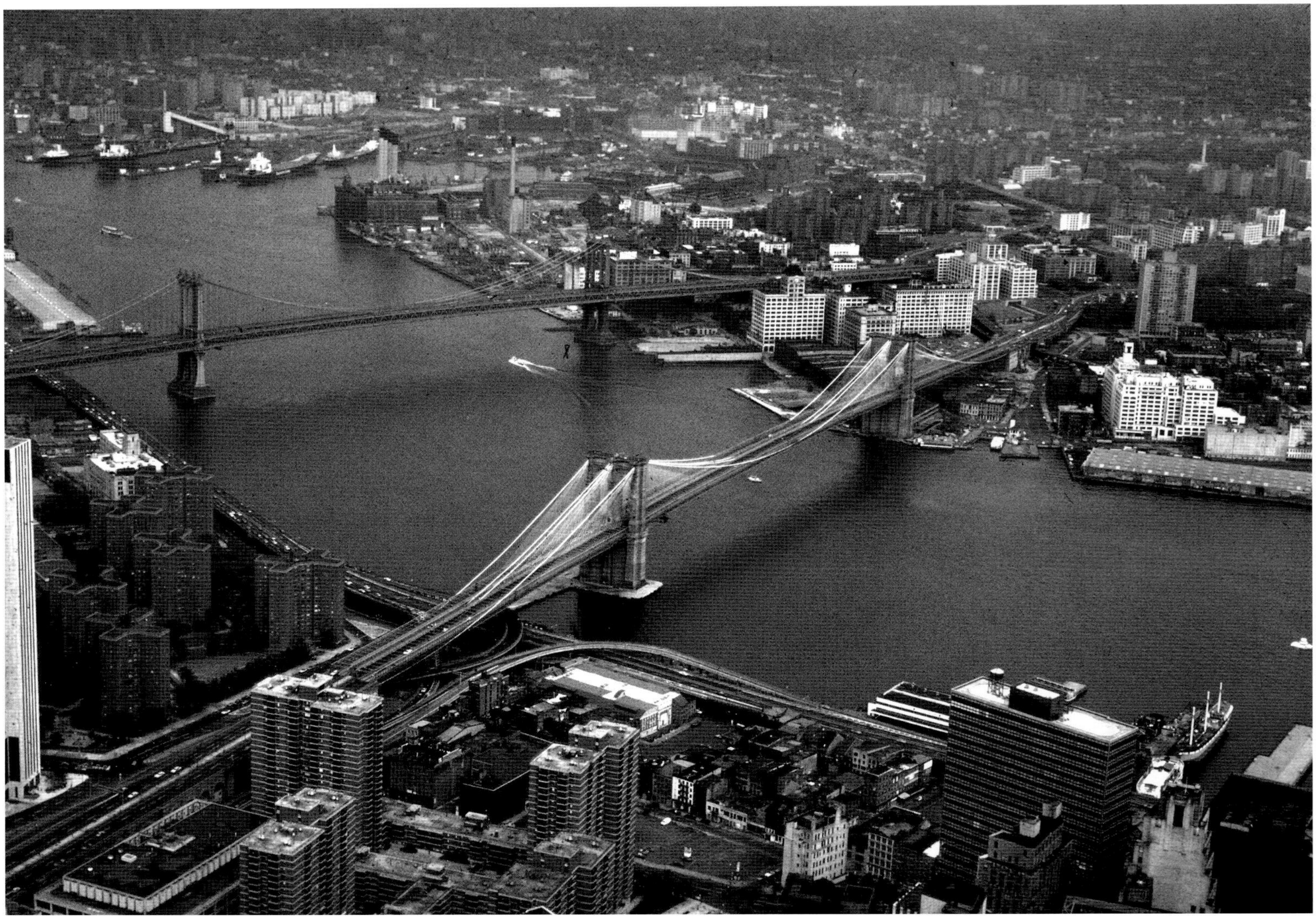

Bridges viewed from towers: This photo, taken in 2001 from the top of the World Trade Center towers, features a view of the East River bridges.

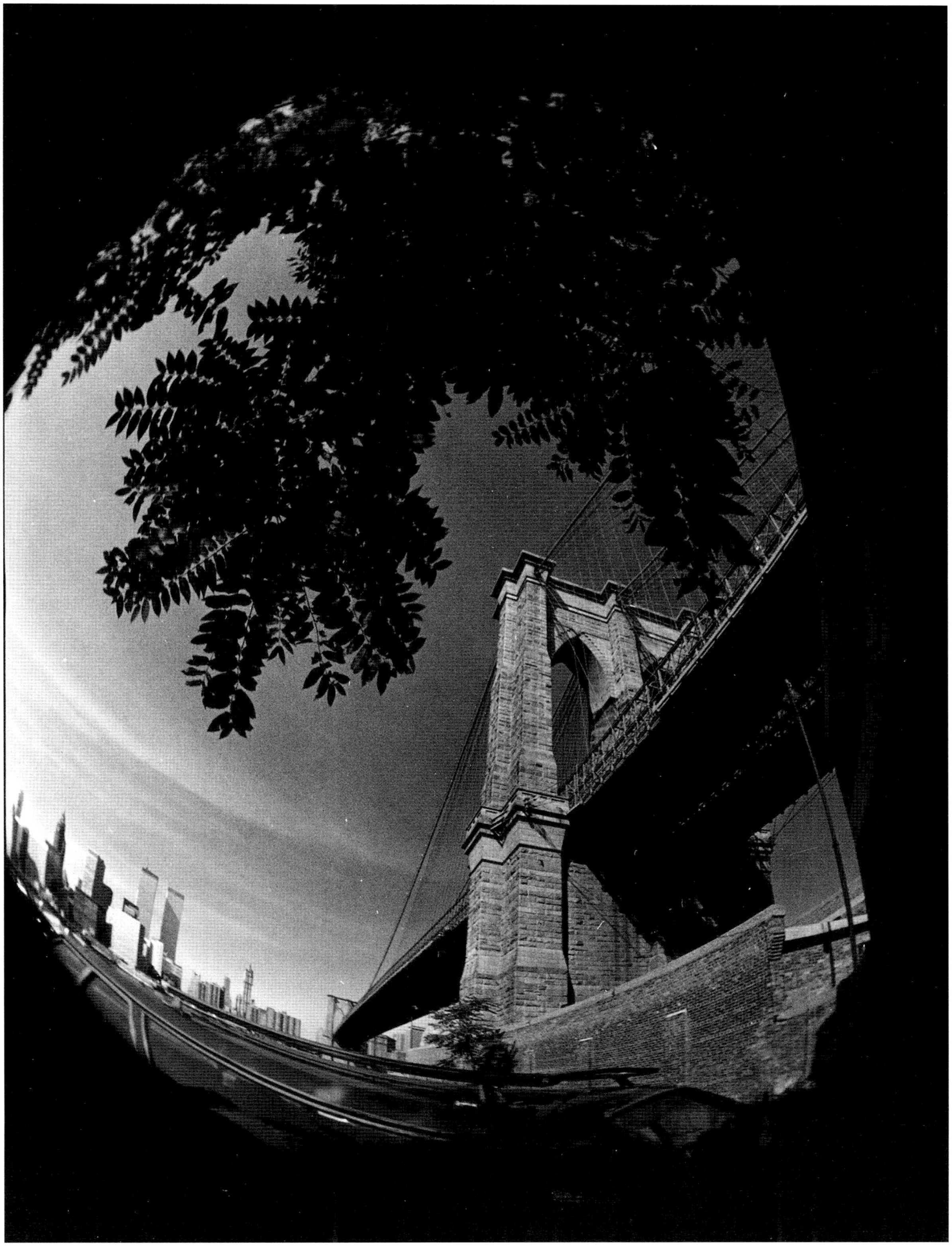

A different view: The Brooklyn Bridge is seen through a fish-eye lens.

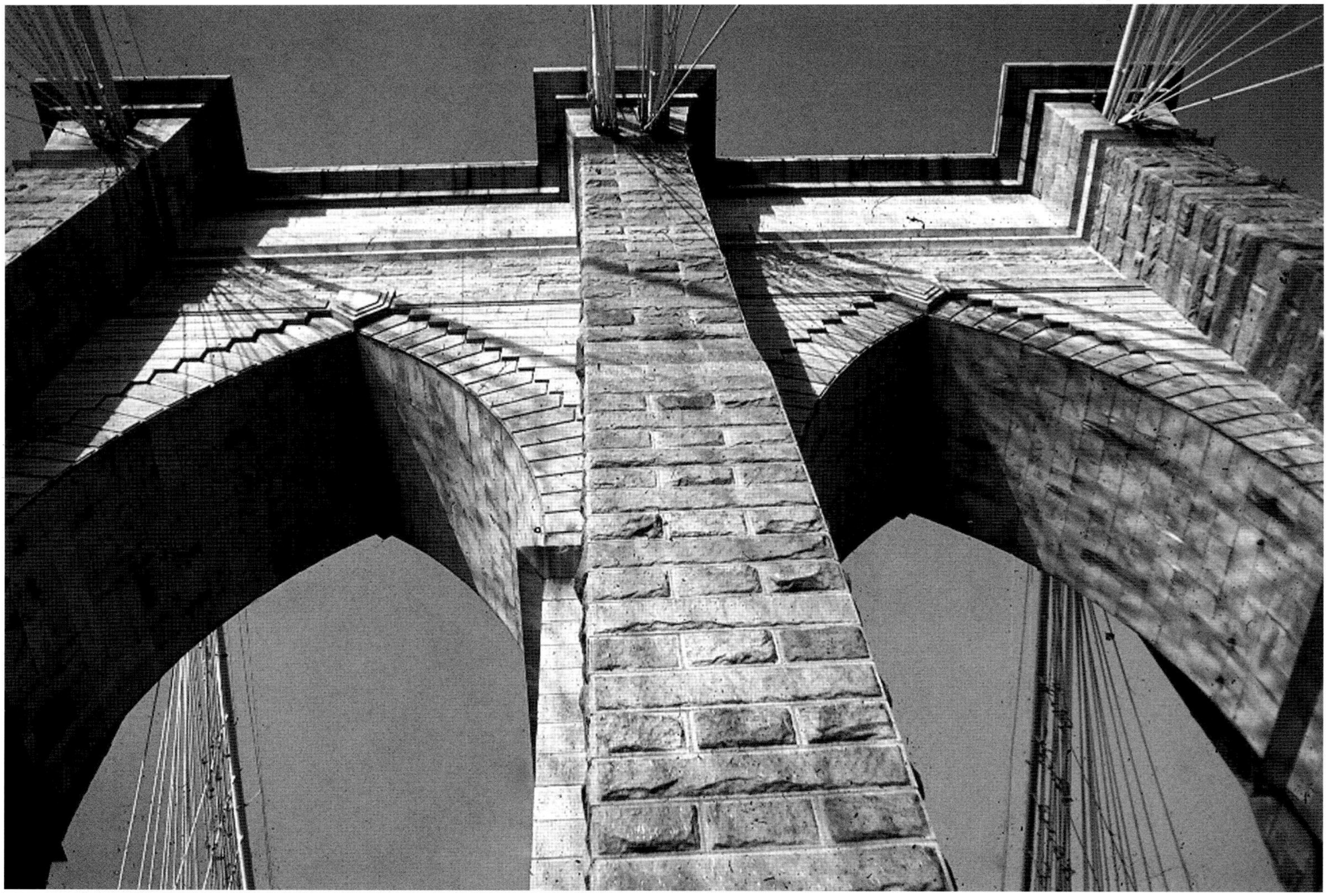

Brooklyn's arches: Arches of the Brooklyn Bridge's New York tower show the carefully crafted stonework.

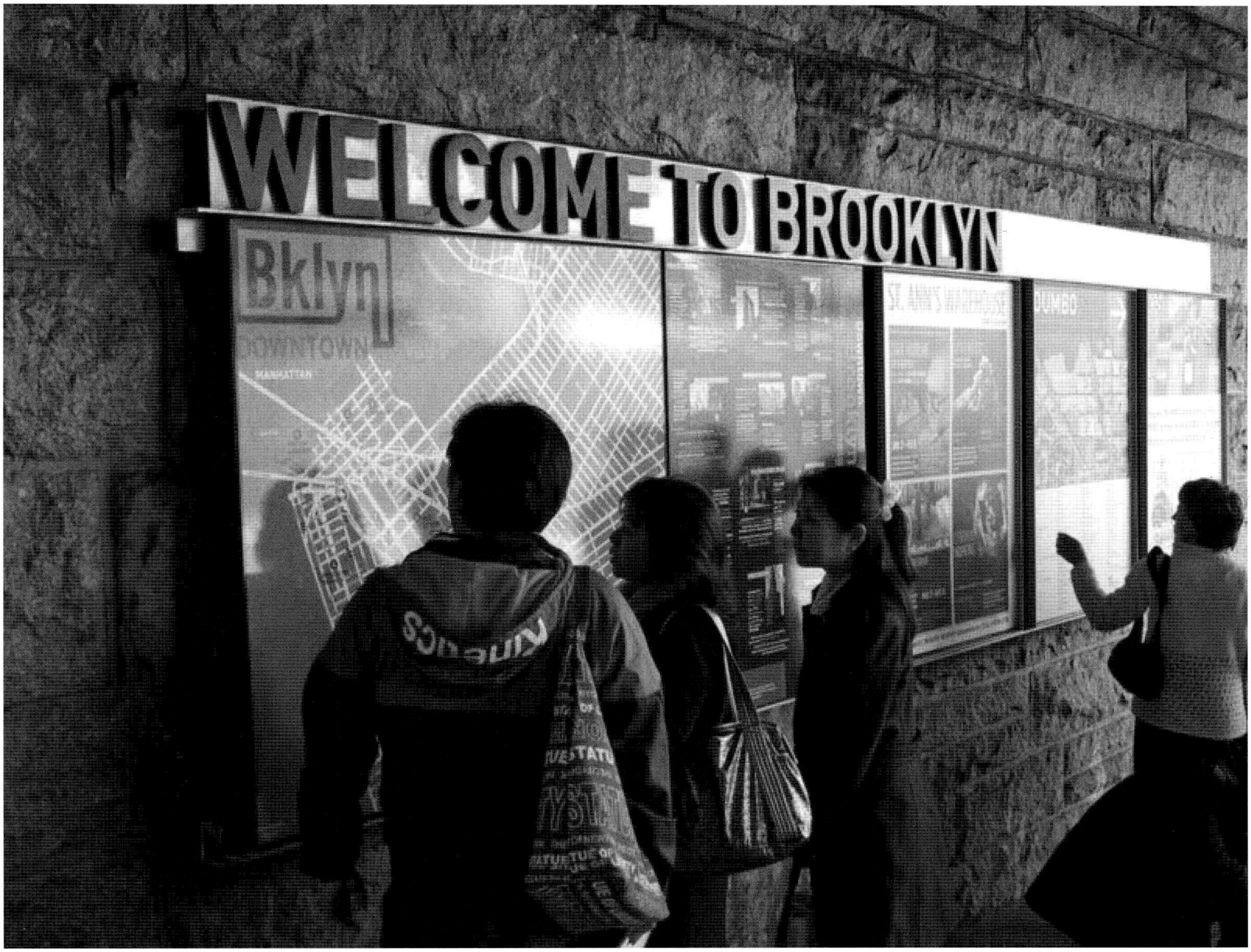

Welcome to Brooklyn: Visitors study a Brooklyn map mounted on the Brooklyn Bridge anchorage before climbing the stairs to walk across its span.

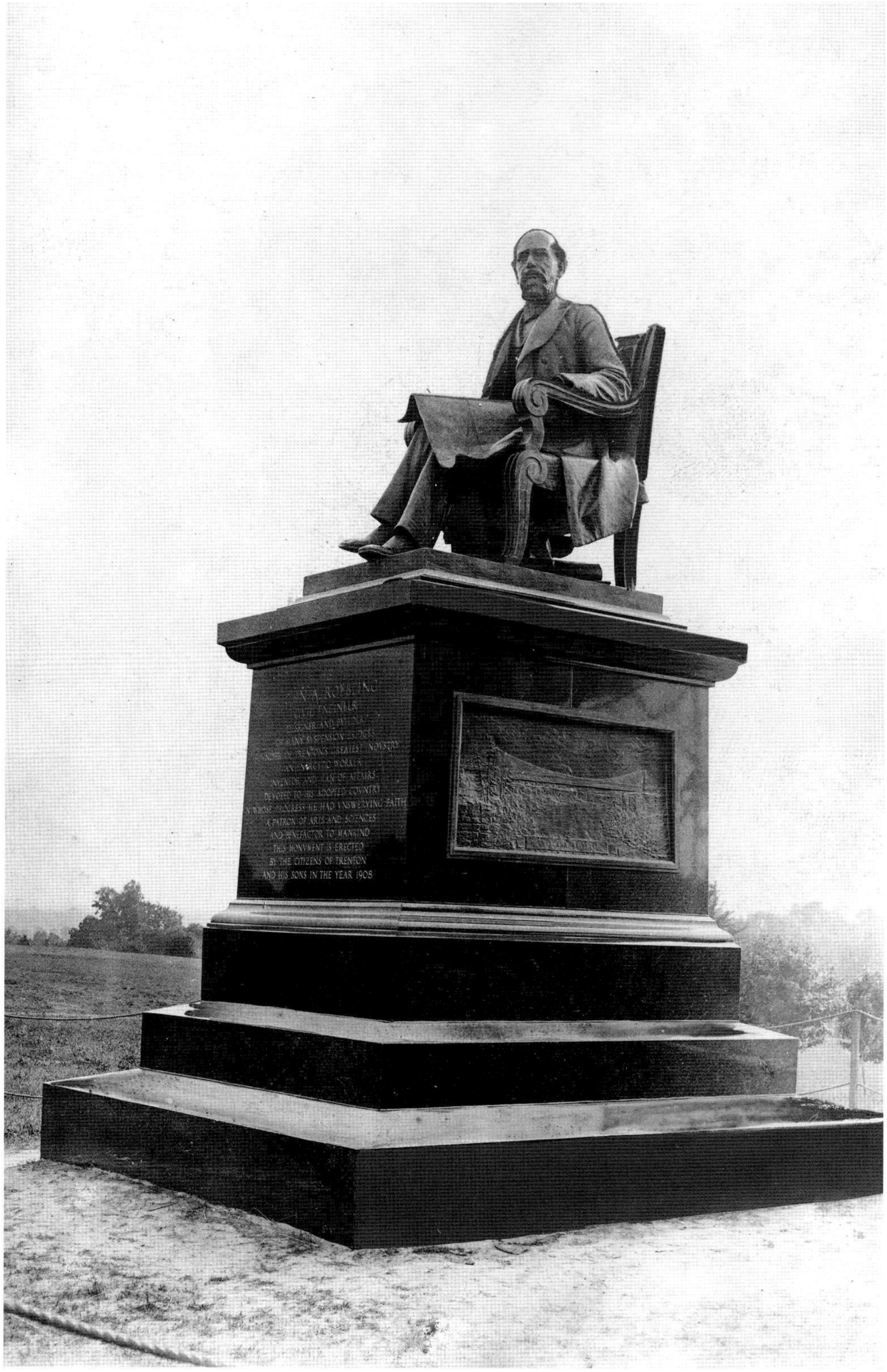

Statue of John Roebling: Outside the Roebling Wire Works stands this statue of the founder, created by William Couper and erected in 1910. A model of the Brooklyn Bridge decorates one side of the monument base. The other side depicts one of Roebling's first projects, the suspension bridge at Niagara Falls. Although Roebling never lived to see the construction of his bridge, his name remains forever tied to his magnificent creation.

The Brooklyn Bridge: A Timeline

1636 Dutch settle in Brooklyn
1642 First ferry across East River operated by Cornelius Dircksen
1788 First proposal for a bridge across the East River
1802 New York State legislature petitioned to build East River bridge
1806 John Roebling born in Mühlhausen, Thuringia (Prussia, later Germany)
1807 John Stevens petitions for a floating bridge
1811 Thomas Pope designs his "Rainbow Bridge" from New York to Brooklyn
1814 Robert Fulton introduces the first steam ferry, *Nassau*, to cross from Fulton Street, Manhattan, to Fulton Street, Brooklyn
1826 John Roebling earns his civil engineering diploma from the Royal Polytechnic Institute of Berlin
1829 Proposal for a chain suspension bridge submitted to the Brooklyn Common Council
1831 Roebling immigrates to U.S. at 25 years old; establishes utopian farming community in Butler County outside Pittsburgh, Pa.
1834 Town of Brooklyn becomes City of Brooklyn
1835 South Ferry (Atlantic Avenue) begins route across East River
1836 Roebling marries Johanna Herting
1837 Having failed as a farmer, Roebling works for Commonwealth of Pennsylvania surveying for railroad routes; Washington Roebling born
1841 John Roebling opens wire rope factory in Saxonburg, Pa.
1842 First successful use of Roebling's wire rope on Allegheny Portage Railroad; Ferdinand Roebling born
1845 Roebling builds first suspension bridge on Allegheny Aqueduct for Main Line Canal in Pittsburgh
1846 Roebling builds Smithfield Street Bridge in Pittsburgh; Covington-Cincinnati Bridge chartered
1848 Delaware Aqueduct Bridge built between Port Jervis, N.Y., and Lackawaxen, Pa.
1849 Roebling moves his factory to the Trenton Wire Rope Plant in New Jersey; Charles Roebling born
1855 John Roebling builds bridge at Niagara Falls
1856 John Roebling is commissioned to build Covington-Cincinnati Bridge over Ohio River
1857 New York State considers a bill to build an East River bridge as designed by Roebling
1865 John Roebling and Wilhelm Hildenbrand draw up plans for the East River Bridge; Colonel Washington Roebling resigns from Union forces, marries Emily Warren, and joins his father in Cincinnati
1866 Washington supervises construction of Covington-Cincinnati Bridge; cables for bridge are completed in Ohio; New York passes bill for construction of bridge; East River frozen over in 1866 and 1867
1867 Covington-Cincinnati Bridge opens; William Kingsley speaks to Senator Henry Murphy about potential for East River bridge; New York Bridge Company incorporated, with John Roebling named chief engineer and draftsman Wilhelm Hildenbrand named design engineer
1868 Brooklyn Common Council authorizes $3 million loan for construction of bridge; New York authorizes half that amount
1869 New York Common Council and Army Corps of Engineers approve bridge construction; surveys begin; Congress approves and President Grant signs bill to build; John Roebling dies after accident on ferry slip; his son Washington is appointed Chief Engineer
1870 Brooklyn named the fastest-growing city in the country; ground broken and cleared for Brooklyn tower foundation; Brooklyn caisson launched; fire discovered in Brooklyn caisson; Washington Roebling gets caisson disease
1871 Brooklyn foundation completed; New York caisson launched in Greenpoint and towed to position
1872 Washington stops construction; New York tower foundation completed; two cranes topple from Brooklyn tower; strike by workers for 50 cents more pay; first worker dies of bends
1873 Brooklyn anchorage begun; Roebling purchases Brooklyn Heights home overlooking bridge construction
1874 Brooklyn Bridge goes public with sale of stock

1875 New York anchorage begun; Brooklyn anchorage and tower completed

1876 First wire for cable stretched from Brooklyn tower to bridge; New York tower completed; E. F. Farrington first to cross towers in boatswain's chair; Roebling resigns in dispute, but resignation not accepted; suit brought against New York mayor to cease construction because cables are run in "navigable waters" and might cause shipping damage, is defeated

1877 Washington Roebling moves to 110 Columbia Heights; first cable crosses East River; temporary footbridge opened to public

1878 Roebling discovers faulty wires supplied by J. Lloyd Haigh; Henry Supple, foreman, whipped from New York tower by snapped cable and killed

1879 Accusation of bribes with steel contract; approaches to bridge started

1881 Emily Roebling leads first walk across the bridge floor

1882 Brooklyn Mayor Seth Low unsuccessfully tries to force Roebling to resign; legislature directs bridge trustees to complete construction

1883 Promenade completed; lights tested on bridge; Brooklyn and Manhattan train terminals completed; Emily Roebling rides first carriage across the bridge; New York and Brooklyn Bridge officially opened on May 24 at 2:00 P.M.; Memorial Day riot on bridge: 12 killed; first cable cars cross bridge; Lexington Avenue elevated line from Brooklyn Bridge to Broadway starts run

1884 P. T. Barnum marches 21 elephants across bridge to prove its safety

1885 Robert Odlum becomes first person to jump from bridge on a bet but does not survive; 20 suicides in the 1890s

1888 One cent toll for pedestrians on footpath lifted at midnight

1892 Record crowd of a quarter million people cross bridge to celebrate 400th anniversary of Columbus' landing

1895 Clara McArthur becomes the first woman to jump from bridge: she survives

1897 Trolley service begins over bridge

1898 Kings County Elevated replaces cable cars; electric trolleys starts service; Brooklyn becomes a borough in NYC consolidation

1903 Emily Roebling dies at 59; proposal for a second deck denied; Williamsburg Bridge opens

1907 Half a million people cross the Brooklyn Bridge on an average day

1908 Cable cars removed from bridge; subway runs from South Ferry, New York, to Borough Hall, Brooklyn

1909 Manhattan Bridge opens

1915 "Brooklyn Bridge" named formally by NYC Board of Aldermen; Roebling's Wire Factory in Trenton burns, is rebuilt

1921 Trolley car collision on bridge injures 21 persons; NYC mayor proposes Brooklyn Bridge be rebuilt, not necessary says Roebling

1922 Traffic banned on bridge until 1925 to study bridge's safety

1926 Washington Roebling dies in Trenton at 89

1933 New York City celebrates 50th anniversary of bridge

1942 Last East River ferry to Greenpoint ceases service; Johnny Weissmuller (Tarzan) climbs Brooklyn Bridge for *Tarzan's New York Adventure*

1944 Elevated train service over bridge discontinued; train terminals demolished

1950 Trolley service on bridge discontinued; Brooklyn Bridge closes for $7 million modernization

1954 Brooklyn Bridge opened after modernization with two new 3-lane highways

1956 Brooklyn Bridge loan paid in entirety: cost in 1883 $25 million plus $50 million in interest over 73 years

1959 Cloverleaf traffic pattern connected to new Brooklyn-Queens Expressway

1964 Brooklyn Bridge declared National Historic Landmark

1967 "Necklace of lights" installed along bridge suspension cables

1973 Bridge roadway closed for 90th anniversary celebration

1983 President Ronald Reagan leads procession over bridge for centennial celebration; performance space, Art in the Anchorage, opens

1984 Time capsule buried in Brooklyn Anchorage; bike lane added to Brooklyn Bridge

1999 Rehabilitation of Brooklyn Bridge

2001 Groundbreaking for Brooklyn Bridge Park between Manhattan Bridge and Atlantic Avenue; vaults in Brooklyn Anchorage closed

Notes on the Images

These notes, listed by page number, attempt to include all aspects known of the images. Each image is identified by the page number, photograph's or drawing's title or description, photographer or artist and collection, archive, and call or box number when applicable. Although every attempt was made to collect all data, in some cases complete data may have been unavailable due to the age and condition of some of the images and records.

ii **Revised Traffic Patterns**
Brooklyn Public Library, Brooklyn Collection
BRID 0110

vi **Suspension Bridge Song Sheet**
Library of Congress
LC-USZ62-7924

x **Rainbow Bridge**
Brooklyn Public Library, Brooklyn Collection
WILC 0190

2 **John A. Roebling**
Brooklyn Public Library, Brooklyn Collection
PORT 0751

3 **The Cincinnati-Covington Bridge**
Library of Congress
LC-DIG-ppmsca-08780

4 **Sectional View of Foundation**
Library of Congress
LC-USZ62-124608

6 **Ferryboat Brooklyn**
Library of Congress
LC-USZ62-52280

7 **Washington A. Roebling**
Brooklyn Public Library, Brooklyn Collection
PORT 0757

8 **Brooklyn Bridge from Columbia Heights**
Brooklyn Public Library, Brooklyn Collection
BRID 0030

9 **Emily Roebling**
Library of Congress
LC-USZ62-120162

10 **Elevation of the Brooklyn Bridge**
Brooklyn Public Library, Brooklyn Collection
BRID 0112

12 **Brooklyn Pier**
Library of Congress
LC-USZ62-124608

14 **Henry Cruse Murphy**
Brooklyn Public Library, Brooklyn Collection
PORT 0099

15 **Stock Certificate**
Brooklyn Historical Society
bk_eph_ix_transp_bkeltra

16 **Hugh McLaughlin**
Brooklyn Public Library, Brooklyn Collection
PORT 0594

17 **William Tweed**
Library of Congress
LC-USZ62-22467

18 **Henry Ward Beecher**
Library of Congress
LC-DIG-cwpbh-03066

19 **Inside the Brooklyn Tower Caisson**
Library of Congress
LC-USZ62-124944

20 **Foundation of Brooklyn Tower**
Brooklyn Public Library, Brooklyn Collection
WILC 0189

21 **Rising Elevation**
Brooklyn Public Library, Brooklyn Collection
WILC 0203

22 **Bird's-eye View of Manhattan Map**
Library of Congress
ppmsc 05738

24 Brooklyn Ferry vs. Brooklyn Bridge
Queens Borough Public Library, Long Island Collection
rugen_841

25 Engraving
Brooklyn Public Library, Brooklyn Collection
BRID 0111

26 Brooklyn Tower Nearing Completion
Library of Congress
LC-USZ62-108448

27 Close-up of Both Towers
Library of Congress
LC-USZ62-108446

28 Cables from Brooklyn to Manhattan
Brooklyn Public Library, Brooklyn Collection
BRAI 0242

30 Brooklyn Bridge Construction
Library of Congress
LC-USZ62-108447

31 Brooklyn Approach
Library of Congress
LC-USZ62-102100

32 Catwalk of Brooklyn Bridge
Brooklyn Public Library, Brooklyn Collection
BRID 0039

33 Farrington Crossing East River
Brooklyn Public Library, Brooklyn Collection
BRID 0041

34 Farrington Plaque
Author's Collection
closeup of Farrington's ride

35 Construction
Brooklyn Public Library, Brooklyn Collection
WILC 0258

36 Cable-spinning Machine
Queens Borough Public Library, Long Island Collection
illustration_wrapping_the_cables

37 Hanging Cables
Brooklyn Public Library, Brooklyn Collection
WILC 0216

38 Construction Progress
Brooklyn Public Library, Brooklyn Collection
WILC 0225

39 Key Figures
Brooklyn Public Library, Brooklyn Collection
BRID 0019

40 Looking Toward New York Tower
Brooklyn Public Library, Brooklyn Collection
WILC 0224

41 Finishing the New York Approach
Library of Congress
LC-USZ62-82882

42 Removing the Footbridge
Library of Congress
LC-USZ62-82880

43 Lashing the Stays
Library of Congress
LC-USZ62-102102

44 Completing the Wooden Walkway
Brooklyn Public Library, Brooklyn Collection
WILC 0181

45 Painting the Bridge
Brooklyn Public Library, Brooklyn Collection
BRAI 0333

46 Plaque
Author's Collection
plaque erected by cities of New York and Brooklyn

48 The Great East River Suspension Bridge
Library of Congress
LC-DIG-pga-00755

49 Invitation to the Opening Ceremonies
Brooklyn Public Library, Brooklyn Collection
BRID 0068

50 Roebling Invitation
Brooklyn Historical Society
bk_eph_bkbridgeexhib_ro

51 Seth Low
Brooklyn Public Library, Brooklyn Collection
PORT 0094

52 The President Crossing the Suspended Highway
Library of Congress
LC-USZ62-108304

53 President Chester Arthur
Library of Congress
LC-USZ62-13021

54 Abram Hewitt
Library of Congress
LC-USZ62-108370

55 Fireworks on the Brooklyn Bridge
Brooklyn Public Library, Brooklyn Collection
BRID 0015

56 Bridge Toll Tickets
Brooklyn Historical Society
bk_eph_tickets001

57 Tragedy on the Bridge
Brooklyn Public Library, Brooklyn Collection
BRID 0015

58 After the Excitement
Library of Congress
LC-USZ62-79054

59 Bridge and Masts
Brooklyn Public Library, Brooklyn Collection
BRAI 0316

60 Bird's-eye View
Library of Congress
LC-DIG-pga-03205

62 Robert Odlum's Jump
Author's Collection
Img011

64 Brooklyn Bridge Arches
New York State Archives
NYSA_A3045-78_D47_NB17

65 Busy Times
Queens Borough Public Library, Long Island Collection
illustration_brooklyn_bridge_4

66 New York Terminal to Brookln Bridge
Library of Congress
LC-USZ62-108303

67 Cable Car on Bridge
Library of Congress
LC-USZ62-108306

68 Tracks for Cable Car
Library of Congress
LC-USZ62-108445

69 Steam Engine Crossing Bridge
Brooklyn Public Library, Brooklyn Collection
BRAI 0338

70 Supports Under Bridge
Brooklyn Public Library, Brooklyn Collection
WILC 0174

71 Entrance to Pedestrian Path
Brooklyn Public Library, Brooklyn Collection
WAY 0005

72 Parallel Routes to Brooklyn
Brooklyn Public Library, Brooklyn Collection
WAY 0004

73 Winter Crossing
Brooklyn Public Library, Brooklyn Collection
BRID 0072

74 Crossing over the Bridge
Library of Congress
LC-USZ62-59953

76 Bridge Trolleys
Brooklyn Public Library, Brooklyn Collection
POST 0048

77 Road to Fulton Ferry
Brooklyn Public Library, Brooklyn Collection
THOM 0020

78 From New York to Brooklyn
Library of Congress
LC-USZ62-100441

79 Walking to Brooklyn
Library of Congress
LC-DIG-det-4a18594

80 Glimpse of New York City
Library of Congress
LC-USZ62-108443

81 Rush Hour Crowds
Library of Congress
LC-USZ62-42820

82 Exterior of Brooklyn Terminal
Brooklyn Public Library, Brooklyn Collection
BRID 0050

83 Terminal Interior
Brooklyn Public Library, Brooklyn Collection
BRID 0044

84 Pedestrians on Parade
Library of Congress
LC-USZ62-79043

85 Daily Scene at New York Terminal
Queens Borough Public Library, Long Island Collection
illustration_brooklyn_bridge_3

86 Fulton Elevated Trains
Queens Borough Public Library, Long Island Collection
rugen_604

87 Elevated Train at Fulton Ferry
Brooklyn Public Library, Brooklyn Collection
TRAN 0363

88 East River Docks
Library of Congress
LC-USZ62-100333

90 Walking from the Fulton Ferry
Brooklyn Public Library, Brooklyn Collection
BRAI 0310

91 Strolling on the Promenade
Library of Congress
LC-USZ62-97318

92 Bridge Pedestrians
Library of Congress
LC-USZ62-133262

93 To the Other Side
New York State Archives
NYSA_A3045-78_D47_NB

94 Crossing Over in 1904
Library of Congress
LC-USZ62-56629

95 The Bridge in 1901
Library of Congress
LC-USZ62-134655

96 Brooklyn Navy Yard
Brooklyn Public Library, Brooklyn Collection
BRAI 0178

97 Bridge and Boat
Queens Borough Public Library, Long Island Collection
illustration_bridge

98 The Bridge in 1896
Library of Congress
LC-USZ62-119639

99 The Bridge Viewed from Manhattan
Library of Congress
LC-USZ62-104633

100 Bird's-eye View
Library of Congress
LC-USZ62-76996

101 Ferry Docks in Brooklyn
Queens Borough Public Library, Long Island Collection
wjm_288

102 Heavy Shipping
Brooklyn Public Library, Brooklyn Collection
BRID 0105

103 Excursion Boat at the Bridge
Brooklyn Public Library, Brooklyn Collection
FROG 0014

104 South Street Activity
Library of Congress
LC-USZC4-2655

105 A Network of Masts
New York State Archives
NYSA_A3045-78-D47_NA

106 Brooklyn Docks
Library of Congress
LC-USZ62-22027

107 Brooklyn Point of View
Library of Congress
LC-D43-T01-600012-X

108 Bridge at Night
Library of Congress
LC-USZ62-67653

109 Lower Manhattan from Bridge Top
Library of Congress
LC-USZ62-69823

110 River Traffic
Library of Congress
LC-USZ62-95565

112 Skyscrapers as Seen from Brooklyn
Library of Congress
LC-USZ62-94926

113 Brooklyn Bridge Upriver
Library of Congress
LC-USZ62-103008

114 Down Under the Bridge
Library of Congress
LC-USZ62-103006

115 Traffic Jam
Library of Congress
LC-USZ62-134654

116 Looking North Toward Williamsburg Bridge
Library of Congress
LC-USZ62-79050

118 Busy Harbor
Library of Congress
LC-USZC2-6211

119 Navy Yard Exercise Yard
Library of Congress
LC-USZ62-91182

120 Electric Light Reflections
Library of Congress
LC-USZ62-92007

121 Brooklyn Bridge Newsie
Library of Congress
LC-DIG-nclc-03669

122 Big Boats, Too
Library of Congress
LC-DIG-ggbain-11042

123 Brooklyn Bridge Subway Station
Brooklyn Public Library, Brooklyn Collection
POST 0244

124 Roebling Factory Blaze
Library of Congress
LC-DIG-ggbain-18233

125 Remains of Roebling's Factory
Library of Congress
LC-DIG-ggbain-18230

126 Fulton Fish Market, 1916
New York State Archives
NYSA_A3045-78_Dn_NY5

128 Classic Pose
Library of Congress
LC-USZ62-74616

129 South Street Looking South
Library of Congress
LC-USZ62-107489

130 The Bridge Overlooking Brooklyn
Library of Congress
LC-USZC4-6727

132 Commerce in Brooklyn Heights
Brooklyn Public Library, Brooklyn Collection
NEIG 0270

133 Trolley Crossing the Bridge at Night
Queens Borough Public Library, Long Island Collection
illustration_brooklyn_bridge_1

134 Classic Night Shot
Brooklyn Public Library, Brooklyn Collection
WILC 0212

136 Bridge Tower Close-up, 1922
Brooklyn Public Library, Brooklyn Collection
BRID 0124

137 Columbia Heights
Queens Borough Public Library, Long Island Collection
ea_kc_bk_15

138 Serene Majesty
Brooklyn Public Library, Brooklyn Collection
UNDE 0094

139 View Overlooking Brooklyn Waterfront
Library of Congress
LC-USZ62-91181

140 Remains of Fulton Ferry House
Queens Borough Public Library, Long Island Collection
ea_kc_bk_153c

141 Artist's Rendering of the Bridge
Brooklyn Public Library, Brooklyn Collection
BRID 0378

142 Proposed Entrance
Brooklyn Public Library, Brooklyn Collection
BRID 0093

143 Arch Proposal
Brooklyn Public Library,
Brooklyn Collection
BRID 0090

144 View of the Bridge from Manhattan
New York State Archives
NYSA_A3045-78_D47_NA52

145 Rooftop View of the Bridge from Wall Street
Library of Congress
LC-USZ62-062618

146 View of the Bridge at Night
New York State Archives
NYSA_A3045-78_D47_NB

147 Manhattan Riverfront
New York State Archives
NYSA_A3045-78_D47_NA58

148 Cars on the Bridge, Barges Beneath
Brooklyn Public Library,
Brooklyn Collection
BRID 0132

149 Manhattan Skyline Through the Cables
Library of Congress
LC-USZ62-90970

150 Ocean Liner Under the Bridge
Queens Borough Public Library, Long Island Collection
illustration_brooklyn_bridge-5

151 Bread Line
Library of Congress
LC-USZ62-91536

152 Brooklyn Tracks
Brooklyn Public Library,
Brooklyn Collection
TRAN 0042

153 Bridge and Old Fulton Street
Brooklyn Public Library,
Brooklyn Collection
BRID 0138

154 Brooklyn Bridge and Lower New York City
New York State Archives
NYSA_A3045-78_16458

155 Ice Floes
Brooklyn Public Library,
Brooklyn Collection
BRID 0140

156 Flag Be-decked Ship
Brooklyn Public Library,
Brooklyn Collection
BRID 0058

157 Fiftieth Anniversary Announcement
Brooklyn Historical Society
bkbridge_eph_50th_anniversary_01

158 Simulated Crossing
Brooklyn Public Library,
Brooklyn Collection
BRID 0063

159 Official Ceremony
Brooklyn Public Library,
Brooklyn Collection
BRID 0060

160 Modern Vehicular Traffic
Brooklyn Public Library,
Brooklyn Collection
BRID 0011

161 Cadman Plaza and Brooklyn Terminal
Brooklyn Public Library,
Brooklyn Collection
NEIG 0534

162 Brooklyn Bridge Plaza
Brooklyn Public Library,
Brooklyn Collection
NEIG 0524

163 Downtown Manhattan and Bridges
New York State Archives
NYSA_A3045-78_C17242

164 Marathon Runners on the Bridge
Brooklyn Public Library,
Brooklyn Collection
BRID 0074

165 Last Train Across the Bridge
Brooklyn Public Library,
Brooklyn Collection
BRID 0122

166 Demolition of Brooklyn Terminal
Brooklyn Public Library,
Brooklyn Collection
BRID 0054

167 First Pedestrians over Modernized Bridge
Brooklyn Public Library,
Brooklyn Collection
BRID 0073

168 64th Anniversary
Brooklyn Public Library,
Brooklyn Collection
BRID 0012

169 Brooklyn Bridge Song Sheet
Author's Collection
Bk Bridge song autographed

170 New Vehicular Approach
Brooklyn Public Library,
Brooklyn Collection
BRID 0025

171 Closed
Brooklyn Public Library,
Brooklyn Collection
BRID 0026

172 Bridge Painter
Brooklyn Public Library,
Brooklyn Collection
BRID 0155

173 New Expressway
Brooklyn Public Library,
Brooklyn Collection
NEIG 0265

174 Inspecting the Bridge
Brooklyn Public Library,
Brooklyn Collection
BRID 0127

176 Restoration Plaque
Author's Collection
Brooklyn Bridge restoration 1954 plaque

177 Workman on Tower Cables
Library of Congress
HAER NY,31-NEYO,90-38

178 Bridges and Navy Yard
Library of Congress
LC-USZ62-118865

179 Speedboat
Brooklyn Public Library, Brooklyn Collection
BRID 0149

180 Crossing the Bridge in 1976
Author's Collection
Brooklyn Bridge at USA bicentennial

181 The Bridge and Brooklyn Waterfront, 1978
Library of Congress
HAER NY,31-NEYO,90-9

182 Street in DUMBO
Library of Congress
HAER NY,31-NEYO,90-29

184 Flags Aloft
Author's Collection
Brooklyn Bridge Centennial01

185 Bridge Mural
Author's Collection
mural of bridge & stores at South Street Seaport02

186 Centennial Birthday Bash
Author's Collection
Brooklyn Bridge centennial birthday cake04

187 Centennial Fireworks Display
Brooklyn College Library Archives, The Phoenix Newspaper Collection
bklyn bridge 1983 3

188 View from the Fulton Ferry Pier
Author's Collection
B0000055

189 Necklace of Lights
Library of Congress
HAER NY,31-NEYO.90-22

190 Historic Landmark Plaque, 1991
Author's Collection
landmarks plaque

191 Panorama from the Recent Past
Brooklyn Public Library, Brooklyn Collection
BRID 0014

192 Evening View of Bridge and Skyline
Brooklyn Public Library, Brooklyn Collection
BRID 0153

193 Bridges and Towers
Queens Borough Public Library, Long Island Collection
erl_1_31

194 Bridges as Viewed from the World Trade Center
Author's Collection
bridges from World Trade tower

195 Unusual Perspective
Brooklyn Public Library, Brooklyn Collection
BRID 0145

196 Arches Close-up
Author's Collection
twin gothic towers of Brooklyn Bridge02

197 Welcome to Brooklyn
Author's Collection
2009 04 17 029_edited-1

198 Statue of John Roebling
Brooklyn Public Library, Brooklyn Collection
PORT 0750